Women in Engineering and Science

Series Editor

Jill S. Tietjen, Greenwood Village, Colorado, USA

The Springer Women in Engineering and Science series highlights women's accomplishments in these critical fields. The foundational volume in the series provides a broad overview of women's multi-faceted contributions to engineering over the last century. Each subsequent volume is dedicated to illuminating women's research and achievements in key, targeted areas of contemporary engineering and science endeavors.The goal for the series is to raise awareness of the pivotal work women are undertaking in areas of keen importance to our global community.

Cecilia D. Craig • Holly J. Teig
Debra A. Kimberling • Janet L. Williams
Jill S. Tietjen • Vicki S. Johnson

Editors

Women Engineering Legends 1952–1976

Society of Women Engineers Achievement Award Recipients

Springer

Editors
Cecilia D. Craig
San Jose, CA, USA

Debra A. Kimberling
San Diego, CA, USA

Jill S. Tietjen
Technically Speaking, Inc
Greenwood Village, CO, USA

Holly J. Teig
Peoria, IL, USA

Janet L. Williams
Albuquerque, NM, USA

Vicki S. Johnson
Fountain Hills, AZ, USA

ISSN 2509-6427 ISSN 2509-6435 (electronic)
Women in Engineering and Science
ISBN 978-3-032-00223-5 ISBN 978-3-032-00224-2 (eBook)
https://doi.org/10.1007/978-3-032-00224-2

This Springer imprint is published by the registered company Springer Nature Switzerland AG
The registered company address is: Gewerbestrasse 11, 6330 Cham, Switzerland

To all of the SWE Achievement Award recipients, who inspire future engineers through their accomplishments and experiences.

Foreword 1

Throughout history, progress has often been measured by the stories of those who dared to break barriers, redefine expectations, and inspire generations to come. Since our founding in 1950, the Society of Women Engineers (SWE) has honored individuals whose achievements embody these ideals—women and allies who have advanced engineering, launched new technologies, shattered glass ceilings, and created pathways for those who follow in their footsteps.

This book, a tribute to the first 25 years of SWE's Achievement Award recipients, is more than a historical record; it is a testament to perseverance, enduring excellence, and the transformative power of diversity in engineering. In reading the stories of these honorees and their extraordinary contributions, you will recognize how commemorating women in engineering is not just about celebrating individual accomplishments—it is about collectively shaping a future where talent, innovation, and leadership know no gender.

In today's rapidly evolving world, where technology is redefining industries and addressing global challenges, the need for diverse perspectives has never been greater. From tackling climate change and developing sustainable infrastructure to advancing artificial intelligence and healthcare solutions, engineers are at the forefront of shaping society. The inclusion of women in this progress is not just a matter of representation; it is an imperative for innovation.

Yet, we know that barriers remain. While we celebrate how far we have come, we must also acknowledge that work is still to be done. Women remain underrepresented in leadership roles, and systemic challenges persist, continuing to impact their retention and advancement. That is why this moment in history is so significant—because every recognition, every milestone, and every story in this book serves as a call to action and a catalyst for change.

Looking back on our 75 years of championing women in engineering, it is even more important that today we celebrate and publicize the accomplishments of women in engineering. Three-quarters of a century ago, women were often omitted from history books, left off patent applications, and barred from full participation in engineering. And now in 2025, women's accomplishments are being removed from the websites of our government, military, academic institutions, and corporations.

Women in engineering continue to need a strong advocate—now more than ever—and the society is proud to continue our legacy as that voice. If not us, then who?

From Mária Telkes, Ph.D., in 1952 to Ada Pressman in 1976, these voices from the past invite you to reflect on the impact these trailblazers have had and the doors they have opened for the next generation. Let each of these stories serve as a reminder of what is possible when talent is nurtured, when voices are heard, and when excellence is recognized.

To our past, present, and future leaders—this is your legacy. And to the engineers of tomorrow, may this book inspire you to forge ahead with the same determination, curiosity, and boldness that have defined the Society of Women Engineers for decades.

Society of Women Engineers
Chicago, IL, USA

Karen Horting

Foreword 2

The stories and statures of some early recipients of the Society of Women Engineers Achievement Award have withstood the test of time. That their legacies have endured the test of time is likely just as much a reflection of the public's understanding of their general fields of work as it is the importance of their technical contributions to those fields. The Society's first annual "Award for Meritorious Contribution of Engineering," as it was known in 1952, was presented to Mária Telkes, a pioneer in solar energy technologies that have reemerged in the public's consciousness some 75 years later. Barbara Crawford Johnson (1974 Achievement Award recipient) is still remembered for her leadership and technical contributions to several NASA space programs. Perhaps most recognizable among SWE's early award recipients is Grace Murray Hopper (the 1964 honoree), whose mid-century work on programming compilers and languages laid the foundations for the digital world we live in today.

Where their legacies have endured, the contributions and lives of other SWE award recipients have been less celebrated over time. The general public can appreciate the importance of computer programming in their daily lives, but Isabella Karle's electron and X-ray diffraction research is harder to visualize and understand. Désirée le Beau's research in rubber reclamation, while having many practical applications, is perhaps a less exciting news story than a rocket launch. For these and other Achievement Award recipients, the totality of their work has largely been condensed down in history to short profiles, when they are remembered at all. Yet, despite the sometimes opaque nature of their work, their contributions to scientific and technological progress remain important.

Of course, all of SWE's Achievement Award recipients are more than the sum of their award citations. By pulling together formal records and personal correspondence from the Society of Women Engineers and other archives, published works, and personal family collections, this volume expands their technical legacies to remember and include their lives: the environment, people, and experiences that shaped who they were and how they approached their work.

Why do we give technical innovators awards, enshrine their legacies in Halls of Fame, read their biographies, and study their lives decades later? The Society of

Women Engineers established its Achievement Award nearly 75 years ago, in part, to let these women know that they and their work had been *seen*, that their work mattered, that they weren't working in the shadows of their male counterparts. And in doing so, the award sent a message to other women in engineering that they too could be seen, that their own work mattered, too.

Decades later, their work, lives, and legacies still matter. SWE's Achievement Award recipients are role models for the future: three-dimensional characters with rich, interesting, and complicated lives that today and tomorrow's engineers can identify with, be inspired by, and learn from. And inspiring and supporting women to become and thrive as engineers and leaders has always lay at the heart of SWE.

Society of Women Engineers Archivist,
Walter P. Reuther Library
Wayne State University,
Detroit, MI, USA

Troy Eller English

Preface

The early stories of the pioneering engineers from the Society of Women Engineers (SWE) have always piqued my interest. I frequently think back to Elsie Eaves, the civil engineer who wrote to the deans of engineering schools in 1919 seeking to connect with other women studying engineering. Who were these women? How did they obtain their engineering education while so many universities barred women's participation? What was it like to work in the engineering field in that era?

Elsie Eaves was at the top of my list when members of my local San Diego County Engineering Council suggested profiling some early engineering pioneers in 2021. The goal was to stay engaged during the pandemic and promote engineering pioneer profiles as role models for the community. We coined the project "Engineering Legends."

Many of the textbook engineers began to pop up in our brainstorming sessions. Several women were included in the list, from the well-publicized astronauts to the human computers of "Hidden Figures" fame. We even included some well-known SWE pioneers, such as Grace Hopper, Edith Clarke, and Lillian Gilbreth.

A SWE volunteer began pulling together the material for the first Engineering Legend profiles. Each profile included a photo with a brief 70-word citation. Upon review, it struck me that the women's profiles were often vague and frequently leaned toward describing their status as the "first woman" to graduate, participate, or be recognized in a particular area. When I inquired further about the sources the volunteer had used to write the citations, she sent me directly to Wikipedia.

Wikipedia readily proclaimed their "first woman" standings: Eaves had eight, Clarke five. Yet their citations lacked their fundamental technical achievements. Their accomplishments were often difficult to identify, even after reading their entire Wikipedia post. The writing of the Engineering Legend citations was reassigned to me, beginning my journey into the linguistic text analysis of gender bias in media. I was flabbergasted to discover the magnitude of overlooked contributions from accomplished women through the decades.

The media tend to use more gender-specific phrases, like "first woman," and to downplay women's achievements by describing their lives through their relationships with men and family.

One classic example of gender bias in the media was the 2013 *New York Times* obituary of Yvonne Brill, the 1986 SWE Achievement Award recipient. The original citation proudly proclaimed, "She made a mean beef stroganoff, followed her husband from job to job and took 8 years off from work to raise three children." Upon public protest, the *New York Times* republished her obituary. The new opening citation, "NASA rocket and jet propulsion engineer known for inventing fuel-efficient rocket thrusters, and presented with the National Medal of Technology and Innovation..." I presented my findings in the 2023 WE Local Seattle Conference session "Out of Focus: Revealing Women's Achievements" as a primer for evaluators, awards chairs, and citation authors.

The SWE Late Career and Retiree (LCR) Affinity Group has been a welcome home base for advancing new concepts that support the SWE Mission. They accepted the "SWE Legends" project as a sub-workgroup to update the Wikipedia profiles of SWE pioneers. We set off to understand the status of the SWE Achievement Award recipients. As a benchmark, we analyzed the Wikipedia profiles of the first seven SWE Achievement Award recipients. Three candidates had no online presence, three required updates to include their technical achievements, and only one even mentioned the SWE Achievement Award. Ouch. Proof of concept.

It was during the annual SWE Fellow's meeting, at the WE22 Houston conference, that momentum continued to pick up for the SWE Legends project. Jill Tietjen stepped forward volunteering to serve as a mentor for the SWE Legends project. She formally introduced me and the SWE Legends project to the SWE Archivist, Troy Eller English. Together we set off to discover the status of the SWE Achievement Award recipient archive documentation.

The SWE Achievement Award recipients are notable because their documentation is in the SWE Archives. The archives include their achievement award nomination packets, letters of recommendation, curriculum vitae, and on occasion an acceptance speech. If you are lucky, you might find an oral history, a transcription, an updated newspaper clipping, a SWE article, a resume, or an obituary that their family sent in. Without the SWE Archives, this book would not have been possible.

The goal of the SWE Legends Phase One was to gain experience editing Wikipedia profiles, to assess the archive needs, and to create viable training materials for future editors. The SWE Archivist created a full listing of the achievement award recipients, complete with their citations online, and compiled the content in each candidate's boxes and folders. We presented our findings at the WE23 Los Angeles conference session entitled "Bringing SWE Legends Into Focus."

Jill Tietjen is the series editor for the Springer Women in Engineering and Science series and is always on the lookout for ideas, editors, and authors for the next women in engineering book. Ceal Craig chatted with Jill after the Awards Reception at the WE24 Chicago conference. Ceal planted the seed for taking the Legends project to a new level with a book on SWE's Achievement Award recipients. After sharing this idea with me, we rapidly focused on the first 25 recipients. We recruited more editors from the Late Career and Retirees Affinity Group, and within 24 hours, the concept of this book to capture the SWE Achievement Award recipients came to life.

This book is divided into sections in increments of 5 years. Our goal is to help set the stage for the technological, cultural, and societal influences our SWE Achievement Award recipients faced growing up and within their careers. The Afterword provides a brief glimpse of the cultural and societal influences that influenced each of the book editors. Enjoy!

SWE Legends Project Leader
San Diego, CA, USA

Debra A. Kimberling

Acknowledgments

No one writes a book alone. Many, many people provide support, guidance, direction, and input. We apologize in advance for any omissions in this list; however, they are unintended.

To Diane Darby, whose creative talent provided the beautiful graphics in each section of this book. We also extend our gratitude for her endless patience with changes and updates.

To Karen Horting and Troy Eller English for their support, encouragement, and contributions to this volume that have lent context and deeper meaning to our endeavor.

To Marie Aloia, Dr. Deborah Fisher, and Linda Reed, guest chapter authors, for their diligent research, writing, and revision assistance.

To Mary James, our editor at Springer, who continues her amazing support for this book and the entire Springer Women in Engineering and Science series. Other staff members at Springer, including A. Thiyagarajan, and Brian Halm, helped us work through nuances and requirements.

To the SWE Fellows community and the SWE Late Career and Retiree (LCR) Affinity Group for responding to our request for their personal memories of studying, living, and working during the years of this volume, along with their own stories of the first 25 Achievement Award recipients.

To the Society of Women Engineers members who have been our role models, mentors, friends, and colleagues for decades.

To all SWE members and SWE Legends Wikipedia writers who expressed their interest, gratitude, and energy in signing up to become volunteer research historians.

To the SWE Awards committee members for their active interests, passions, and drive to highlight some of SWE's most accomplished women in engineering. Thank you for ensuring the way to store and update their information within the SWE Archives.

To the SWE Late Career and Retiree Affinity Group for actively encouraging the SWE Legends project. Thanks for your willingness to create a welcoming home, to support the members, and help them advance the project visibility within the community and SWE.

To all of the SWE Achievement Award recipients and their families, including the pioneers we recognize in this book. Through your persistence and drive, we captured each unique story using SWE Archive documents. Thank you for telling us about your accomplishments beyond receiving your award, updating us on your activities in retirement, and providing a glimpse into your life stories. Thank you to those who have already provided your written or oral histories. Thank you to your families for providing updates and the historical context so we could pull the full story together, from the birthplace of inspiration to the legacy you left behind.

To our co-editors, SWE colleagues, and friends for sharing a passion and creating this book. You have accelerated the SWE Legends project and are helping to ensure that women are written into history.

To all our families who helped us complete this important project, colleagues who helped us along the way, and to all the co-editors and authors, thank you for the patience and grace you have shown. This has been a learning journey for us all.

This book is only possible due to our SWE foremothers' foresight, wisdom, and strategic direction in creating the SWE Archives in 1957, now stored at the Walter P. Reuther Library at Wayne State University as of 1993.

Contents

About the Editors and Authors

Cecilia D. Craig, PhD, F.SWE Ceal's early memories of her civil engineer father were of surveying with him and building birdhouses for the backyard. He died in his early 30s when she was in fourth grade; her younger sister had died the year before. Now a widow with no job, her mother began raising two children (Ceal and a younger brother) alone in the early 1960s. Ceal's mom took college classes, hoping to be a teacher. Various illnesses hit her mom in those first few years, and Ceal had to change schools in fourth grade and then fifth grade, living with an aunt and then her grandmother for months while her mother was in the hospital. Ceal's mom remarried when Ceal was in eighth grade, and then two more brothers arrived. These experiences molded Ceal into an independent person, often taking on home leadership roles until she left home and her birth state of Minnesota to start college at 17.

A male chemistry teacher in high school suggested she try for a National Science Foundation (NSF) summer program. She did and left Minnesota for Ohio for many weeks studying FORTRAN, electronics, statics, and other engineering things, along with six other young women and 40-plus young men, all between their junior and senior year in high school: The summer of 1969. They all watched Apollo 11 land on the moon on a small black and white television in the basement of one of the dorms at The Ohio State University (OSU). This summer was pivotal for Ceal and her career aspirations.

Thus, in her high school senior year, Ceal said, "I'm going to study engineering." Her parents were supportive; most teachers were supportive. She applied to many institutions, was accepted by all, and all were out of state. OSU offered a small scholarship, and Ceal left for Columbus after her youngest brother was born. Studying engineering that first year was rough. Mathematics, a subject that has always been easy, was hard. She made it through. That summer, she was a counselor for that same NSF program at OSU. She dated her future husband, an aeronautical engineering student, that summer, and they married that December, her sophomore year. As he finished graduate school and later she earned a master's in engineering, this partnership provided her stability and support for the next 51 and a half years until he died.

After 35 years working in high-tech companies, Ceal moved into the STEM education field. She was also swept into robotics via *For Inspiration and Recognition of Science and Technology* (FIRST) programs. Working with young people on robotics programs inspired her to pursue a Ph.D. in Education at the age of 61, studying the influence of FIRST on young women's career decisions.

Dr. Craig has focused her later years on exploring the influences of engineering experiential activities on young women's career decisions. Her passion is researching and supporting activities that inspire young women to pursue careers in engineering, physics, and computer science. She has published two books: Editor and chapter author, *Mentoring in STEM Through a Female Identity Lens: Heroes Make a Difference for Women* (Vernon Press), and *Sinking Underwater, a ghost town's amazing legacy* with Anita Goldwasser.

Holly J. Teig, F.SWE enrolled at the University of North Dakota (UND) in the fall of 1984, determined to earn an engineering degree. Teig had done well in high school math, science, and accounting classes but also had been active in the arts—choir, band, and theater—leading her high school guidance counselor to advise her to pursue theater arts rather than a STEM career. Wanting financial stability, Teig scanned the salary comparison by major list at the library and discovered engineering degrees were at the top. She chose mechanical engineering, liking the hands-on nature of how things worked, discovered during her upbringing on her parents' small farm. Teig joined the Society of Women Engineers

(SWE) freshman year, finding a great support network. She also enrolled in the UND engineering co-op program, alternating class semesters with engineering work in industry, allowing her to pay for her remaining schooling while gaining engineering experience. In 1989, Teig graduated with a B.S. in Mechanical Engineering.

Teig worked 30 years for Caterpillar Inc., beginning with roles in manufacturing research and production support. A business unit downsizing 5 years into her career created a fork in the road. She switched focus, joining the logistics and supply chain division, and discovered a passion for process improvement projects. Teig progressed from a technical contributor into the supervisor and manager ranks, gaining customer service and financial skill sets. She traveled around the world to work with global colleagues and customers on inventory management and process improvement, something she never imagined doing back in high school. Mid-career, she joined a multi-company team that was transforming the supply chain infrastructure—from software to processes to physical layout—creating a legacy of modernization that launched breakthrough technology to the industry and was the capstone of her Caterpillar career.

SWE has been a constant in Teig's life in the 40+ years since her collegiate days. Teig is a SWE Fellow and life member and a founding member of the SWE Central Illinois section. Teig is currently Treasurer of the SWE Board of Trustees overseeing the SWE Endowment Fund Inc, ensuring women pursuing engineering aspirations have financial support to pursue their education and be successful in the STEM workforce. Teig has been an honored recipient of the SWE Emerging Leader in Procurement/Logistics award, SWE Central Illinois Pillar of Support award, and Peoria 25 Women in Leadership award. She was a frequent presenter at industry conferences on supply chain technology and logistics, including presentations entitled: *Engineering the Supply Chain* and *Rise of the Industrial Drone*.

Since 2021, Teig has been a certified University of Illinois Extension Master Gardener, providing community garden education and support. A lifelong gardener and environmentalist, she volunteers at multiple community garden sites, presents on topics for the public, and supports horticulture hotline questions. In 2023, Teig received the Illinois Outstanding Master Gardener recognition. Teig enjoys getting outside with her husband Randy (also a UND mechanical engineer) in the garden, hiking, golfing, or traveling. She also spends time in her creative space sewing, writing, reading, or scrapbooking.

Debra A. Kimberling, F.SWE became an engineer by accident. Even though she graduated in the top 1% of her class in a large suburban high school, she enrolled in a community college in the Fall of 1976. She had no idea of which career to pursue. It was a college math professor, for whom she was grading papers, who recognized her talents and suggested she consider a career in engineering. After some research on what engineers do and their job prospects, she changed majors and transferred to Arizona State University (ASU). She graduated with a B.S. in Engineering Science from ASU (Tau Beta Pi).

Kimberling accepted a position at the Naval Ocean Systems Center in San Diego after listening to their career presentation hosted by SWE at ASU. Her first job at the center was working on submersibles. After 3 years at the center, she decided to pursue her master's degree at the University of Washington. She majored in aerospace engineering because she had enjoyed the structural analysis side of her work. After graduation, she became an Associate Professor of Mechanical Engineering Technology at Purdue University and then an engineer at General Dynamics (GD) Space Systems Division and Solar Turbines, where she remained for 22 years.

Growing up, Kimberling's father had ensured she had practical hands-on skills. He taught her how to use a saw, change tires, and do basic car maintenance, among other things, just in case. Her mother taught her persistence. The family moved often because of her father's naval career. After failing a third grade math test at a new school, her mother sat her down at the kitchen table every afternoon for 2 months until she started getting A's on her tests. She has been comfortable with math ever since. When her college professor suggested engineering, she had the confidence she would succeed.

Kimberling had been either the only woman or one of two women in her classes and workplace departments. It was at GD that she became SWE active, ultimately Section President, then Region Director. When she moved to Solar Turbines, she was thrilled to finally work at a company where she could introduce SWE initiatives. She implemented a Bring Your Child to Solar Day to over 1,200 students through the years, and possibly more importantly she introduced the company to SWE. Kimberling started the first Affinity Group at her company to help carry out the SWE mission.

Positive role models of women in engineering benefit everyone: girls, boys, parents, educators, and the general public. They have a way of inoculating people against negative stereotypes. In 2010, Kimberling began the SWE Parent and Educator Programs to introduce why so few women in STEM research and look for additional ways to put engineering role models in the community.

Kimberling received the SWE Distinguished Service Award in part for introducing the SWE Legends Wikipedia project, which is the source of inspiration for this book. She is a SWE Fellow and a founder of the SWE Late Career and Retiree Affinity Group. Kimberling has been a San Diego SWE member for 35 years and has received numerous recognitions for her community service and leadership. Today she is repackaging the SWE Legends into presentations and is extremely grateful for her SWE friends, mentors, and role models.

Engineering has been a rewarding career for Kimberling. She has helped launch communication satellites into orbit, written a design manual, and developed fuel tools to serve the engineers and sales folks at her companies. In her spare time, she enjoys hiking and is in an improv troupe.

Janet L. Williams, F.SWE is an "accidental" engineer. She graduated near the top of her high school class in Northern California, but as a first-generation college attendee, she lacked career guidance and mentorship from parents or counselors to lead her to a STEM career. She graduated from the University of California, Davis with a BA in Spanish and moved to Kansas City, where she found an administrative job at a prominent engineering consulting firm.

Based on the demonstrated strength of her analytical and communication skills, she began helping engineers with technical tasks. She soon realized that she could do the work that engineers did and went back to college. After several years of part-time and full-time coursework at the University Missouri, Kansas City, she obtained her BS in Mechanical Engineering, one of two women granted that degree in 1983.

Jan worked for 2 years doing facilities projects in a plant that manufactured mechanical and electrical components for the U.S. nuclear weapons enterprise. This experience led to a 33-year career at Sandia National Laboratories in Albuquerque, New Mexico. During her career she obtained an MS in civil engineering with an emphasis in construction management. She progressed through mechanical design, facility construction project management, site planning, and ultimately in program management for infrastructure and facility investment to maintain specialized capabilities for the nuclear weapon enterprise. Her career included special assignments at the National Nuclear Security Administration in Washington, DC and the Atomic Weapons Establishment in the United Kingdom. She also spent significant time recruiting and hiring diverse candidates for Sandia Labs, in addition to her other assignments.

Throughout her career, Jan has served in numerous volunteer positions at local and national levels in the Society of Women Engineers (SWE). She is passionate about STEM outreach because she herself did not get exposure to STEM careers as a possible career path. Often the only woman in meetings throughout most of her career, she is distressed by the painfully slow progress in getting more women and minorities into the engineering workforce. She has participated in many programs over the years to introduce young people, especially girls and minorities, to engineering. SWE recognized her long-time service when she was named Fellow in 2009 and received SWE's Distinguished Service Award in 2017. She continues to serve the society in many roles.

Since her retirement, she has devoted much of her time to STEM advocacy at the local and national levels. Her efforts include attending legislative meetings and introducing state legislation to establish a centralized STEM network that will ensure every child has access to STEM learning and knowledge of potential STEM careers. Realizing that women in STEM are often unrecognized and underappreciated for their contributions, she has made it a personal project to nominate deserving women in engineering for awards within and beyond the Society of Women Engineers. She is also part of a team of women working to improve the accuracy and equitable treatment of Wikipedia pages for women in engineering. Another current project is speaking in public about women in STEM and how they have influenced history and made our world a better place through their persistent advocacy and amazing innovations. In short, Jan is dedicated to ensuring that women's technical contributions are recognized and that their stories are heard.

Jill S. Tietjen, P.E., F.SWE entered the University of Virginia in the Fall of 1972 (the third year that women were admitted as undergraduates after a suit was filed in court by women seeking admission) intending to be a mathematics major. But midway through her first semester, she found engineering and made all of the arrangements necessary to transfer. In 1976, she graduated with a B.S. in Applied Mathematics (minor in Electrical Engineering) (Tau Beta Pi, Virginia Alpha) and went to work in the electric utility industry.

Galvanized by the fact that no one, not even her Ph.D. engineer father, had encouraged her to pursue an engineering education and that only after her graduation did she discover that her degree was not ABET-accredited, she joined the Society of Women Engineers (SWE) and for more than 45 years has worked to encourage young women to pursue science, technology, engineering, and mathematics (STEM) careers. In 1982, she became licensed as a professional engineer in Colorado.

Tietjen started working jigsaw puzzles at age two and has always loved to solve problems. She derives tremendous satisfaction seeing the result of her work—the electricity product that is so reliable that most Americans just take its provision for granted. Flying at night and seeing the lights below, she knows that she had a hand in this infrastructure miracle. As an expert witness, she worked to plan, license, and put into rate base new power plants.

Her efforts to nominate women for awards began in SWE and have progressed to her acknowledgment as one of the top nominators of women in the country. Her nominees have received the National Medal of Technology and the Kate Gleason Medal; they have been inducted into the National Women's Hall of Fame and state Halls, including Colorado, Maryland, and Delaware, and have received university and professional society recognition. Tietjen believes that it is imperative to nominate women for awards—for the role modeling and knowledge of women's accomplishments that it provides for the youth of our country.

Tietjen received her MBA from the University of North Carolina at Charlotte. She has been the recipient of many awards, including the Distinguished Service Award from SWE (of which she has been named a Fellow and is a National Past President), the Distinguished Alumna Award from the University of Virginia, the Distinguished Alumna Award from the University of North Carolina at Charlotte, and the Distinguished Engineering Alumni Award (special category as a non-alum) from the University of Colorado at Boulder. She has been inducted into the Colorado Women's Hall of Fame and the Colorado Authors' Hall of Fame and elected to the National Academy of Construction. Tietjen sits on the board of Georgia Transmission Corporation and served for 11 years on the board of Merrick & Company.

Her publications include the bestselling and award-winning book *Her Story: A Timeline of the Women Who Changed America* for which she received the Daughters of the American Revolution History Award Medal and *Hollywood: Her Story, An Illustrated History of Women and the Movies* which has received numerous awards. Her award-winning book *Over, Under, Around and Through: How Hall of Famers Surmount Obstacles* was released in 2022. The memoir of Dr. Antonia Novello, *Duty Calls: Lessons Learned from an Unexpected Life of Service*, which she co-wrote, was released in 2024. She is the series editor of the Springer Women in Engineering and Science series, has authored two volumes in the series, served as co-author for a few volumes, served as co-volume editor for four more volumes, and written chapters for a number of volumes.

Vicki S. Johnson, PhD, F.SWE graduated from high school in 1977 with a private pilot's license and a determination to become an aerospace engineer. She didn't know any engineers and didn't have any idea what aerospace engineers did. She knew that Cessna 172s seemed very safe; her goal was to get a degree and go to work for Cessna Aircraft Company. Johnson chose the University of Missouri—Rolla for college, thanks to a Curator's scholarship and the availability of a cooperative engineering (co-op) program.

Becoming a co-op student at NASA-Langley Research Center (LaRC) in Hampton, Virginia, led to five work periods while earning her undergraduate degree. Johnson's graduating class in aerospace engineering had 13 members; Johnson was the only woman. The co-op program also led to a full-time job at NASA-LaRC when she graduated in 1982, along with the opportunity to obtain a master's degree in flight sciences part time through George Washington University and a Doctor of Philosophy from the University of Kansas where she was the first woman to earn a PhD in aerospace engineering.

Johnson left NASA-Langley after 12 years and went on an odyssey of jobs in various sectors. Johnson served as senior program officer at the Aeronautics and Space Engineering Board of the National Research Council and then an employee of Universities Space Research Association (USRA) as Program Manager of the NASA/USRA/Universities Advanced Design Program. Johnson's stint in academia was as associate professor of aerospace engineering at Embry-Riddle Aeronautical University (ERAU) with 75% release time to be Director of Student Success Programs. Johnson moved to Wichita, Kansas, in 2000 to work for the National Institute for Aviation Research. The Boeing Company in Wichita was her first foray into industry. Johnson spent 12 years at Cessna Aircraft Company/Textron Aviation (it only took 25 years to achieve her initial career goal of working for Cessna!). While at Cessna, Johnson earned a Master of Business Administration in the management of technology. She retired from Spirit AeroSystems in Wichita, Kansas, after a 42-year career.

Upon arrival in Wichita in 2000, Johnson became a member of the KU Aerospace Engineering Advisory Board where she was the lone woman until 5 years ago.

In 2016, Johnson was invited to join the KU School of Engineering Advisory Board which she has chaired since 2021.

Johnson joined the Society of Women Engineers (SWE) while at ERAU, was very involved in Wichita for 15 years, and then got re-involved in SWE with the development of the Late Career and Retiree (LCR) Affinity Group (AG) in 2019. Johnson became an SWE Fellow in 2020 and received the Distinguished Service Award in 2024. Johnson was invited to join the group writing this book because of her involvement with the LCR AG.

Johnson enjoyed her career. The work was generally challenging and satisfying. Her earning power led to a good retirement. Mentoring young engineers has made good use of her experience. And documenting the excellent engineering skills and accomplishments of some of the first 25 SWE Achievement Award recipients has been very rewarding. Johnson was especially interested in how engineers in the first half of the twentieth century accomplished their work. They were amazing!

Marie Aloia In 1972, Marie Aloia majored in math and physics at Montclair State University because they were her favorites. However, she had no career plan at the time. As the first one in her family to go to college, she had no guidance and worked two or three part-time jobs to help pay for college.

After graduation, she took a technician job at Exxon Research and Engineering to earn money for graduate school. She saw engineering for the first time and fell in love with it there. Her manager encouraged her to enroll at the New Jersey Institute of Technology (NJIT). After five years of night classes, she earned her second bachelor's degree in chemical engineering and a few promotions along the way. Having found a love for engineering later, Exxon's educational speakers' group became a passion for her because they visited middle schools where the students' parents were not likely to be engineers, bringing career talks and a "Mr. Wizard kit" of science demos. She eventually ran this program, training other speakers, visiting schools, and judging science fairs.

In the mid-1990s, a coworker invited her to a SWE meeting, and she has been an active member ever since. With SWE, Marie found new opportunities to reach out and mentor young people, especially girls like her, in science and engineering. SWE also helped Marie confirm her identity as an engineer. She left Exxon for the more lucrative high-tech world from 1998 to 2003. When the *dot.com* bubble burst, she was laid off. Marie decided it was time for a completely different job. After many years of classroom visits and many students' thank-you letters that read "I wish you were my science teacher," why not try teaching for real?

In September 2003, Marie became an alternate route teacher of chemistry and physics at Bayonne High School, Bayonne, New Jersey. Taking advantage of her background in industry, her director asked her to start a competitive science research program and then an engineering program which became her primary teaching role. She brought in many new STEM programs, including FIRST Tech Challenge Robotics and Lemelson-MIT InvenTeam, to include as many of her students as possible. Her students won top awards at local and national competitions. An NJIT professor who remembered her and heard she was a teacher invited her to join the Research Experience for Teachers (RET) program that summer, an invitation exended for the next 10 years. She brought a few students each summer on American Chemical Society Project SEED stipends to share her RET research projects. Many students asked her for college and graduate school recommendation letters. Her students even wrote recommendation letters for her. Each year, National Engineers Week Girl Day was celebrated at the school with SWE speakers and an evening of engineering activities, with high school girls mentoring younger girls. Marie received many awards, including the 2013–14 Bayonne School District Teacher of the Year. Now, even in retirement, Marie continues to have fun with STEM, mentoring students, judging science fairs, and refereeing for robotics.

Deborah J. Fisher-Rebstock, PhD was raised in a highly educated family, being the third generation Dr. Fisher. She was the granddaughter of a Canadian medical doctor and the daughter of a Presbyterian minister. Both her parents were theologians, and dinner discussions were often centered around esoteric topics in philosophy. She graduated from high school in 1969 in central Texas in a place and time when young ladies who went to college were supposed to study the arts and work on their "MRS degree." She graduated with a BA in modern languages in 1973 from Texas A&M University, which had only become co-educational during her junior year. She had no interest in marriage and having children, like all her friends. It was during the oil embargo and jobs were scarce, but it was also a time when the government was promoting affirmative action. While interviewing for scarce jobs, she learned that companies were desperate for female engineers. She didn't know what an engineer was, but she was determined to get a job in a non-arts field that paid well.

She was mentored by a family friend who was a professor at Texas A&M University, where she received her Master's in Civil Engineering in 1976. She married a PhD graduate and followed him to work for Halliburton in Oklahoma, designing pneumatic conveying systems and ASME-coded pressure vessels. She then followed him back to Texas to work for Foster Wheeler Engineering, where she moved up, managing design and construction of hundreds of millions of dollars in petrochemical projects. Her professors were not sure that she could ever be licensed as a PE without a bachelor's degree, but she surprised them by passing both the EIT and FE exams, becoming licensed in both Texas and New Mexico. When her husband wanted her to move a third time, she said no and divorced him. The economy fell in the mid-1980s and her engineering office closed, so newly single and unemployed, she returned to grad school, obtaining a PhD in Civil Engineering in 1988 from the University of Texas at Austin.

During grad school, she married a second time to an engineer and followed his career by taking a job at the University of Houston as Director of the Engineering Management program in the Industrial Engineering

Department. An undergrad female student told her about SWE, and she became the faculty advisor for the student section. Surprisingly, when she got an offer to become the first AGC Chaired Professor at the University of New Mexico (UNM) and the first female in the Civil Engineering Department in 1994, her second husband moved for her. She received tenure and promotion there and became the faculty advisor for the SWE student section. She also helped plan the National SWE Conference in Albuquerque in 1997.

Wanting to become a mom, she and her second husband adopted a one-year-old girl from Russia, but being the first female in the department, and a new mom, she felt like she hit a glass ceiling. Even though she received the SWE Distinguished Engineering Educator award in 2005, she chose to leave UNM and to do consulting, co-author a book, and teach part-time for Embry Riddle Aeronautical University Worldwide online. Dr. Fisher officially retired in 2016 and has been active in volunteering with the Central New Mexico Section of SWE ever since, doing outreach, advocacy, and faculty advising. She is proud to be a thirty-year member of SWE.

Linda A. G. Reed, F.SWE entered Michigan State University in the Spring of 1977 as a non-major freshman. Despite her interest in science, her high school counselors did not suggest engineering. She graduated in 1981 with a Bachelor of Arts in Advertising Communications. She began work at an advertising agency and quickly transitioned to one of the firm's clients, PacTel Cellular, one of the first cellular companies in the country. She managed their advertising program and outside agencies.

During her time at PacTel, she was exposed to the engineering side of the cellular business, and this sparked her earlier interest in science and technology. She designed helicopters in high school and even developed the concept for small disks that contained directions to the shopping mall—all before compact disks and computer disks existed.

She returned to school in 1990, at California State University, Long Beach (CSULB), intending to study mechanical engineering and work in the field of prosthetic design. After realizing that seeing and thinking in three dimensions was difficult (before CAD programs made things easier), she shifted to electrical engineering. During her four years at CSULB, she became involved with the SWE student section.

Graduating in 1994 with a Bachelor of Science in Electrical Engineering and a minor in Mathematics, she joined TRW as a manufacturing engineer. She also became more involved with SWE, eventually becoming the LA Section Secretary, Section Representative and President, and 50th Anniversary Banquet Chair. She is the recipient of the Loring Nicholson Award for volunteerism. At the national level, she served on the SWE magazine Editorial Board.

Reed remembers sitting in her neighbor's living room in the middle of the night, watching Neil Armstrong step out of the lunar lander to set foot on the moon. That inspired a love of space. When an internship at TRW became available while she was studying at CSULB, she snatched it and converted it to full-time employment upon graduation. She was then working for the company that built the lunar landing module descent engine. Hardware she has been involved with still orbits our planet, and some have even left the solar system.

Reed is a Fellow of the Society of Women Engineers and the Institute for the Advancement of Engineering. She is a certified 6 Sigma Master Black Belt. Her love of process engineering has carried over to her hobbies of wine making and olive oil pressing. She now writes fiction and has published three novels. Her first, a time-travel fantasy, explored the ideas of time travel, generational ancestry, and family ties. Her second was a middle-grade fantasy that allowed her to flex her creative wings. The last book brought her back to science, technology, and math, with a story about a little girl who wants to direct movies in space but has difficulty in class when it comes to science. She has to win a science fair to secure a place at Space Camp, and the book includes an actual science fair program on the fundamentals of flight.

Reed and her husband live in San Miguel, CA. After years living in cities, they love the country life: quiet, stars, and fresh air.

Chapter 1
Setting the Stage: Pre 1952

From the Birth of the Profession of Engineering through the Founding of the Society of Women Engineers

Vicki S. Johnson and Jill S. Tietjen

1.1 Introduction

What were the lived experiences of the 25 women who were honored as the first Society of Women Engineers' (SWE) Achievement Award recipients? The short answer: very different from the lived experiences of young people starting their careers today. Some of these pioneering engineers were born in the nineteenth century, before electricity. Some of these trailblazers never had a computer to use. Others made the transition to the early, much bulkier computers than today's exceedingly powerful smartphones but did much of their work using a slide rule. Most were usually one of the few or the only woman in engineering throughout their education and their career.

The focus of this book is documenting the lives and accomplishments of the women who made outstanding technical contributions recognized by the first 25 years of the SWE Achievement Awards (1952–1976). This chapter sets the stage leading up to 1952: the year that the first Achievement Award was presented. Then, each 5-year section discusses the state of technology, the state of women, the state of SWE, and the state of culture and laws during that time frame. Each Achievement Award recipient is profiled in her own chapter.

V. S. Johnson (✉)
Fountain Hills, AZ, USA

J. S. Tietjen
Technically Speaking, Inc, Greenwood Village, CO, USA

C. D. Craig et al. (eds.), *Women Engineering Legends 1952-1976: Society of Women Engineers Achievement Award Recipients*,
https://doi.org/10.1007/978-3-032-00224-2_1

The advancement of technology in the late nineteenth and early twentieth centuries was rapid and intense. Electrification was a key technology that enabled so many others. Women began to have the ability to become educated as engineers, and the World Wars (especially World War II) provided many job openings for women. The opportunities were not as widespread after the men returned from the war, but women persisted and achieved. SWE was established in 1950 to provide a mechanism for women to work together to increase the opportunities for and success of women in Science, Technology, Engineering, and Math (STEM) careers. The Great Depression resulted in legislation like the National Fair Labor Standards Act and Social Security that provided improved working conditions and retirement income for both men and women. Meanwhile, women were busy at work and being activists to improve conditions for all women.

1.2 State of Technology Prior to 1952

A vast array of technologies was developed in the nineteenth and first half of the twentieth centuries that enabled improvement in nearly every sector of human activity. The world became a healthier, safer, and more productive place primarily because of these engineering achievements.

Many of these technologies are identified in the National Academy of Engineering's (NAE) Top Engineering Impacts of the twentieth century released in 2000 (Science Daily 2000). The NAE identified electrification as the top achievement since it powers almost everything in modern society. That list forms the basis for this section on the state of technology related to electrification, communications and media, human health and comfort, and the technologies used at work. The first three categories affected the quality of life both at home and at work; the last category shows what tools the SWE Achievement Award recipients had available to accomplish their work and make their contributions.

1.2.1 Electrification

The most important engineering achievement of the twentieth century, electricity and the technologies it powers, has significantly improved the standard of living and quality of life of people in the US and around the world. In this one century, humankind advanced from light bulbs to AC current to nuclear power.

Lightbulb Thomas Edison is credited with inventing the lightbulb in 1879; what he and his colleagues did was improve the filament. His final bamboo filament design would burn for 1200 h compared to 14.5 h for the uncoated cotton thread carbonized filament. Edison also worked on improving the systems that made use of the lightbulb practical: distribution of direct current electricity from a central generator through a conduit, electrical power generation, and the electric power meter to enable billing of individual customers for their actual power use (US Department of Energy 2013a).

Alternating Current Power System Nikola Tesla invented alternating current which overcame two of the disadvantages of direct current: limited transmission distances; and difficulty in increasing or decreasing the voltage. Tesla held 40 US patents for his work with current and the equipment which generates, transmits, and uses it including an 1888 patent for electric transmission of power. George Westinghouse was determined to supply America with the Tesla system, so he bought all of Tesla's patents (American Physical Society 1988). Tesla had long admired the apparent power in waterfalls and achieved his lifelong dream of extracting energy from them by designing the first hydroelectric power plant at Niagara Falls in 1895 (US Department of Energy 2013b).

Vacuum Tubes Invention of early vacuum tube types started in 1875. The first practical vacuum tube was invented by John Ambrose Fleming in 1904. The first commercial electron tube manufacturing started at RCA in 1920. Uses of the vacuum tube included amplifying a weak current, rectifying an alternating current to a direct current, and generating oscillating radiofrequency power for radio and radar (Bellis 2019). Two disadvantages of the vacuum tube were its bulky size and high-power usage (Effectrode Thermionic n.d.).

Transistors The use of transistors got off to a slow start because the early versions invented in 1947 were fragile and could not withstand high temperatures. Designing with transistors was much more complicated than designing with vacuum tubes. In the meantime, vacuum tubes were becoming smaller, and factories knew how to make them reliably and at low cost. The discovery of the transistor resulted in the Nobel Prize in Physics being awarded to Walter Brattain, John Bardeen, and William Shockley in 1956. It was the mid-1950s before the transistor came into widespread use (Vedin n.d.).

Nuclear Technologies In 1942 Dr. Enrico Fermi achieved the first controlled nuclear chain reaction with a uranium device moderated with graphite. The chain reaction took place in Chicago Pile 1, the first demonstration reactor. The Atomic Energy Act was signed by President Harry S. Truman in August 1946. Nuclear energy began to be used for commercial power production in 1955 (PBS American Experience n.d.).

1.2.2 Communications and Media Inventions

It is now hard to envision a time where the only means people had to communicate when they weren't together was by paper and writing implements. That all changed with the advent of the telegraph, the telephone, and television.

Telegraph The wired electric telegraph was first demonstrated in the US in 1837 by Samuel Morse. The wired telegraph era in the US was initiated on May 24, 1844 (McGillem 2024). The system was limited and expensive because of the required wires. Guglielmo Marconi invented the wireless telegraph in 1896. His first successful transmission across the Atlantic Ocean (2100 miles) took place in December 1901 (Lemelson-MIT 2025).

Telephone Italian innovator Antonio Meucci is credited with inventing the first phone in 1849, and Charles Bourseul in France developed a phone in 1854. Alexander Graham Bell was awarded the first US patent for the phone in 1876. By 1878, the first telephone exchange was in operation. Almost 49,000 telephones were in operation a short 3 years later. By 1920, there were 5.8 million telephones. The transcontinental telephone line began operating in 1915. By 1948, the 30 millionth phone was connected in the US (Elon University 2005).

Television Scottish inventor John Logie Baird made a public display of moving visuals for the first time in 1925 on a mechanical scan television which employed a cathode ray tube. Picture quality was not good, and the receivers were too complex to become popular in homes. An electronic television system designed by Philo Taylor Farnsworth was first demonstrated in San Francisco in 1927. Electronic television uses images and sounds transmitted by radio waves, microwaves, or infrared rays to produce images on a video screen. The first color television was developed between 1946 and 1950 by a RCA Laboratories' research team. It was 1953 before advertisements for color television started airing on TV (TCL Guides 2023a).

1.2.3 Transportation

Horses and buggies, and later trains, were the primary means of transportation for much of humankind's recent history. But in the twenty-first century, we had automobiles, planes, and the fuels needed to power them.

Automobiles The first successful gas-powered car was produced by Charles Edgar Duryea and his brother Frank in 1893. The first American car manufacturing company was established by the Duryea brothers (Library of Congress 2024). In the US, by 1898 more than 100 companies had been established to manufacture automobiles. In 1908, Henry Ford revolutionized the automobile by making the Model T cheap, versatile, and easy to maintain. By the mid-1920s manufacturing and selling cars had become an important part of the economy (Cromer et al. 2025). The automobile became the world's major mode of transportation (Science Daily 2000).

Airplanes The Wright Brothers (Orville and Wilbur) are credited with making the first powered flight on December 17, 1903. Once the Wright Brothers demonstrated that controlled, powered flight was possible, military and civil aviation developed quickly. World War 1 drove progress in airplane capabilities. An airmail system was established by the US Post Office in 1918 to stimulate aircraft production and generate trained pilots. Correspondence arrived on the West Coast from New York in 2 days by air instead of 5 days by rail. The Contract Air Mail Act of 1925 privatized the mail service. The Air Commerce Act of 1926 established an organization to enforce procedures for the licensing of aircraft, engines, pilots, and other personnel. The 10-passenger Boeing Model 247 airliner was introduced in 1933. The legendary 21-passenger Douglas DC-3 entered service in 1936. By 1939, DC-3 transports were carrying 90% of the world's airline traffic. The 55-passenger Boeing Stratocruiser offered luxury travel in the late 1940s and early 1950s. Even before jet aircraft were introduced, piston-engine aircraft had taken over from railroads and steamships as the principal mode of transportation for long-distance trips (Crouch et al. 2024).

Petroleum and Gas Technologies Automobile fuel demand overtook kerosene demand in 1920. To increase gas production, refineries began to use thermal cracking. High-octane products could be made by using catalytic cracking which was invented in the 1920s. Fuel demand surged during World War 1, and increasing use of gasoline engines kept fuel demands increasing after the war. Demand also rose for petrochemicals (non-fuel products made from petroleum). Petrochemical items developed in the first half of the twentieth century included Styrofoam in 1941, polyvinylidene chloride (PVC), plastic cling wrap, women's nylon stockings, and isopropyl alcohol. World War II also drove an acceleration in petrochemical and refining technologies. Higher performance fuels were developed to meet the Allies' needs for aviation, truck, jeep, motorcycle, and tank fuels as well as providing lubricants for Allied weapons and machinery. Achievement Award recipient Dr. Margaret Hutchinson Rousseau (1955) was a chemical engineer contributing to development of high-octane aviation fuel. At the same time, the petrochemical industry produced synthetic rubber for vehicle tires, TNT, and insulation for radar and underwater cables (American Fuel and Petrochemical Manufacturers 2020).

1.2.4 Human Health and Comfort

Advances in food safety, human health, and human comfort all resulted from the advent of refrigeration, air conditioning, X-rays and antibiotics, resins and polymers, and a variety of kitchen appliances including the microwave.

Refrigeration The invention of artificial refrigeration in 1748 by William Cullen demonstrated that quickly evaporating a liquid into a gas had a cooling impact. Key enabling technologies included German scientist Carl von Linde developing a revolutionary method for liquefying gases in the late 1800s and American Jacob Perkins creating the first vapor compression device in 1834. By 1900, there was widespread commercial refrigeration, especially in breweries and meatpacking factories. The first home electric refrigerator was built in 1913 by American Fred W. Wolf. It was a refrigeration unit on top of an icebox. The first house refrigerator with a self-contained compressor was developed by William C. Durant in 1918; this was the beginning of mass production of domestic refrigerators (TCL Guides 2023b). The use of home electric refrigerators was limited by the availability of electric power to homes. President Roosevelt signed the Rural Electrification Act (REA) into law in 1936 because 90% of the farms across the nation lacked access to electricity. The use of home refrigerators was also limited by cost (from $500 to $1000) (Parker & Sons 2025). Freon was developed in the 1930s, providing a safer alternative to the toxic gases originally used in the vapor compression process; this increased home refrigeration use even further (TCL Guides 2023b).

Air Conditioning Willis Carrier installed the first air conditioning system on July 17, 1902, which he would later patent. The breakthrough thought that spurred his invention was recognition that air could be dried by passing it through water to create fog. This allowed manufacturing air with specific amounts of moisture in it (Carrier n.d.). Carrier designed a system that used cooling coils to control humidity. His patent "Apparatus for Treating Air" could either heat water to humidify the air or cool water to dehumidify the air. The public first experienced comfort cooling in 1903. The term "air conditioning" was coined in 1906. A patent was issued to H.H. Schultz and J.Q. Sherman for a window air conditioning unit in 1931. First available in 1932, the units were not popular due to their high cost. Frigidaire started marketing central air conditioning systems for houses in 1931. It took until 1947 for low-cost air conditioners to become widely available (US Department of Energy 2015).

X-rays The discovery of X-rays was first reported by W.C. Röntgen in December 1895. X-rays were first used on patients in January 1896 to visualize internal body structure. After observing skin erythema, X-rays were used against skin lesions (radiotherapy) beginning in June 1896 (Tubiana 1996).

Antibiotics The first antibiotic, Salvarsan, was developed in 1910. Penicillin (discovered in 1928) started the golden age of natural product antibiotic discovery that peaked in the mid-1950s. Penicillin was approved for clinical use in 1943. Antibiotics were used to treat infectious diseases. They also made many modern medical treatments possible (e.g., cancer treatment, organ transplants, and open-heart surgery) (Hutching et al. 2019). Mass production and usage of antibiotics led to vast health improvements (Science Daily 2000).

PTFE Polytetrafluoroethylene (PTFE), a resin, was discovered in 1938 when DuPont chemist Dr. Roy Plunkett was trying to invent a replacement for Freon gas. He left a batch of gas overnight. When he arrived the next morning, he found the gas had polymerized spontaneously, leaving a slippery, waxy solid which was impenetrable by corrosive acids, had excellent performance in extreme high and low temperatures, and did not dissolve in solvents. PTFE was entered into the Guiness Book of World Records under the heading "Lowest Friction." PTFE had a coefficient of static and dynamic friction of 0.02, which is equivalent to wet ice on wet ice.[1] In addition to being used in Teflon pans, PTFE contributed to the Manhattan Project by providing material for gaskets, packings, and linings to handle the corrosive uranium hexafluorides in the first atomic bomb (1945) (AFT Fluorotec 2016).

Microwave Ovens The microwave oven was invented by Percy Spencer, a self-educated Maine farm boy and American Appliance Company (later Raytheon) employee. The patent for the microwave oven was awarded on October 8, 1945. The first oven was almost 6 ft tall, had to be water cooled, weighed more than 750 pounds, and cost more than $50,000 in today's currency. It was first installed on the nuclear-powered cargo ship *Savannah*. In January 1947, a Radar Range was offering cooked hot dogs in New York City at the Grand Central Terminal. The public was afraid of microwave radiation, so the oven didn't take off at first (Conover 2015).

1.2.5 Technologies at Work

The work tools for engineers at the beginning of the twentieth century included slide rules, drafting equipment, and probably abaci. With the advent of computing machines, typewriters, and teletype machines, significant advances in productivity were possible.

Slide Rule The slide rule was invented in 1622 by Anglican minister William Oughtred. He placed two scales side by side and slid them to multiply and divide directly by reading the distance relationships. The slide rule was very important to the Industrial Revolution, and the Industrial Revolution had a dramatic impact on the slide rule as demonstrated by the number of different designs. Between 1800 and 1899 there were 250 slide rule types and manufacturers. There were more than 90 designs in the first 10 years of the twentieth century. The slide rule was used to design the thrust profile of the Redstone Rocket and the approaches to the Golden Gate Bridge. The slide rule was the hand computing device for engineers until the advent of somewhat affordable hand calculators in the early 1970s (IEEE Pulse 2019).

Drafting/Drawing Drafting has beginnings traceable to 2000 BC. The first drafting table was patented in 1905. Drafting by hand required planning because making changes was difficult and time consuming. Basic drafting tools included a drafting table, pencils, several types of paper, erasers, compasses, T-squares, and rulers. Computer-aided drawing started in the late 1950s (Brandywine CAD 2022).

[1] Wet ice is frozen water at 32 °F or less (as opposed to dry ice which is solid carbon dioxide at temperatures less than 100 °F). Wet ice on wet ice has the lowest possible static and dynamic coefficients of friction.

Fig. 1.1 (Left to right) Friden Model C10, Marchant Model DRB8, and Monroe "Executive" Model L-160X computing machines. (Courtesy of John Wolff)

Computing Machines Author Margot Shetterly wrote in her book *Hidden Figures*, referring to the engineering work done in the mid-1930s and 1940s, "The work of most of the women, like that of the Friden, Marchant, or Monroe computing machines they used, was anonymous" (Shetterly 2016). In those days, the computers wore skirts! The Friden Model C10 was used between 1935 and 1949 (John Wolff's Web Museum 2024a). The Marchant Model DRB8 was used in the 1920s and 1930s (John Wolff's Web Museum 2024b). The Monroe "Executive" Model L-160X was used beginning in 1929 (Fig. 1.1) (John Wolff's Web Museum 2024c).

Electric Typewriter The first electric typewriter (the Electromatic 01) was produced by IBM in 1935 (Batchelor Business Machines n.d.).

Teletype A teletype (teleprinter) is a communications device that allows operators to send and receive text-based messages using a typewriter-style keyboard and printed paper output. The term "teletype" originated as a trademarked term for a brand of teleprinters created by the Teletype Corporation in 1928. Teletype Corporation's products became so ubiquitous that "teletype" evolved into a generic term synonymous with "teleprinter," especially in the field of computers (Edwards 2021).

1.3 State of Women

Understanding women in engineering in the US leading up to 1952 requires examining the history of engineering education, the attempts to professionalize those working as engineers, and the development of professional societies. Looking across time by major world events, it is also useful in understanding how women progressed in the profession of engineering.

1.3.1 Engineering Education

As the engineering discipline evolved in Europe in the 1700s and 1800s, formal schools were established to teach the needed curriculum. European universities, however, did not admit women. Most European universities had been founded to teach theology, medicine, and law; these professions were all closed to women. Thus women, with very few exceptions, were denied access to a university-level education. The École Nationale des Ponts et Chaussées (National School of Bridges and Roads) was established in 1747 in France as the first formal school of engineering (Ambrose et al. 1997). In the nineteenth century, the practice and focus of engineering was significantly expanded with the development of Newtonian mechanics and the development of the steam engine. These developments led to an even greater need for formal education (Kass-Simon & Farnes 1990).

Male French engineers assisted the colonies during the American Revolution, stayed to construct the new nation, and provided the foundation of engineering faculty for the US Military Academy (USMA) at West Point (Barker 1994). When it was established in 1802 to educate engineers, the USMA did not admit women. Rensselaer Polytechnic Institute, established in 1824 in Troy, New York, now the oldest surviving non-military engineering school in the US, also did not admit women (Ambrose et al. 1997; Turner 1954). Prior to the Civil War, there were only six schools of engineering across the entire US; none admitted women (Turner 1954). The women who called themselves engineers during this time were not formally educated in an engineering degree program.

While the US population was centered in the East, all state universities west of the Mississippi River admitted women from 1871 onward (Turner 1954). But even then, the number of women formally studying engineering in the late 1800s was very few. The first woman to graduate from an engineering degree program in the US was Elizabeth Bragg, who completed a BS in civil engineering at the University of California at Berkeley in 1876 (Kass-Simon & Farnes 1990). Kate Gleason was the first woman to enter Cornell's Sibley College of Engineering in 1884, but did not stay to complete her studies, as she was called back to help the family business (Barker 1994). Elmina Wilson, purported to be the second female engineer by education, graduated in 1892 from Iowa State College with a civil engineering degree and was the first female instructor at that school (Rossiter 1992). In 1893, Bertha Lamme graduated from The Ohio State University with a degree in mechanical engineering with an option in electricity. She was the first woman to graduate with a degree in a field other than civil or architectural engineering (Kass-Simon & Farnes 1990).

When the Society for the Promotion of Engineering Education (later named the American Society for Engineering Education) was formed in 1893, only the three women noted above were recorded as having received engineering degrees in the US. However, as women began to enter the educational system, graduate, and then try to find work as engineers, a backlash developed (Rossiter 1992).

1.3.2 Professionalization

Professionalization of engineering began in the late nineteenth and early twentieth centuries. This term referred to upgrading the membership or image of a profession by excluding or diminishing the influence of persons who could be perceived to be "amateurs." This movement deliberately created barriers between engineers with college degrees and relevant professional experience and those other "engineers" who had learned their jobs by experience and lacked the "professional" credentials. It led to great concerns about the perceived prestige of the organization (Rossiter 1992).

Since most engineering schools did not admit women (and thus women could not get the desired "professional" credentials), the most significant impact of professionalization was to exclude women (Lebold & LeBold 1998). As women were finally able, at least in small numbers, to gain an engineering education and engineering employment, they also endeavored to join the engineering societies. These organizations, however, did not welcome women and developed a strict hierarchy of requirements for each of several levels of membership. They were already in the midst of upgrading themselves and the entrance of women into the profession was not seen as a positive development by most of the men involved in the leadership positions of these organizations (Rossiter 1992).

The lack of standards in engineering education at the time also spurred the emergence of professionalization. Engineering educational standards in the late 1800s and early 1900s were not yet at a level necessary to earn a college education. Consequently, engineers were not invited to serve on national scientific advisory boards, nor were engineers recognized as part of the established scientific community until 1916. And it was not until 1932 that the Engineering Council for Professional Development (ECPD), now known as ABET, was created to provide accreditation of engineering degree programs, in partial response to reports sponsored by the engineering societies (Barker 1994).

1.3.3 Professional Societies

By the end of the nineteenth century, civil and mechanical engineering were firmly established as engineering disciplines, with electrical and chemical engineering following closely behind (Lebold & LeBold 1998). Engineering societies were forming. These types of organizations are deemed the hallmark of a profession. They define intellectual style and norms of conduct and generally act to promote the interests of their members. The early engineering societies placed a high value on free enterprise, individualism, hard work, ambition, and success intertwined with the concepts of prestige, status, and professionalism (Barker 1994).

The "Founder" societies, the five original engineering societies that founded the United Engineering Society in 1904 (which later became the United Engineering

Foundation) included the American Society of Civil Engineers (ASCE), the American Institute of Mining Engineers (AIME) (now called the Society for Mining, Metallurgy and Exploration—SME), the American Society of Mechanical Engineers (ASME), the American Institute of Electrical Engineers (AIEE—a predecessor organization to the Institute of Electrical and Electronics Engineers—IEEE), and the American Institute of Chemical Engineers (AIChE) (AIME 2000a; United Engineering Foundation 2000a, b). A brief look at their history and their lack of admittance of women shows the impact that professionalization and the associated membership requirements had on the recognition for and advancement of women in the engineering profession.

ASCE, America's oldest national engineering society, was founded in 1852. Twelve founders met at the Croton Aqueduct in New York City on November 5, 1852, and agreed to incorporate as the American Society of Civil Engineers and Architects (later the ASCE) (ASCE 2000a, b). Emily Warren Roebling, probably the first woman field engineer, became the first woman to address the ASCE in 1892, when she argued that her husband should not be replaced as the formal director for the construction of the Brooklyn Bridge (Kass-Simon & Farnes 1990). In 1909, Nora Blatch de Forest, a graduate of Cornell University in the top five of her class, was admitted as a "junior member" of the ASCE but was unable to advance any higher. When her junior membership expired in 1916, ASCE refused to promote her to full membership, despite her meeting the stated requirements, and instead, dropped her from the rolls. She brought a lawsuit against the Society but did not prevail (Lebold & LeBold 1998). Elsie Eaves became an associate member of ASCE in 1927 and later the first female member (1957), first female life member, first female Fellow, and the first woman elected to honorary membership (1979) (Kass-Simon & Farnes 1990; SWE Newsletter 1959).

AIME was founded in 1871 by 22 mining engineers in Wilkes-Barre, Pennsylvania (AIME 2000a, b). The first woman member was Ellen Henrietta Swallow Richards, who became the first woman member of any engineering society when she was elected a full member of AIME in 1879. Richards was aided by her MIT degree, her publications in mineral chemistry, and the fact that her husband was a vice president of the organization (Lebold & LeBold 1998). In 1917, the Woman's Auxiliary to the AIME was established and is still active today as the WAAIME (AIME 2000b; Society for Mining, Metallurgy and Exploration n.d.).

By 1880, 85 engineering colleges had been established in the US and most offered a full mechanical engineering curriculum. Thirty engineers met in New York City in February 1880 and decided to form the ASME. In April, a formal organizational meeting was held with 80 engineers at the Stevens Institute of Technology in Hoboken, New Jersey. The first annual meeting of the organization was held in November 1880 (ASME n.d.). The first woman member, Kate Gleason, was admitted to full membership of the ASME in 1914 (Ingels 1952). Lydia Weld, a 1903 naval architecture graduate of MIT, became an associate member of ASME in 1915. She was allowed to become a full member in 1935, when the ASME finally granted full membership status to women (Rossiter 1992).

By 1884, 25 prominent figures in electrical technology signed a "call" to establish an American electrical engineering national society, mindful that civil, mining, and mechanical engineers had already established their own national societies. Twenty-five electrical engineering practitioners met in the headquarters of the ASCE on April 15 to devise an organizational structure for what became the AIEE. The first general meeting was held on May 13, also at ASCE headquarters (Institute of Electrical and Electronics Engineers 2000a). In 1926, Edith Clarke (1954 SWE Achievement Award recipient), who would later become one of the first AIEE female fellows, was the first woman to address the AIEE (Read & Witlieb 1992). As late as 1942, there were only 3 women in the AIEE and more than 17,000 men (Read & Witlieb 1992). The Fellow grade was established in 1912 for engineers who had demonstrated outstanding proficiency and had achieved distinction in their profession. It was not until 1948, however, that the first women were elected AIEE Fellows (Rossiter 1995). These three distinguished women were Edith Clarke (1954 SWE Achievement Award recipient), who significantly contributed to knowledge of and modeling for electric utility systems; Vivien Kellems, the founder of Kellems Company, a manufacturer of cable grips and shell lifters; and Mabel Rockwell (1958 SWE Achievement Award recipient), who significantly contributed to electrical control systems (Goff 1946; Institute of Electrical and Electronics Engineers 2000b; Society of Women Engineers 1999).

AIChE was founded in 1908 at the Engineers' Club in Philadelphia by 19 men. Chemical engineering was just coming into its own and was somewhere between chemistry and mechanical engineering. The founding of AIChE helped to establish chemical engineering as a separate discipline. At the time, the founding members of AIChE believed that about 500 people were practicing chemical engineering across the country (AIChE n.d.). The first female member of AIChE was Margaret Hutchinson Rousseau (1955 SWE Achievement Award recipient). She was the first woman to receive a Sc.D. in chemical engineering from MIT in 1937 and made significant contributions to the field, yet she was not admitted to AIChE until 1945 (Rossiter 1995).

In addition to the Founder Societies, an engineering honor society, Tau Beta Pi, was established. The engineering equivalent of Phi Beta Kappa (which had been founded at the College of William and Mary in 1776), Tau Beta Pi, was established at Lehigh University in 1885. Membership in Tau Beta Pi was limited to men until 1969. "Women's Badges" had been authorized in 1936 as an alternative to membership for women. Only 619 Women's Badges had been awarded by 98 chapters until women were admitted to full membership in 1969 (84 years after the founding of the organization) (Tau Beta Pi 1999a, b; Phi Beta Kappa n.d.).

Additional engineering societies came into existence in the early to mid 1900s. However, these organizations also excluded women from membership or often relegated them to lower membership status. Marie Luhring was elected as an associate member of the American Society of Automotive Engineers in 1920. That same year, Ethel H. Bailey became the first full female member of the organization (Ingels 1952). The American Society of Safety Engineers (in 2025, the American Society of Safety Professionals), founded in 1911 in New York City with 62 members as the

Fig. 1.2 Lillian Moller Gilbreth. (Photograph courtesy of Society of Women Engineers Records, Walter P. Reuther Library, Archives of Labor and Urban Affairs, Wayne State University)

United Society of Casualty Inspectors, admitted its first female member, Vera Burford, in 1946 (American Society of Safety Engineers n.d.; Rossiter 1995). The first time a woman was elected as a junior member of the Society of Naval Architects and Marine Engineers was 1946 (Department of Labor 1948). The first female fellow of the Illuminating Engineering Society, Gertrude Rand, a researcher on the way color perception is affected by illumination and on color blindness, was elected in 1954 (Read & Witlieb 1992; Rossiter 1995).

Lillian Gilbreth (Fig. 1.2), "The First Lady of Engineering" and the co-founder of the field of industrial engineering, was made an honorary member of the Society of Industrial Engineers in 1921 (as a personal favor to her husband, Frank Gilbreth), but not admitted to regular membership (Lebold & LeBold 1998; Rossiter 1995). She was, however, the first woman elected to the National Academy of Engineering (NAE), an event that occurred in 1965, only 1 year after the founding of the NAE in 1964 (Institute of Industrial Engineers n.d.; National Academy of Engineering n.d.).

Once women were admitted to full membership, it would still be decades before these professional societies elected their first woman president. That occurred in 1986–1987, when Nancy Fitzroy (1972 SWE Achievement Award recipient) was elected president of ASME, becoming the first woman to lead any major engineering society. Next was IEEE. Martha Sloan became its first female president in 1993. In 2000, Sheila Widnall (1975 SWE Achievement Award recipient) became the first female president of the American Institute of Aeronautics and Astronautics. In 2003, Patricia Galloway became the first woman president of ASCE as did Diane Dorland for AIChE. Barb Filas served as the first female president of SME in 2005

(AIAA n.d.; American Institute 2024; ASME 2014; IEEE Computer Society n.d.; SME n.d.; University of Arizona 2023;Wastewater Digest 2003).

1.3.4 World War I

During World War I (April 1917–November 1918), women were encouraged to participate in the workforce and support the war effort. This "war to end all wars" has also been called the first engineers' war, as the results of military applications of scientific and technological advances appeared as tanks, airplanes, and submarines (Barker 1994). Because of the shortage of male engineers during World War I, women had the opportunity to work in factories and offices where they were instrumental in keeping the manufacturing industry working (Rossiter 1992). They were drawn into tool design and chemical research, and designed buildings and automobiles (Goff 1946).

However, because the US was involved in the conflict for only 19 months, the manpower emphasis was on the emergency training of mechanics to service the new military technology and not the education of engineers with degrees (Barker 1994). Some women became wartime mechanics through special training; some of those women, because of this introduction, graduated with engineering degrees in the 1920s. Approximately 120 women, some without any college degree, worked during the war in engineering jobs. The end of the war signaled the end of this opportunity. However, prior to 1920, at least 45 women are known to have graduated from college with engineering degrees (Turner 1954). It is ironic that war, an activity that women tend to oppose, increases women's economic progress, as men are drawn out of the civilian workforce (Goff 1946).

Many historians believe that the vote for women's suffrage, which was finally ratified in 1920, was out of gratitude for women's efforts during the war (Lebold & LeBold 1998). However, despite that "gratitude," most women were not allowed to keep the jobs they had filled during the war when it was over. The few women who did remain in engineering after World War I must have been nearly invisible, because according to the *American Men of Science* directory for 1921, there were no women at all in engineering, although the 1920 Census reported 41 women out of a total of 130,000 engineers (Lebold & LeBold 1998; Schneider & Schneider 1993).

Elsie Eaves was a 1920 civil engineering graduate from the University of Colorado who embarked on what would be an illustrious career. Eaves were instrumental in establishing a national organization for women engineers: the Society of Women Engineers and Architects. In 1919, she and several of her colleagues at the University of Colorado wrote to engineering schools across the country, asking for information on women engineering students and graduates. They found 63 women enrolled in engineering at 20 universities—including 43 at the University of Michigan alone! The Michigan women had organized a group of their own, the T-Square Society, in 1914. Most schools, however, did not yet admit women to their engineering programs. One professor responded to Eaves' letter: "I would state that we have not now, have never had, and do not expect to have in the near future, any women students registered in our engineering department." In the end, there were too few women engineers around the country and they were too scattered geographically to keep the organization alive, so it folded (Kindya 1990).

Women engineers made little headway in the 1920s and 1930s. Despite the economic boom in the 1920s and the accompanying optimism and expansion, women in the engineering profession did not seem to benefit from the strong economy. Out of 197 fellowships awarded in engineering during one period of the 1920s, one went to a woman. However, the number of women engineering graduates did grow by 113 from 1920 to 1929, reaching 158. And 53 engineering colleges (an increase from 35 in 1920) now admitted women (Turner 1954).

Those women who were able to graduate with a degree and find employment still faced significant obstacles in advancing in their careers. After graduation, a man would start as a junior engineer and could generally expect, if his performance was satisfactory, to be promoted to senior engineer and then project engineer. Then, he would move on to management, with each step representing increased status, power, and salary. Women engineers, however, were usually limited to desk work, sometimes bordering on clerical work, and did not have access to plant or field work, which denied them the necessary experience to move up the ladder. Many higher level positions required travel to remote locations, behavior not acceptable for "gentle women." Women often needed advanced training—not generally required of men. Although a few women were able to hold positions with professional responsibility, many others operated on the fringes of the profession as writers, editors, secretaries, librarians, industrial teachers, or laboratory assistants—all more "acceptable" positions for women than being an actual engineer (Barker 1994).

1.3.5 Great Depression

Men and women alike were forced to cope when the stock market crashed in September 1929, and the Great Depression followed. Many women had to sacrifice personal ambition and accept a life of economic inactivity (Rossiter 1992). Thus many women engineers either voluntarily or involuntarily ceased their careers after 1929. Jobs in general were scarce, and what jobs were available were unlikely to be given to women.

However, as college enrollments dwindled, colleges began to look for women as students. Twenty-seven more engineering schools began to admit women and the number of female engineering graduates increased by 156 during 1930–1939. With a new emphasis on graduate education for engineers as a result of technology developments during World War I, six schools that excluded women at the undergraduate level allowed them to pursue graduate engineering degrees. Prior to 1942, 18 women had received master's degrees in engineering across the entire US. One woman did receive an engineering PhD in 1920, but no more women earned that capstone degree again until three women did so in the 1930s. However, women still accounted for only about 3 out of every 1000 engineers. And no women were on the engineering faculty at any of the 20 largest doctoral universities in the country in 1938 (Barker 1994; Turner 1954).

By 1938, the percentage of women in engineering represented significantly less than one half of one percent of all engineers. Although it was estimated that about 1000 women engineers and architects were trained in the US, the 1938 edition of *American Men of Science* reported 8 women, representing 0.2% of its approximately 3500 engineers (Lebold & LeBold 1998; Rossiter 1995). The 1940 Census listed 730 women employed as engineers (Table 1.1), less than 0.3% of the total; but many of those are thought not to have the education credentials. And in 1938, there were zero women engineers among the almost 20,000 engineers employed in the Federal classified civil service (Department of Labor 1948).

The highly visible women we know about today were either linked with a male relative who was making a significant impact in the engineering field or were so outstanding on their own that their accomplishments could not be ignored. Female engineering students were still rarities in a man's world and had to deal with both implicit and explicit constraints in their education and in their jobs, constraints which often relegated them to the margins of the profession. In addition, the number of women in engineering was so small that no feminized branches of engineering even had a chance to develop (Barker 1994). Most women were still not admitted to the professional engineering societies in their field of expertise, and a nationwide organization for women engineers had not yet been established.

Table 1.1 Distribution by branch of engineering of men and women employed as professional engineers, 1940

Branch of engineering	Total	Number		Percent		Percent women are of total
		Men	Women	Men	Women	
All employed professional engineers	245,288	244,558	730	100.0	100.0	0.3
Civil engineers	80,362	80,171	191	32.8	26.2	0.2
Mechanical engineers	82,443	82,255	188	33.6	25.8	0.2
Electrical engineers	53,267	53,103	164	21.7	22.5	0.3
Industrial engineers	9283	9209	74	3.8	10.1	0.8
Mining and metallurgical engineers	8813	8739	74	3.6	10.1	0.8
Chemical engineers	11,120	11,081	39	4.5	5.3	0.4

Fig. 1.3 Rosie the Riveter. (Courtesy US Department of Defense)

1.3.6 World War II

World War II, in similar fashion to World War I, presented opportunities for women to assist in the war effort. This movement of women back into the workforce is often personified by famed metalworker and poster woman *Rosie the Riveter* (Fig. 1.3). Although World War I had seen the serious use of technology for airplanes, submarines, tanks, bombs, and other equipment, technological development and organized scientific research in the ensuing 20 years was such that the machines of World War II were more destructive and more powerful. Consequently, they needed engineers and technicians to design, build, and maintain them. Many of the men with the skills to fill these engineering and technical jobs were now in the military (Barker 1994).

Women were encouraged to enter the workforce because the men were gone. Indeed, they began to be invited into the more prestigious positions, not just those invisible jobs discounted as "women's work" (such as silk hose manufacturing) (Barker 1994). People were needed at drawing boards and in engineering shops to keep planes, tanks, and other essential materials rolling off the assembly line—everywhere that the war effort needed support. The problem for the government was that with the men gone, the two reserve labor forces of the country, women and Blacks, were the only ones available to fill these positions. However, there were not enough appropriately trained workers in these labor pools. In engineering, the shortages were so great that women were vigorously recruited during World War II (Bix 1999; Rossiter 1995).

The Office of War Information (OWI) and the War Manpower Commission began issuing literature and propaganda glorifying women as scientists and engineers to bolster the war effort. In 1942, a film, *Women in Defense*, narrated by Katharine Hepburn from a script written by Eleanor Roosevelt, urged women to go to work in government or on scientific projects. In 1942, the movie *Madame Curie,* starring award-winning actress Greer Garson, was released, further glorifying women's contributions in science. New training programs for scientists and engineers were established and, by 1943, women (and to some extent, Blacks) were being specially recruited and trained for jobs in industry. Bright high school students were sought out and urged to major in science in college through the Westinghouse National Science Talent Search (established 1943) and the Bausch and Lomb Science Talent Search (established 1944). Both programs had women among their early winners, and the Westinghouse program required that the percentage of women winners be proportional to female entrants. Books and articles were released during the 1942 through 1945 period urging women to pursue careers in science and engineering. Edna Yost's book *American Women of Science* (1943), which lauded women's past and current contributions to science, painted a bright picture for women in these fields (Rossiter 1995).

To fill the personnel gap, the US government began running crash courses in science, engineering, and management for both men and women after the war began in 1941. Some specific courses were targeted at helping women become engineering aides and engineering cadettes. Many engineering schools set up special engineering training courses for women sponsored by the War Department, the Signal Corps, the Ordinance Department, and the Air Force. Twenty-nine institutions that had heretofore excluded women including the Carnegie Institute of Technology, Columbia University's School of Engineering, and Rensselaer Polytechnic Institute, admitted women engineering students for the first time between 1940 and 1945 to train them in support of the war effort (Barker 1994; Bix 1999; Department of Labor 1948; Rossiter 1992, 1995).

Engineering Aides, Engineering Cadettes and Engineers

In 1942 and 1943, the call went out for more women engineers. Elsie Eaves, of the *Engineering News-Record,* warned women that the word "engineer" was being applied to two very different career paths: individuals with engineering degrees, and those who effectively apprenticed without college coursework. Those few women with degrees in the field could expect to be hired into professional positions. But those non-engineering college female graduates with a few additional courses in drafting and machine testing, and those women without college degrees, were hired into sub-professional jobs as engineering aides or engineering cadettes. Sometimes, they were temporary assistants to men who had been promoted from lower positions within the organization. Nonetheless, there were plenty of jobs for both kinds of "engineers" at this point. Even the US government changed its longtime policy and began to hire women engineers in 1942 (Rossiter 1995).

The demand was so great for the subordinate type of personnel (those classified as aides or cadettes) that a special program was set up by the federal government as early as October 1940 in anticipation of entering World War II. The US Office of Education administered this defense training program that was funded by a special appropriation. It was one of the first federal government efforts to increase and train scientific manpower. Manpower shortages were expected to be severe if the US entered the war—the aircraft industry on Long Island and in New Jersey alone would require about half of all the engineering graduates nationwide. The first Emergency Defense Training (EDT) program course was offered in December 1940, and by June 1941, more than 100,000 people (almost all men) had received training in engineering subjects (Barker 1994; Rossiter 1995).

At first, the Engineering, Science, and Management War Training program (ESMWT) offered training only at 4-year technical schools. Colleges were responsible for determining local training needs and developing courses to meet those needs, but not to provide a complete engineering education nor necessarily any academic credit for the courses (Barker 1994). As the war continued and personnel shortages intensified, the program and the corresponding training was extended to a total of 227 colleges and universities, including several women's colleges, where courses were taught in elementary engineering, mathematics, chemistry, physics, and safety engineering. The ESMWT opened doors of opportunity to many women and some minorities (Rossiter 1995). In a report issued in December 1942, the ESMWT reported classes were offered at Clarkson, Cornell, the University of Rochester, Syracuse University, Rensselaer Polytechnic Institute, and Union College. Union College had, in fact, given courses to 145 women out of 817 students, including a special course in electricity and mechanics for "girl high school graduates employed on testing work in the Engineering Department of the General Electric Company" (Barker 1994).

In 1943 and 1944, 21.8% of the enrollees were women and minorities. By 1945, more than 1.8 million people had been trained, including 280,000 women (15.7%) and 25,000 Blacks (men and women, 1.4%) (Rossiter 1995). About one-fifth of these women were enrolled in engineering drawing, while significant numbers were learning aeronautical engineering, inspection and testing, mapping and surveying, and engineering fundamentals; the rest were registered for other scientific and management courses (Barker 1994).

Other departments of the federal government conducted their own programs to recruit and train women for technical positions. The Office of the Chief of Ordnance trained women with high school mathematics for civil service appointments as junior engineering aides during a 3-month intensive course at the University of Michigan. Female engineering aides for the Frankford and Picatinny Arsenals were trained at similar programs sponsored at Rutgers University, Drexel Institute, Temple University, and the University of Pennsylvania. The Signal Corps and the Army Air Forces provided technical training to hundreds of women who were often then assigned clerical duties, which resulted in a high turnover rate (Barker 1994).

Some companies had other training requirements or couldn't wait for the results of the government programs to bear fruit. General Electric recruited women with degrees in math or physics and then gave them on-the-job training so that they could handle computations in GE's machine-tool department. However, so few women existed with these types of credentials that the companies began reaching down further, pursuing women still in college (Bix 1999).

Aircraft companies started special programs for engineering aides and engineering cadettes. The Goodyear Aircraft Corporation developed a 6-month program in aeronautical engineering at the University of Cincinnati to prepare women as "junior engineers" (Barker 1994).

The Vega Aircraft Corporation needed aircraft design engineers starting in 1941. In cooperation with Lockheed Aircraft Corporation and the California Institute of Technology (Caltech), their engineering shortage was temporarily solved by hiring engineers trained in other disciplines—civil, mechanical, or electrical—from other industries and training them for aircraft work with 8 weeks of college-level instruction at Caltech, followed by 8 weeks of an on-the-job apprenticeship. Other technical and clerical employees with some experience took evening classes specially arranged with the University of California in aircraft engineering-related subjects. But soon, all available candidates for these programs had been trained. Next, Vega established a full-time training program at the University of California at Los Angeles (UCLA) to teach drafting for women. These women were not already employees but were guaranteed jobs if they finished the course. By 1943, Vega established a full-time 52-week program at CalTech for 2 groups of 20 full-time employees. The 40 students, all women, ranged in age from 18 to 49 years old, and already worked in the engineering department at Vega. They were able to take a full college course in engineering, stripped of all non-essentials, and earn the chance to be upgraded to higher level jobs (Barker 1994).

In 1942, the Curtiss-Wright airplane company developed a plan for training CW Cadettes, young women with at least 2 years of college education, including 1 year of college-level mathematics. Seven colleges, some of which had never previously enrolled women—Cornell, Penn State, Purdue, Minnesota, Texas, Rensselaer Polytechnic Institute, and Iowa State—gave over 600 students a 10-month crash course of engineering, math, job terminology, aircraft drawing, engineering mechanics, airplane materials, theory of flight, and aircraft production. After their training, these cadettes were assigned to airplane plants to work in research, testing, and production (Barker 1994; Bix 1999; Rossiter 1992).

The Engineering Cadette Program in the School of Electrical Engineering at Purdue University sponsored by Radio Corporation of America (RCA) helped convince skeptical faculty that women could excel technically and academically. Cadettes accounted for 20% of the engineering staff in some RCA plants, allowing women to move beyond being tokens (Barker 1994).

By 1945, the number of women engineers listed on the National Roster had increased from 144 in 1941 to 395 (an amazing increase of almost 300% in 4 years), mostly at the bachelor's level. And many more engineering colleges and universities now at least had seen women as students (Rossiter 1995). The number of women enrolled as engineering undergraduates had increased to 1800 by 1945 (Barker 1994).

Faculty

Women also were encouraged to become faculty members at colleges and universities, temporarily replacing men who had been called into government service or into the military. In fact, a significant number of female scientists, including mathematicians and chemists, who were not working on other war projects, moved into the ranks of college and university faculty. In 1942, 2412 women scientists (including 50 women in engineering out of a total of 5394 engineering teachers) were on science faculties. By 1946, the number had increased to 7746 (a more than 200% increase in 4 years) and the number of women engineering faculty, including some women who taught at all-male institutions, had increased to 53 (Barker 1994; Rossiter 1995).

These women cadettes, engineering aides, engineers in industry, and faculty all made significant contributions to the war effort, and to the advancement and acceptance of women in technical fields. All in all, about 45,000 women were trained for engineering jobs during World War II (Barker 1994).

1.3.7 After World War II

When the war ended, millions of veterans—primarily men—came home expecting to be gainfully employed. Many of the women who had been employed found themselves no longer welcome in the workforce. In fact, these women were now expected to go home and attend to their families or "go back to the kitchen" and not "steal" work from returning GIs. The various special educational programs discussed in the previous section were shut down. And because many engineering schools still did not admit women, educational opportunities for women who wanted to pursue an engineering career were once again limited. By January 1946, four million women had left the labor force (Barker 1994; Rossiter 1992).

By the time World War II ended, however, women were not quite as eager to accept the inevitable as they were when World War I ended. They began generating statistics and scientific data to prove they were just as capable as men and were being treated unfairly. Still, even for those who managed to hold on to their positions, they were unable to advance as men did. Women were still channeled into less challenging work, kept in lower ranks, and paid lower salaries by employers who minimized their contributions (Rossiter 1995).

A very significant impact of the war's end was the GI Bill, which provided funds for veterans' education. Schools that had been underutilized during the war were suddenly faced with a deluge of male students. It is estimated that 7.8 million veterans chose to take advantage of the GI Bill's educational provisions. Female enrollment at some coeducational schools had reached 50% or higher during the war, but as the demand for slots increased, many institutions introduced or reintroduced maximum quotas on female enrollment. Some colleges refused to accept applications from out-of-state women. Women were told that space was not available and

Table 1.2 Women earning undergraduate engineering degrees 1948–1953

Year	Number of degrees
1947–1948	191
1949–1950	171
1951–1952	60
1952–1953	37

that they would have to attend other institutions. For a few years, women who were already in the educational pipeline, women veterans, and wartime women engineering aides getting engineering undergraduate degrees, contributed to a brief expansion in the number of women getting engineering degrees. Through the 1950s the percentage of women in colleges dropped to just 25–35% of the total college undergraduate enrollment. The number of women graduating with engineering degrees declined as well (Table 1.2) (Rossiter 1995).

Discrimination against women was widespread; women were systematically pushed out of science and engineering at the undergraduate level, at the graduate level, and as faculty. Men even took over women's dormitories and many universities added temporary housing to accommodate the returning male GIs. Male veterans replaced female staff and faculty as well as female engineering students. There were no discrimination laws in place (these would not be enacted until the 1960s or later) to prevent any of this behavior. The nation was also extremely grateful to its veterans and felt there was, in many cases, a duty to make their adjustment to civilian life as easy as possible. Objections or protests were few and far between and not effective. In addition, society as a whole still seemed to feel that the proper role for women revolved around the home and family. These ambitious women in engineering, who were certainly atypical, threatened presumed male and female spheres (Rossiter 1995).

The *Outlook for Women in Architecture and Engineering*, a Bulletin of the Women's Bureau of the US Department of Labor published in 1948, painted a bleak picture for women in engineering:

> Advancement for women in engineering is conceded to be difficult. They seldom follow the usual line from junior engineer to senior engineer to project engineer, nor are they often transferred to non-engineering work in sales, purchasing or administration. Usually limited by custom to office work, as compared with field or plant work, women engineers rarely find opportunities to obtain the rounded experience necessary for normal progression. The fact that many of the positions representing advancement often require field work or travel to remote locations further reduces their chances. However, a few women have broken through these bounds (Department of Labor 1948).

Women engineers, who had been welcomed with open arms during the war years, were now told to abandon the idea of an engineering career. The Women's Bureau reported that there were 950 women in engineering as of 1946–1947. These women constituted only 0.3% of the total 317,000 engineers in the US. Of course, this was much higher than the eight women reported to be practicing engineering in the 1949 edition of *American Men of Science,* which despite its title, reported on all individuals practicing in the sciences. The US Census of 1950 counted 6475 women

Table 1.3 Membership in the Founder Societies (Department of Labor 1948)

Organization	Total membership	Number of women members
American Institute of Electrical Engineers	24,526	14
American Society of Civil Engineers	21,100	23
American Society of Mechanical Engineers	20,060	33
American Institute of Mining and Metallurgical Engineers	12,600	26
American Institute of Chemical Engineers	5788	5

employed as engineers, of whom only 41% had 4 years or more of college, and 17% had not graduated from high school. Probably about 3600 of these women were truly engineering professionals. There were now an adequate number of women in engineering to finally constitute enough of a critical mass to move the cause of women in engineering forward, both to seek more women as engineers and to improve career opportunities for those already in the field (Barker 1994; Department of Labor 1954; Kindya 1990; Rossiter 1995).

In fact, women seeking careers or advancement in their existing careers and students studying engineering started to band together for mutual support and job-hunting help. In the late 1940s, the war-support effort and the associated defense industry had made cutbacks across the board. Employment for many engineering disciplines was at a low. Many professional societies still excluded women, either outright or by refusing them full membership, and thus had very few women among their membership (Table 1.3).

But, this time, efforts for mutual support among women engineers were enabled by a nationwide infrastructure that had not existed in 1919. Now there was the telephone, the automobile, the superhighway, and the railroad (Kindya 1990). Female engineers, fueled by isolation and discrimination, and the new national infrastructure, founded the Society of Women Engineers in 1949–1950 (Rossiter 1992).

1.4 State of SWE

SWE traces its founding back to a group of female students at Drexel University in Philadelphia and a gathering of Cooper Union and City College of New York graduates and students in New York City. The Drexel students had begun meeting in 1948, calling themselves the "Philadelphia District of the Society of Women Engineers." The New York Group had been meeting informally for several years and had even had meetings with the Drexel students. In November 1949, the New York women officially started the New York Organization with a designated treasurer and bylaws chair. Within a few months, the New York Organization would elect a president as well. Women from Boston and Washington, DC also considered themselves part of the "National Organization" (Hayes 2000; Kindya 1990; Rossiter 1995; Society of Women Engineers 2015a).

The Philadelphia District's newsletter announced in November 1949 that the National Organization would hold a convention in the spring or fall of 1950. The meeting at Green Engineering Camp of The Cooper Union was held May 25, 1950 and more than 60 women attended from the 4 districts of Philadelphia (including the Drexel students), New York, Boston, and Washington, DC. The women who attended the Camp Green meeting represented two major constituencies: (1) the pioneers—those who had been trained and were working as engineers prior to or during World War II; and (2), the engineering aides who had returned to college after the war to get their engineering degrees. Many topics were discussed and decided at this meeting, including whether the organization would be composed of just graduate engineers or would also include student members. That meeting is generally referred to as the official founding of SWE. The first president, Beatrice Hicks (1963 SWE Achievement Award recipient), was elected, membership requirements were established, and dues were set. The initial board of directors consisted of five officers, three directors-at-large and one director from each section. A nominating committee was established (Hayes 2000; Kindya 1990; Rossiter 1995; Society of Women Engineers 2015a).

During the rest of 1950, the organizational structure of SWE was refined. Nominating, membership and finance, and convention committees were established. The logo developed by the Drexel students was refined by the emblem committee, and a publications committee was established that included Elsie Eaves on its editorial board. SWE would put together an official publication, it would be called *Journal of the Society of Women Engineers* and released quarterly. It was also decided that SWE would not participate in controversial political matters and that non-profit status would be sought (Hayes 2000).

The first issue of the *Journal* was published in the spring of 1951. That issue reported that Lillian Moller Gilbreth, who was a staunch supporter of SWE, had accepted honorary membership. Although technical articles were published in the first few issues, by mid-1952, the *Journal* turned its focus to reports of section activities, publicity received by the sections, notes from members, and listings of new members. Not until the 1980s would SWE's publication (renamed *US Woman Engineer* by then) feature technical articles again (Hayes 2000).

In 1951, the SWE membership approved the organization's bylaws, and the Pittsburgh Section was chartered. In addition, a convention was held in New York City. SWE was incorporated in February of 1952 and registered as a corporation in the District of Columbia (Hayes 2000). As contained in the Articles of Incorporation, the objectives of the organization were (Hayes 2000; Kindya 1990):

- To inform the public of the availability of qualified women for engineering positions; to foster a favorable attitude in industry towards women engineers; and to contribute to their professional advancement.
- To encourage young women with suitable aptitudes and interest to enter the engineering profession, and to guide them in their educational programs.
- To encourage membership in other technical and professional engineering societies, participation in their activities, and adherence to their code of ethics.

Ninety-two women joined the fledgling organization, cognizant that for their ranks to grow, they needed to support and encourage one another (Hayes 2000).

By 1952, SWE had established the Achievement Award. This award, the highest tribute given by SWE, honors a woman engineer who has made outstanding contributions in a field of engineering over a significant period of time. The inaugural recipient was Mária Telkes, for her many innovations that utilized solar energy, as evidenced by more than 20 patents, many for solar-powered devices (Society of Women Engineers 2015b). Finally, there was an organization willing to accept women engineers, recognize them for their accomplishments, and encourage other women to be engineers.

Significantly, in 1952, SWE was invited to participate in the "Centennial of Engineering" to celebrate 100 years of the ASCE. This lent credibility to the organization since all the major professional societies would be represented. The Chicago and Detroit Sections were chartered in 1952 and the organization became truly national in 1953 when the Los Angeles Section was chartered (Hayes 2000).

Today, the Society of Women Engineers has been in existence for 75 years. Women engineers still have not achieved gender equity, but some progress has been made. The Achievement Award continues to be bestowed annually to a woman engineer with significant lifetime contributions. As part of the organization's 75-year celebration, this book profiles and celebrates the first 25 recipients of that illustrious award and their enduring legacy to engineering and humankind (Table 1.4).

Table 1.4 Legacy of the first 25 SWE Achievement Award recipients

Year	Milestone	Recipient
1919	First woman to receive a Master's in Electrical Engineering from Massachusetts Institute of Technology (MIT)	Edith Clarke
1922	First professionally employed woman electrical engineer in the US	Edith Clarke
1926	First woman to address what is today the Institute of Electrical and Electronics Engineers	Edith Clarke
1926	First woman to have a paper appear in the AIEE Transactions	Edith Clarke
1927	First woman in Canada to obtain a BS in Electrical Engineering [University of Toronto]	Elsie MacGill
1929	First woman to receive a Master's degree in Aeronautical Engineering in North America, and possibly the world [University of Michigan, Ann Arbor]	Elsie MacGill
1937	First woman to earn a Doctorate of Science degree in chemical engineering from MIT	Margaret Hutchinson Rousseau
1938	First woman member of the Engineering Institute of Canada [transitioning from Fairchild to Canadian Car]	Elsie MacGill
1938	Likely the first woman to carry the title "Chief Aeronautical Engineer" [Canadian Car and Foundry]	Elsie MacGill
1938	First woman to design and manufacture an aircraft from the ground up [Canadian Car and Foundry]	Elsie MacGill

(continued)

Table 1.4 (continued)

Year	Milestone	Recipient
1940	First woman to graduate from Rice University with a PhD in physics	Marguerite Rogers
1942	First woman engineer to be hired by Western Electric	Beatrice Hicks
1943	First woman to run her own aeronautical engineering design consulting firm [Toronto]	Elsie MacGill
1945	First woman 'active' member of the American Institute of Chemical Engineers	Margaret Hutchinson Rousseau
1945	First woman faculty member of the chemistry department at the University of Michigan	Isabella Karle
1945	First woman Director of Research for the Midwest Rubber Reclaiming Company; the first woman to hold such a position in the rubber industry	Désirée Le Beau
1946	First woman engineering graduate at the University of Illinois	Barbara Crawford Johnson
1946	First scientist to discover a new isotope of calcium	Dorothy Martin Simon
1947	First woman to chair a United Nations committee [The Stress Analysis Committee of the International Civil Aviation Organization]	Elsie MacGill
1947	First full-time woman professor of electrical engineering in the US	Edith Clarke
1948	One of the first three women Fellows of the Institute of Electrical and Electronics Engineers	Edith Clarke
1948	One of the first three women Fellows of the Institute of Electrical and Electronics Engineers	Mabel MacFerran Rockwell
1948	Awarded first degree in Biophysics (MS) [Cornell University]	Alice Stoll
1949	First woman to graduate from Rensselaer Polytechnic Institute with a bachelor's degree in chemical engineering	Nancy D. Fitzroy
1949	Likely first woman Professional Engineer in San Bernardino County, CA	Rebecca Sparling
1950	Co-founder and first President of the Society of Women Engineers	Beatrice Hicks
1950	One of the ninety-two charter members of the Society of Women Engineers	Laurence Delisle Pellier
1951	First woman manager, Director of Research and Development at Aerovox	Elise Harmon
1952	Inaugural recipient of the Society of Women Engineers Achievement Award	Mária Telkes
1956	First woman to ride the centrifuge [US Naval Air Development Center]	Alice Stoll
1961	First woman engineer appointed by the US Secretary of Defense to the Defense Advisory Committee on Women in the Services	Beatrice Hicks
1961	First woman engineering faculty member at the University of Washington	Irene Peden

(continued)

Table 1.4 (continued)

Year	Milestone	Recipient
1961–1967	First woman appointed to the Los Angeles Board of Airport Commissioners	Laurel Van der Wal
1962	First woman to receive a PhD in engineering from Stanford University	Irene Peden
1964	The first undergraduate alumna appointed to the MIT faculty	Sheila Widnall
1964	The first woman appointed to the MIT School of Engineering faculty	Sheila Widnall
1964	The first and only woman to receive the Ford postdoctoral fellowship when appointed to the MIT faculty	Sheila Widnall
1964	First woman corporate officer at AVCO as Vice President over six divisions	Dorothy Martin Simon
1965	First woman to receive an honorary doctorate (DEng) from Rensselaer Polytechnic Institute	Beatrice Hicks
1968	First woman engineering professor at Stanford University	Irmgard Flügge-Lotz
1969	First computer science "Man of the Year" award from the Data Processing Management Association	Admiral Grace Murray Hopper
1969	First woman to be elected President of any section of the Instrument Society of America	Ada Pressman
1970	First American woman engineer or scientist to live and work in Antarctica	Irene Peden
1970	First woman Fellow of the American Institute of Aeronautics and Astronautics	Irmgard Flügge-Lotz
1971	First woman to be invited by the American Institute of Aeronautics and Astronautics to speak as the 1971 von Kármán lecturer	Irmgard Flügge-Lotz
1973	First woman and first person from the US to be named a Distinguished Fellow of the British Computer Society	Admiral Grace Murray Hopper
1973–1985	First woman trustee at Worcester Polytechnic Institute, where she set up the Dorothy M. Simon Endowed Fund for Fire Safety Studies	Dorothy Martin Simon
1974	First woman to be Chief Control Systems Engineer for Bechtel Corporation	Ada Pressman
1975	First person to be registered in California as a Control System Engineer	Ada Pressman
1975	First Navy nominee to win the Federal Women's Award	Marguerite Rogers
1975	First woman elected to the board of directors of the Institute of Electrical and Electronics Engineers	Irene Peden
1975	First woman honored with University of Illinois Distinguished Alumni	Barbara Crawford Johnson
1978	First woman named to the advisory board for the US Merchant Marine Academy	Irene Peden
1979	First woman to be Engineering Manager for Bechtel Corporation	Ada Pressman
1979	First woman chair of the faculty at MIT	Sheila Widnall

(continued)

Table 1.4 (continued)

Year	Milestone	Recipient
1983	First woman recipient of the prestigious Founders Award of the American Institute of Chemical Engineers	Margaret Hutchinson Rousseau
1983	First "three-pin" SWE President—Fellow, Past President, Achievement Award	Ada Pressman
1986	Upon retirement, the oldest active duty commissioned US Navy officer in the entire fleet and in the history of the Navy	Admiral Grace Murray Hopper
1986	First woman to be national president of the American Society of Mechanical Engineers	Nancy D. Fitzroy
1988	First American woman to receive an Honorary Fellow of the Institution of Mechanical Engineers of the United Kingdom, ASME's sister organization in the U.K.	Nancy D. Fitzroy
1991	First individual woman to be awarded the National Medal of Technology	Admiral Grace Murray Hopper
1993	First woman to receive the Bower Award and Prize for Achievement in Science from the Franklin Institute	Isabella Karle
1993–1997	First woman appointed Secretary of the Air Force and first woman to serve as the head of a US military branch	Sheila Widnall
1996	USS Destroyer Hopper was the second destroyer named for a woman who had served in the US Navy	Admiral Grace Murray Hopper
1997	First woman to receive the Thomas Alva Edison Medal of the Institute of Electrical and Electronics Engineers	Esther M. Conwell
2000	First woman to serve as President of the American Institute of Aeronautics and Astronautics	Sheila Widnall
2011	ASME created the 'Nancy DeLoye Fitzroy and Roland V. Fitzroy Medal', recognizing pioneering contributions to the frontiers of engineering that have led to a breakthrough in existing technology	Nancy D. Fitzroy
2014	First woman recipient of the Davies Medal for Engineering Achievement from Rensselaer Polytechnic Institute School of Engineering	Nancy D. Fitzroy
2017	AIChE created the "Margaret Hutchinson Rousseau Pioneer Award for Lifetime achievement by a Woman Chemical Engineer"	Margaret Hutchinson Rousseau
2019	First woman to receive the Guggenheim Medal in its 90-year history	Sheila Widnall
2021	First woman honored on the Philadelphia Women's Walk of Fame	Admiral Grace Murray Hopper

1.5 State of Culture and Laws

In the 1800s, families lived primarily on farms or in small towns, and women worked at home caring for the home, cooking, sewing clothes, and raising children. Between 1800 and 1900 the US workforce switched from 75% agricultural to two-thirds industrial and service (Wells 2018). During the Industrial Revolution (1876–1900), women did work in factories; their wages were lower than men's wages, and their working conditions were generally poor (Library of Congress 2025). It wasn't until 1839 in Mississippi when states started to give married women some control over property and earnings (Howe 2005). In 1848 at the Seneca Falls Convention, the first demand for equal rights for women was formalized by developing the Declaration of Sentiments which mirrored the Declaration of Independence (Schneider 2023).

The era of progressive reform (1890s–1920s) saw women increasingly active in social reforms including child labor, education, and women's health. Settlement houses like Hull House in Chicago (founded by Jane Addams) provided education, health care, and social services to immigrants and the working class. Women's clubs provided platforms for women to engage in politics and community activism by advocating for public health, education, and social services (Lange 2015).

Until the mid-twentieth century, the access to contraceptives was banned, and abortion was very restricted, being illegal in most states. Margaret Sanger founded the American Birth Control League (which later became Planned Parenthood) in 1921. Sanger fought for access to contraception and challenged laws like the Comstock Act which banned the distribution of contraceptives and information about birth control (Michals 2017).

By 1920, more people were living in cities than on farms and in small towns. Families transitioned from the extended family on the farm to the nuclear family (father, mother, and children) in the city. Americans also began to have longer life expectancies due to better health care and sanitation along with the development of successful public health programs (Social Security Administration 2025).

The Roaring Twenties (1920s) brought the "flapper" as a symbol of women's liberation. Flappers were young women who wore shorter dresses, bobbed their hair, smoked, drank, and rejected the Victorian ideals of femininity (Pruitt 2021). The Great Depression (1929–1939) put a damper on all life. World War I (1914–1918) and especially World War II (1939–1945) were pivotal in shifting gender roles. Millions of men went off to war and were replaced by women who worked in factories, served as nurses, and contributed to the war effort in many ways. Working outside the home and earning wages was a turning point in women's roles in society (McDermott 2023).

Some of the laws and activities to improve the state of work for all Americans are given in Table 1.5.

Table 1.5 Culture and legal highlights: pre-1952

Date	Event
1913	**National Safety Council** founded by industry to pool resources about on-the-job safety. Companies were focused on trying to make the workplace safer. Safety glasses and hard hats are just two examples of required safety equipment that resulted from this focus (Aldrich 2001)
1910–1921	**Workmen's Compensation laws** adopted by most states. Wisconsin was first in 1911. By 1920, 44 states had Workmen's Compensation laws. These laws provided payment for medical expenses and time off due to accidents at work. Prior to this, workers had to sue the company and win the lawsuit in order to have their medical expenses and time off work paid (Aldrich 2001) The Federal Employees' Compensation Act was passed in 1916 to cover government employees (Guyton 1999; Nordlund 1991)
1920	**19th Amendment** granting women the right to vote was a monumental legal victory for gender equality. However, it did not bring true equality, especially for women of color due to poll taxes, literacy tests, and other discriminatory practices that kept them from voting for many years after the amendment's passage (National Archives 2025a)
1935	**Social Security Act** created a social insurance program to pay workers aged 65 or over a continuing income after retirement based on workers contributing to their own future retirement benefit by making regular payments into a fund (Michals 2017)
1935	**National Labor Relations Act** guaranteed employees the right to join labor unions and bargain with employers through those unions. By the end of the 1930s, more than 800,000 women belonged to unions. Provisions of the NLRA were later expanded under the Taft-Hartley Labor Act of 1947 (National Archives 2025b)
1938	**Fair Labor Standards Act** included provisions that established a minimum wage ($0.25/h initially), restricted the work week to 44 h without paying overtime, and prevented children younger than 16 from entering the workforce (Grossman 1978; Payne-Patterson & Maye 2023)

Author's Note Portions were adapted from the first three chapters of *Engineering Women: Re-visioning Women's Scientific Achievements and Impacts* by Jill S. Tietjen, P.E., the first volume in the Springer Women in Engineering and Science series.

References

AFT Fluorotec. 2016. The History of PTFE. https://www.fluorotec.com/news/blog/the-history-of-ptfe/. September 27. Accessed February 2, 2025.

Aldrich, Mark. 2001. History of Workplace Safety in the United States, 1880-1970. EH.Net Encyclopedia, edited by Robert Whaples. August 14. URL https://eh.net/encyclopedia/history-of-workplace-safety-in-the-united-states-1880-1970/ Accessed February 9, 2025.

Ambrose S, Dunkle K, Lazarus B, Nair I, Harkus D. 1997. Journeys of Women in Science and Engineering: No Universal Constants. Philadelphia, Pennsylvania: Temple University Press.

American Fuel and Petrochemical Manufacturers. 2020. The refining and petrochemical industries: 170 years of innovation. https://www.afpm.org/newsroom/blog/refining-and-petrochemical-industries-170-years-innovation. October 5. Accessed February 2, 2025.
American Institute of Aeronautics and Astronautics (AIAA). n.d. AIAA Presidents. https://aiaa.org/wp-content/uploads/2024/12/aiaa-presidents-history.pdf. Accessed April 19, 2025.
American Institute of Chemical Engineers (AIChE). n.d. A brief history of AIChE. www.aiche.org/welcome/history.htm. Accessed August 20, 1999.
American Institute of Chemical Engineers (AIChE). 2024. Institute Milestones. https://www.aiche.org/about/institute-milestones. Accessed March 31, 2024.
American Institute of Mining Engineers (AIME). 2000a. Historical Highlights. www.idis.com/aime/history.htm. Accessed December 15, 2000.
American Institute of Mining Engineers (AIME). 2000b. WAAIME. www.idis.com/aime/WAAIME.HTM. Accessed December 19, 2000.
American Physical Society. This Month in Physics History. 1988: Tesla Patents "Electric Transmission of Power". May. https://www.aps.org/archives/publications/apsnews/200305/history.cfm. Accessed January 14, 2025.
American Society of Civil Engineers (ASCE). 2000a. ASCE Profile. www.asce.org/aboutasce/profile.html. Accessed December 15, 2000.
American Society of Civil Engineers (ASCE). 2000b. 150 Years of Civil Engineering. www.asce.org/150/1506years.html. Accessed December 15, 2000.
American Society of Mechanical Engineers (ASME). n.d. The History of ASME International. www.asme.org/history/asmehist.html. Accessed December 15, 2000.
American Society of Mechanical Engineers (ASME). 2014. ASME's First Female President Honored by Alma Mater. https://www.asme.org/topics-resources/society-news/asme-news/asmes-first-female-president-honored-alma-mater. May 29.
American Society of Safety Engineers. n.d. Society History. www.asse.org/hsoci.htm. Accessed December 19, 2000.
Barker, Anne M. 1994. Women in Engineering During World War II: A Taste of Victory. November 21. Unpublished. Rochester Institute of Technology.
Batchelor Business Machines. n.d. History of the IBM Typewriter: IBM Typewriters Milestones. https://ibmtypewriters.com/history-of-the-ibm-typewriter-categories/ibm-typewriters-milestones.html. Accessed November 18, 2024.
Bellis, Mary. 2019. The History of Vacuum Tubes and Their Uses. https://www.thoughtco.com/the-history-of-vacuum-tubes-1992595. February 3. Accessed January 24, 2025.
Bix, Amy Sue. 1999. 'Engineeresses' 'Invade' Campus: Four Decades of Debate over Technical Coeducation. Proceedings of the 1999 International Symposium on Technology and Society—Women and Technology: Historical, Societal, and Professional Perspectives, 195-201, New Brunswick, NJ. July 29-31.
Brandywine CAD. 2022. The Evolution of Drafting. https://www.bcad.com/the-evolution-of-drafting/. August 16. Accessed November 18, 2024.
Carrier. n.d. Willis Carrier - The Inventor of Modern Air-Conditioning. https://www.carrier.com/commercial/en/us/about/willis-carrier/. Accessed January 24, 2025.
Conover, Emily. 2015. 1945: First Patent for the Microwave. APS - APS News - This Month in Physics History. https://www.aps.org/apsnews/2015/10/first-patent-for-microwave. October 1. Accessed February 2, 2025.
Cromer, Orville C., Foster, Christopher G., Purdy, Ken W. 2025. automobile. Encyclopedia Britannica. https://www.britannica.com/technology/automobile. January 31. Accessed February 2, 2025.
Crouch, Tom D., Boyne, Walter James and Bilstein, Roger E. 2024. history of flight. Encyclopedia Britannica. https://www.britannica.com/technology/history-of-flight, October 30. Accessed February 1, 2025.
Department of Labor. 1948. The Outlook for Women in Architecture and Engineering. Bulletin of the Women's Bureau No. 223-5., Washington, DC: US Government Printing Office.

Department of Labor. 1954. Employment Opportunities for Women in Professional Engineering. Women's Bureau Bulletin No. 254. Washington, DC: US Government Printing Office.

Edwards, Benj. 2021. What Are Teletypes, and Why Were They Used with Computers? How to Geek. ttps://www.howtogeek.com/727213/what-are-teletypes-and-why-were-they-used-with-computers/. May 17. Accessed November 18, 2024.

Effectrode Thermionic. n.d. Vacuum Tubes and Transistors Compared, Effectrode Thermionic. https://www.effectrode.com/knowledge-base/vacuum-tubes-and-transistors-compared/. Accessed February 16, 2025.

Elon University.2005. 1870s-1940s: Telephone. https://www.elon.edu/u/imagining/time-capsule/150-years/back-1870-1940/. Accessed January 25, 2025.

Goff, Alice C. 1946. Women Can Be Engineers. Ann Arbor, Michigan: Edwards Brothers, Inc.

Grossman, Jonathan. 1978. Fair Labor Standards Act of 1938: Maximum Struggle for a Minimum Wage. US Department of Labor. https://www.dol.gov/general/aboutdol/history/flsa1938. Accessed February 9, 2025.

Guyton, Gregory P. 1999. A Brief History of Workers' Compensation. The Iowa Orthopaedic Journal. National Library of Medicine. https://pmc.ncbi.nlm.nih.gov/articles/PMC1888620/. Accessed February 17, 2025.

Hayes, Dianne Williams, 2000. Five Decades of The Society of Women Engineers. Society of Women Engineers.

Howe, LeAnne. 2005. Betsy Love and the Mississippi Married Women's Property Act of 1839. Mississippi History Now. https://mshistorynow.mdah.ms.gov/issue/betsy-love-and-the-mississippi-married-womens-property-act-of-1839. June. Accessed February 17, 2025.

Hutching, Matthew I., Truman, Andrew W., Wilkinson, Barrie. 2019. Antibiotics: past, present and future. Current Opinion in Microbiology. Volume 51, Pages 72-80. https://doi.org/10.1016/j.mib.2019.10.008. https://www.sciencedirect.com/science/article/pii/S1369527419300190. Accessed January 24, 2025.

IEEE Computer Society. n.d. Martha Sloan. IEEE Computer Society. https://www.computer.org/profiles/martha-sloan. Accessed March 31, 2024.

IEEE Pulse. 2019. The Slide Rule, https://www.embs.org/pulse/articles/the-slide-rule/. May 28. Accessed November 18, 2024.

Ingels, Margaret. 1952. Petticoats and Slide Rules. Western Society of Engineers. September 4.

Institute of Electrical and Electronics Engineers. 2000a. The Origins of IEEE. www.ieee.org/organizations/...ical_articles/history_of_ieee.html. Accessed December 15, 2000. (In 2025, https://www.ieee.org/about/ieee-history.html)

Institute of Electrical and Electronics Engineers. 2000b. Welcome to the IEEE Fellow Program: Our History. www.ieee.org/about/awards/fellows/fellows.htm. accessed December 19, 2000.

Institute of Industrial Engineers. n.d. www.iienet.org/historg.htm. Accessed August 20, 1999.

John Wolff's Web Museum. 2024a. The Friden Calculating Machine Company. https://www.johnwolff.id.au/calculators/Friden/Friden.htm. Accessed November 18, 2024.

John Wolff's Web Museum. 2024b. The Marchant Calculating Machine Company. http://www.johnwolff.id.au/calculators/Marchant/Marchant.htm. Accessed November 18, 2024.

John Wolff's Web Museum. 2024c. The Monroe Calculating Machine Company. https://www.johnwolff.id.au/calculators/Monroe/Monroe.htm. Accessed November 18, 2024.

Kass-Simon, G. and Farnes, Patricia, Editors. 1990. Women of Science: Righting the Record. Bloomington, Indiana: Indiana University Press.

Kindya, Marta Navia. 1990. Four Decades of the Society of Women Engineers. New York, New York: Society of Women Engineers.

Lange, Allison. 2015. Female Reformers in the Progressive Era. National Women's History Museum. https://www.crusadeforthevote.org/progressive-era-reformers. Fall. Accessed February 17, 2025.

LeBold, William K. and Dona J. LeBold. 1998. Women Engineers: A Historical Perspective. ASEE Prism. March. Volume 7, Number 7, pp. 30-32.

Lemelson-MIT. 2025. Guglielmo Marconi - Wireless Telegraphy - Computing and Telecommunications. https://lemelson.mit.edu/resources/guglielmo-marconi. Accessed February 1, 2025.

Library of Congress. 2024. Who invented the automobile? https://www.loc.gov/everyday-mysteries/motor-vehicles-aeronautics-astronautics/item/who-invented-the-automobile/. December 5. Accessed February 2, 2025.

Library of Congress. 2025. Industrial Revolution 1876-1900. US History Primary Source Timeline, Rise of Industrial America, 1876-1900. https://www.loc.gov/classroom-materials/united-states-history-primary-source-timeline/rise-of-industrial-america-1876-1900/. Accessed February 9, 2025.

McDermott, Annette. 2023. How World War II Empowered Women: How did women's service during World War II inspire their fight for social change and equality? History. https://www.history.com/news/how-world-war-ii-empowered-women. August 15. Accessed February 17, 2025.

McGillem, Clare D. 2024. telegraph. Encyclopedia Britannica. https://www.britannica.com/technology/telegraph. September 16. Accessed January 14, 2025.

Michals, Debra. 2017. Margaret Sanger (1879-1966). National Women's History Museum. www.womenshistory.org/education-resources/biographies/margaret-sanger. Accessed February 17, 2025.

National Academy of Engineering. n.d. About the NAE. www.nae.edu/nae/naehome.nsf/weblinks/NAEW-4NHMQM?opendocument. Accessed December 19, 2000.

National Archives. 2025a. 19th Amendment to the US Constitution: Women's Right to Vote (1920). https://www.archives.gov/milestone-documents/19th-amendment. Accessed February 17, 2025.

National Archives. 2025b. National Labor Relations Act (1935). https://www.archives.gov/milestone-documents/national-labor-relations-act. Accessed February 9, 2025.

Nordlund, Willis J. 1991. The Federal Employees' Compensation Act. Monthly Labor Review. https://www.bls.gov/opub/mlr/1991/09/art1full.pdf. September. Accessed February 17, 2025.

Parker & Sons. 2025. When Did Homes First Get Electricity? A Historical Insight. https://www.parkerandsons.com/blog/when-did-homes-first-get-electricity-a-historical-insight. Accessed on January 24, 2025.

Payne-Patterson, Jasmine, and Maye, Adewale A. 2023. A History of the Federal Minimum Wage: 85 years later, the minimum wage is far from equitable. Economic Policy Institute. https://www.epi.org/blog/a-history-of-the-federal-minimum-wage-85-years-later-the-minimum-wage-is-far-from-equitable/. August 31. Accessed February 9, 2025.

PBS American Experience. n.d. Timeline of Nuclear Technology: Meltdown at Three Mile Island. https://www.pbs.org/wgbh/americanexperience/features/three-nuclear-technology/. Accessed February 1, 2025.

Phi Beta Kappa. n.d. The Phi Beta Kappa Society: A Short History of Phi Beta Kappa. www.pbk.org/history.htm. Accessed December 19, 2000.

Pruitt, Sarah. 2021. How Flappers of the Roaring Twenties Redefined Womanhood. History. https://www.history.com/news/flappers-roaring-20s-women-empowerment. April 16. Accessed February 17, 2025.

Read, Phyllis J., and Bernard L. Witlieb. 1992. The Book of Women's Firsts. New York, New York: Random House.

Rossiter, Margaret W. 1992. Women Scientists in America: Struggles and Strategies to 1940. Baltimore, Maryland: The Johns Hopkins University Press.

Rossiter, Margaret W. 1995. Women Scientists in America: Before Affirmative Action 1940-1972. Baltimore, Maryland: The Johns Hopkins University Press.

Schneider, Dorothy and Schneider, Carl F. 1993. The ABC-CLIO Companion to Women in the Workplace. Santa Barbara, California: ABC-CLIO.

Schneider, Jennifer. 2023. 175th Anniversary of the Seneca Falls Convention, Smithsonian American Women's History Museum. https://womenshistory.si.edu/blog/175th-anniversary-seneca-falls-convention. July 13. Accessed February 17, 2025.

Science Daily. 2000. National Academy of Engineering Reveals Top Engineering Impacts of the 20th Century: Electrification Cited as Most Important. https://www.sciencedaily.com/releases/2000/03/000303075501.htm. March 3. Accessed January 21, 2025.

Shetterly, Margot. 2016. Hidden Figures: The American Dream and the Untold Story of the Black Women Mathematicians Who Helped Win the Space Race. New York, New York: William Morrow. p. 83.

SME. n.d. SME Past Presidents. https://www.smenet.org/What-We-Do/SME-Past-Presidents. Accessed March 31, 2024.

Social Security Administration. 2025. Historical Background and Development of Social Security. https://www.ssa.gov/history/briefhistory3.html. Accessed February 9, 2025.

Society for Mining, Metallurgy and Exploration. n.d. The Woman's Auxiliary to the American Institute of Mining, Metallurgical, and Petroleum Engineers, Inc. a division of the Society for Mining, Metallurgy and Exploration (SME). http://www.smenet.org/waaime/. Accessed April 5, 2015.

Society of Women Engineers. 1999. www.swe.org/SWE/Awards/achieve3.htm. Accessed September 1, 1999.

Society of Women Engineers. 2015a. The SWE Story. http://societyofwomenengineers.swe.org/membership/history/2454-the-swe-story.Accessed May 24, 2015.

Society of Women Engineers. 2015b. Individual Awards Overview: Achievement Award. http://societyofwomenengineers.swe.org/awards/individual-awards. Accessed May 24, 2015

SWE Newsletter. 1959. Elsie Eaves – Pioneer from the West. May, p. 3.

Tau Beta Pi. 1999a. www.tbp.org/TBP/INFORMATION/Info_book/membership. Accessed August 20, 1999.

Tau Beta Pi. 1999b. Tau Beta Pi: Integrity and Excellence in Engineering. www.tbp.org. Accessed August 20, 1999.

TCL Guides. 2023a. Who Invented Television: History of TV. https://www.tcl.com/global/en/blog/playbooks/history-of-tv. June 28. Accessed February 1, 2025.

TCL Guides. 2023b. History of Refrigerator Invention & Who Invented Refrigerator, https://www.tcl.com/global/en/blog/playbooks/history-of-refrigerator-invention. November 19. Accessed January 24, 2025.

Tubiana M. 1996. National Library of Medicine. Wilhelm Conrad Röntgen et la découverte des rayons X [Wilhelm Conrad Röntgen and the discovery of X-rays]. Bulletin de l'Academie nationale de medecine, 180(1), 97–108, https://pubmed.ncbi.nlm.nih.gov/8696882/. Accessed January 24, 2025.

Turner, Edna May. 1954. Education of Women for Engineering in the United States 1885-1952. New York University. Ann Arbor, Michigan: UMI Dissertation Services.

United Engineering Foundation. 2000a. An Overview of the United Engineering Foundation: History. www.uefoundation.org/overview.html. Accessed December 15, 2000.

United Engineering Foundation. 2000b. Founder Societies of the United Engineering Foundation. www.uefoundation.org/fndsoc.html. Accessed December 15, 2000.

US Department of Energy. 2013a. The History of the Light Bulb. https://www.energy.gov/articles/history-light-bulb. November 22. Accessed January 14, 2025.

US Department of Energy. 2013b, Top 11 Things You Didn't Know About Nikola Tesla. https://www.energy.gov/articles/top-11-things-you-didnt-know-about-nikola-tesla. November 18. Accessed February 16, 2025.

US Department of Energy. 2015. History of Air-Conditioning. https://www.energy.gov/articles/history-air-conditioning. July 20. Accessed January 24, 2025.

University of Arizona. 2023. 1978 Mining Alumna in the Business of Ridding Industry Barriers. College of Engineering News. https://news.engineering.arizona.edu/news/1978-mining-alumna-business-ridding-industry-barriers. November 28.

Vedin, Bengt-Arne. n.d. Ericsson. https://www.ericsson.com/en/about-us/history/products/other-products/the-transistor%2D%2Dan-invention-ahead-of-its-time. Accessed January 24, 2025.

Wastewater Digest. 2003. ASCE's First Woman President Assumes Office Saturday. https://www.wwdmag.com/water/news/10908068/asce039s-first-woman-president-assumes-office-saturday. November 14.

Wells, Christopher. 2018. Industrializing Women. National History Education Clearinghouse., teachinghistory.org. https://teachinghistory.org/history-content/ask-a-historian/25037. Accessed February 17, 2025.

Part I
1952–1956

Chapter 2
Setting the Stage: 1952–1956 Before Sputnik and Post WWII Influences

Cecilia D. Craig and Janet L. Williams

2.1 Introduction

Technology changes skyrocketed during 1952–1956, similarly to the early years of the Industrial Revolution in the late 1800s. Miniaturization from the invention of the transistor and other similar electronic devices, fiber, and other material inventions developed for communication products, entertainment, long-distance data transmissions, and home products all drove technology and consumer usage. These technology-fed changes and cultural changes in the post-World War II (WWII) era lit up civil rights actions, and women began to build on the industrial roles they had taken up during WWII.

C. D. Craig (✉)
San Jose, CA, USA
e-mail: Ceal@Research.Druai.com

J. L. Williams
Albuquerque, NM, USA

C. D. Craig et al. (eds.), *Women Engineering Legends 1952-1976: Society of Women Engineers Achievement Award Recipients*,
https://doi.org/10.1007/978-3-032-00224-2_2

Discoveries in transistor technology at Bell Labs culminated in the licensing of several transistor patents in the early 1950s that launched commercial transistor products and marked the dawn of the semiconductor industry.

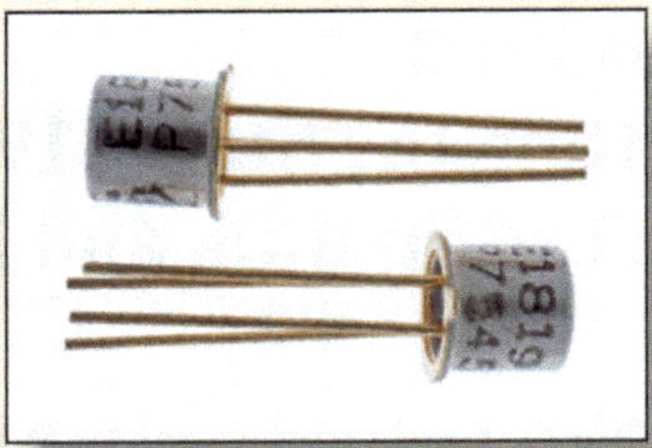

Maria Telkes, PhD received the Achievement Award for her many innovations that utilized solar energy, as evidenced by more than 20 patents, many for solar-powered devices, including an oven and salt water distiller.

The first "compiler" to translate English language instructions to machine code was invented by Grace Murray Hopper in 1952. It led to the first computer software.

Elsie Gregory MacGill received the Achievement Award for her accomplishments in aeronautical engineering including precision manufacturing of HAWKER Hurricanes during WWII for Britain.

1952

1953

- In 1950, 34% of women worked outside the home.
- Most women worked in clerical, retail, or manufacturing positions, or as teachers or nurses.
- Women usually had to have male relatives co-sign to get credit.
- No reliable birth control method was available.

Researchers at Bell Labs demonstrated the first practical silicon solar cell in 1954. They linked several cells together to produce a solar battery, paving the way for solar panel technology.

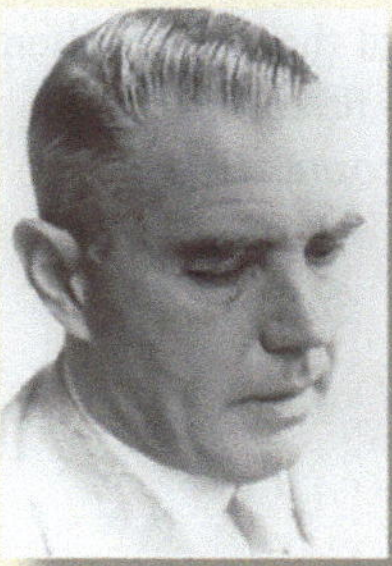

In 1953, Dutch scientist Bram Van Heel first demonstrated image transmission through bundles of optical fibers with transparent cladding to reduce interference.

Edith Clarke received the Achievement Award for her many contributions to the field of electrical engineering for the power industry, including a seminal textbook and a graphical calculator for solving power transmission problems.

1954

The Second Sex by French feminist Simone de Beauvoir was published in the USA, where the term "women's liberation" was used for the first time.

In 1955, Tappan released the first microwave oven for home use. Due to their size and cost, microwave ovens do not become a household fixture until the 1960s, when they could fit compactly on a kitchen countertop.

Margaret H. Hutchinson Rousseau, ScD received the Achievement Award for contributions to chemical engineering, including the development of methods for the first large-scale production of penicillin and refining of high-octane aviation fuel.

1955

- Women represented 2.6% of the members of the 83rd Congress, with three senators and 11 representatives.
- The participation of women in the workplace increased to 36%.

By 1956, the technology for a video tape recorder was widely used in the television industry. It had been perfected five years earlier at Ampex Corporation by Charles Ginsburg.

Elise F. Harmon received the Achievement Award for early printed circuit board manufacturing and circuit miniaturization, and for adapting airplane generator components to enable flight at higher altitudes in WWII aircraft.

Marc Grégoire and his wife, Colette, launched the Tefal corporation in 1956, applying Teflon (PTFE) to cooking pans to create a non-stick surface.

1956

Rosa Parks was arrested in Montgomery, Alabama, for refusing to move to the back of a city bus.

2.2 State of Technology

In the early years of the 1950s, specifically 1952–1956, the electronics revolution began in earnest, driven by new inventions, miniaturization, and a burgeoning middle-class (post-WWII) that had an appetite for new products and could afford them. Industries that had flourished during WWII, like aerospace, materials, and communications, began moving forward quickly to commercialize what had been key technologies during the war. New ideas in energy and medicine were nascent, just beginning to stretch those industry's boundaries.

These highlights encompass electronics hardware, computer hardware and software, media and entertainment, inventions for the home, energy generation, and medicine providing a backdrop for what the SWE Achievement Award recipients would accomplish in the years ahead.

2.2.1 *Hardware Products*

Transistors William Shockley (1954) from Bell Labs wrote in 1954 that the transistor became public on "30 June 1948, [when] Bell Labs announced John Bardeen and Walter Brattain's game-changing invention of the transistor" which was invented in 1947 (Marsh 2024, para. 4). By the time Shockley authored an article in the *American Scientist* (1954), hearing aid product applications using the transistor were being sold. He asserted that transistor electronics was a young industry and predicted significant industry growth in future years (Shockley 1954). Bell Labs, General Electric, Philco, and Fairchild created different types of transistors from 1947 to 1959 that were used in numerous applications. The transistor would transform existing industries and create several new ones. Keyes stated in 2009 that while the transistor "has changed in form, material, and fabrication, nothing has come along in half a century to displace it" (Keyes 2009).

Transistor Radios Chris Haggerty, Texas Instruments (TI) Vice President, asked engineers in May 1954 to quickly develop a transistor radio for stores in time for the Christmas rush. They created a prototype in one week! He then committed 10% of TI's revenue for this venture. With help from other partners to bring this first-time product into production, they made the deadline. Raytheon released a tabletop model with higher sound quality and then Tokyo Telecommunications Engineering Corporation (Sony) released the TR-55 in 1955. In the next few years, Sony popularized the transistor radio by decreasing cost and size, ultimately dominating the market (Marsh 2024).

Miniaturization Many other electronic systems benefited from being made smaller; faster transmission speeds, ease of use, and the smaller physical size lowered costs. Elise Harmon (1956 SWE Achievement Award recipient) developed new printed circuit board fabrication techniques in the 1950s and later led other miniaturization efforts affecting many industries.

2.2.2 *Computer Hardware and Software*

The age of computers began in this period, enabled by the invention of the transistor in 1947, and the advent of nascent electrical and then electronic general-purpose computer systems, in the post WWII years.

John Mauchly and J. Presper Eckert led a team to build the first general-purpose computer, ENIAC (Electronic Numerical Integrator and Computer), in 1948. Initially developed during WWII as an Army project to process ballistic calculations more quickly, the ENIAC was not completed until after the war. The ENIAC was a room-sized computer that

> Contained 17,468 vacuum tubes, 7200 crystal diodes, 1500 relays, 70,000 resistors, 10,000 capacitors and around five million hand-soldered joints. It weighed more than 27 tons, occupied 1800 square feet of space, and consumed 150 kilowatts of power. (National Museum US Army n.d.)

Over the next decades, switching from electrical to electronic hardware drove dramatic reductions in the size of these general-purpose computers.

Initially, *programmers* ran programs by physically making *hardware* wiring changes versus, in later years, making *software* code changes. The first computer *software compiler* was in operation on a UNIVAC computer in the spring of 1952. Grace Murray Hopper (1964 SWE Achievement Award recipient) is recognized as the "Mother" of the first English computer language known as FLOW-MATIC. FLOW-MATIC provided the foundation for COBOL, which stands for COmmon, Business-Oriented Language. She later defined a compiler process in 1955 with two critical principles:

> in the case of a *verb,* the catalog entry will lead to a generator or subroutine and its satellite routines in a volume of the library. In the case of a *noun,* the catalog entry will lead to a standard item design and its satellite routines in a volume of the library (Sammet 1978, 217).

Her work on this new concept was just one of the many computer languages developed in the ensuing years.

To define what became COBOL, the Short-Range Committee (known as the PDQ committee) was formed as a multi-entity group to define a compiler that was portable across hardware platforms. The PDQ committee had 15 government representatives, 11 users and consultants, and 16 company representatives. Formed in May 1959, the Department of Defense facilitated the first meeting after a request from six people (including Hopper, first initiated by Mary Hawes at Burroughs Corporation). Ultimately, the PDQ committee defined the acronym COBOL on September 18, 1959. The compiler language used ideas and knowledge from many sources, in particular: "the three main inputs to COBOL in terms of influencing the work were FLOW-MATIC [Remington-Rand UNIVAC],

AIMACO [developed under Air Force with input from Hopper], and Commercial Translator [IBM]" (Sammet 1978, 209). By 1956, FLOW-MATIC ran in several organizations, such as DuPont, Westinghouse, U. S. Navy, and more (Sammet 1978).

2.2.3 Communications and Media Inventions

Videotape Recorder At Ampex Corporation in 1952, Charles Ginsburg led a team to develop a "broadcast-quality videotape recorder (VTR), US patent number 2,956,114" (Lemelson-MIT n.d., para. 2), first sold in 1956 for $50,000. "The system used a rapidly rotating recording head to apply high-frequency signals onto a reel of magnetic tape. The VTR revolutionized television broadcasting" (National Inventors Hall of Fame n.d.). It quickly gained widespread use in broadcast television, with CBS initiating the adoption process in 1956. This technology invention was a foundation for videocassette recorders (VCR) ubiquitous in the 1980s and beyond (Lemelson-MIT n.d.).

Television Color televisions were first demonstrated in the early 1950s but did not have wide adoption until RCA developed a standard adopted by the US Federal Communication Commission in 1953, still used today: the National Television System Committee (NTSC) color system (SRI n.d.). NBC made the first coast-to-coast broadcast when it telecast the Tournament of Roses Parade on January 1, 1954, and in 1964, provided it in color for the first time (Lloyd 2024).

Fiber Optics Fiber-optic cables are an industry standard today for transmitting telecommunication data. With roots back to the late 1800s, the pivotal concept that made these tubes of light more practical happened in 1954. Dutch scientist Bram van Heel described image transmission using bundles of transparently cladded optical fibers. This idea of cladding the fibers "protected the total reflection surface from contamination and greatly reduced crosstalk between fibers" (Alwayn 2004, para. 3). This cladding improved efficiency and was more practical (Oncea 2024).

2.2.4 In the Home

Non-stick Pans DuPont patented Teflon in 1941 while looking for a freon replacement; Teflon was used on parts for the Manhattan Project. Teflon was later applied to household pans as a non-stick coating. In 1954, Marc Grégoire patented Teflon coating as a non-stick surface for cookware, inspired by his wife, Colette, who had suggested the application after seeing him coat his fishing gear with it to prevent sticking. She had also noticed the effect on their pans, which helped keep eggs from sticking. Together, they launched the Tefal corporation in 1956 applying Teflon to cooking pans to create a non-stick surface (Gormley 2022; Tefal 2024).

Ovens Raytheon engineer Percy Spencer discovered that microwaves accidentally heated food when he visited a place with magnetrons, and a food item in his pocket heated up. Spencer's colleagues remember a more methodical discovery process. Either way, Raytheon moved forward with a product for restaurants, albeit at more than two times the price of $495 offered 12 years later by Amana (a Raytheon acquisition) (Ackerman 2016/2024).

Mária Telkes (1952 SWE Achievement Award recipient) developed a low-cost solar-powered oven that reached 350°F in 30 min; she spent much of her career in solar energy and its uses.

2.2.5 *Energy*

While the concept of solar energy capture had been studied since the 1800s, in the 1950s, Bell Laboratories focused efforts on silicon as a more efficient material than others. However, the cells were still only "6% efficient" (Chu and Tarazano n.d., para. 7). A symposium was organized in 1954 by UNESCO on Wind and Solar Energy to highlight the potential positive social impact. The Association for Applied Solar Energy began that year, too. Solar energy would not move into the mainstream for some decades; however, the world was starting to invest and take notice of its potential (ISES 2020).

2.2.6 *Air and Space*

During WWII, aeronautical technology advanced due to military aircraft needs. In the 1950s, the golden age of air travel began, characterized by increased options, lower costs, faster travel, and newer planes. "In 1955, for the first time, more people in the United States traveled by air than by train" (Smithsonian n.d., para. 1). PanAm, which had a virtual monopoly during WWII for civilian air travel, faced competition from new airlines, like Trans World Airways. The field of aeronautical engineering grew, and engineers designed faster planes and, in later years, ultimately developed vehicles to orbit the Earth.

2.2.7 *Medicine*

In 1952, "Elizabeth Hazen and Rachel Brown's invention, nystatin, the world's first useful antifungal antibiotic, was introduced to the market" (SWE 2001, 38). Engineering and medical professionals were developing lifesaving technologies. An external heart stimulation device kept a patient alive for over 50 h in 1952, leading to further research and, ultimately, to artificial pacemakers in the late 1950s (Geselowitz and Leder 2017).

2.2.8 State of Technology Summary

Technological changes and inventions exploded in this period. Costs decreased; sizes were smaller; and an increasing number of people were using these new gadgets. Computers had found new platforms in both hardware and the nascent field of software. The events described above are only a fraction of the technology changes that occurred yet demonstrate the environment in which Achievement Award recipients found themselves.

2.3 State of Women

In the 1950s, a typical stereotype for women was the MRS. Degree; that is, women attended college simply to meet prospective husbands. The concept of nuclear family—father, mother, and children was promoted as a way to show that "Americans [were] superior to the Communists" (NPR n.d., para. 2). Large families were common and encouraged; single mothers were "undesirable" (para. 4).

Women engineers were small in numbers during the 1950s, after higher numbers during WWII, "drop[ping] to 1,500 in 1952" (SWE 2000, 24). Some educational institutions did not allow women to study engineering, while others had for many decades. This discriminatory practice began to change in the 1950s: "Georgia Tech began admitting women in 1952" (p. 25), and the trend continued into the 1970s when Caltech finally admitted female students. By 1957, the number of women enrolled in engineering programs jumped to 500 in the United States (SWE 2000). Nonetheless, the number of women engineers was small throughout the 1950s. In 1966, 35,826 bachelor's degrees in engineering were awarded but only 146 were earned by women (NSF 2008).

Mainstream America was not supportive of women pursuing technical fields. In April 1956, the popular television series *Father Knows Best* aired an episode entitled "Betty—Girl Engineer". Betty's school provided opportunities for students to intern at a job, and Betty uses her initials to sign up for a civil engineer position. A male college student, Doyle, leading the work crew shows Betty how to use a transit, but keeps saying, "Why don't you go home where little girls belong?" (Weingroff 2023). He gives Betty a hard time, saying it is not right for a girl to want to be an engineer. Betty left the job site, annoyed and angry. Doyle visits Betty at home, bringing chocolates. Her father asks Doyle what he thinks about Betty's idea. Doyle shares "She might be a damned good [engineer]…but what a dirty trick…on some guy," implying she would be taking their job. Receiving little support from her parents, Betty gives up her interest in engineering (Weingroff 2023).

Engineering was a ripe career field at this time. Early advocates, such as Dr. John Russell, an electrical engineering professor at Columbia University, asserted, "The current need [for engineers] will result in far greater opportunities for women to study and practice engineering than in the past" (NYT 1951b March 12, 9 para. 4). Although not common, woman-owned companies often found their gender an asset, not a liability. Dot Merrill from Chicago owned and operated an engineering sales firm by selling "electrical and power generating equipment and often do[ing] the fieldwork in supervising the installation of my own jobs" (NYT 1952, 79 para. 8).

2.4 State of SWE: SWE's Early Years

While SWE as a group of 60 members began 2 years earlier, "on February 13, 1952, at 12:47 p.m. in incorporation Liber No. 78 Folio 241, the Society of Women Engineers was registered as a corporation in the District of Columbia" (SWE 2000, 7). Beginning with a Governing Council headed by President Beatrice Hicks (1963 SWE Achievement Award recipient), the group grew to 415 members by 1955.

The first four sections were located in New York City, Philadelphia, Washington, D.C., and Boston, followed rapidly by Pittsburgh, Chicago, Detroit, and Los Angeles. The organization expanded from 4 sections to 20 sections by 1954, adopting a Board of Directors model for several years, where each section appointed a Director. Over time, this governance model changed as the organization grew, as it needed a stronger financial basis to manage the growth more effectively (SWE 2000).

A Professional Guidance and Education committee was established to provide information that members could use for outreach to high schools and Parent-Teacher Association (PTA) programs. The committee printed a booklet containing speeches that had been delivered at the Centennial of Engineering by SWE members, including a paper from Margaret Ingels (1952) titled "Petticoats and Slide Rules." SWE participated in the 75th Anniversary of the ASME at the Stevens Institute of Technology. SWE thoughtfully established a statistics committee to track statistics on women in engineering which proved very valuable to the organization, industry, and government. In 1956, a headquarters fund was started to provide income for the eventual establishment of a Headquarters office, the scholarship fund was established, and a Board was elected to manage both funds (SWE 2000).

Early on, the group wanted to recognize engineering achievements. Beatrice Hicks, President of SWE, announced (NYT 1951a March 11) "an annual award would be made to women deemed to have made the most significant contribution to engineering, in design, research, education or administration" (p. 47 para. 3). In 1952, Mária Telkes became the first SWE Achievement Award recipient in that first year of SWE's incorporation for *meritorious contributions to the utilization of solar energy*. At that time, she had 20 patents for solar devices, portable solar stills for converting salt water to fresh water (vital during WWII), a chemical process for storing solar energy, and a much-publicized low-cost solar oven aimed for use in poor and arid regions of the world. At the 1952 SWE Convention, Telkes received a pin (the SWE emblem surrounded by a laurel wreath), a certificate citing the award, and life membership (SWE 2000). From 1952 to 1956, five women were recognized with the SWE Achievement Award, marking the beginning of a long legacy of honoring notable and legendary women engineers.

2.5 US Culture and Laws

Table 2.1 illustrates bellwether examples of changes across the spectra of family life, race and gender, and medical advances.

- *Middle-class suburbia arrives:* Economic prosperity rose dramatically post-WWII. The interstate highway system began construction in the US. The rise of suburbia exploded, with cities growing in geographic size as they incorporated these new suburbs over time, expanding their population footprint and driving higher energy use (primarily through utilities and transportation, with cars being the primary mode of transportation for the longer distances associated with suburban living) (Rafferty 2024).
- *Race and gender activism:* Efforts build momentum, resulting in new laws that build a foundation for significant changes in the coming decades.
- *Medical advances accelerate*: Vaccines were developed for significant diseases.

Table 2.1 Culture and legal highlights for 1952–1956

Date	Event
1952–1956	**Levittowns**: Suburbs and housing grew dramatically; Levittown number two established in Pennsylvania in 1952. People were earning more post-WW II; manufacturing was growing, and buying a house became the "American Dream." Levittowns were planned communities sparking the rise of suburbia. Originally, founded in 1941 (first Levittown in New York) with restrictive covenants preventing Black families from purchasing these affordable homes, these planned communities pushed an "emphasis on conformity, and a return to traditional gender roles" (Sheidlower 2020, para. 6). While these restrictions were legally removed with a court case in 1948, the next highlight had the larger influence (Britannica 2024a; Rafferty 2024; Sheidlower 2020)
1954	***Brown v. Board of Education***: This unanimous U. S. Supreme Court ruling upended public schools, ending an extended history of segregation by race in the US. The National Association for the Advancement of Colored People (NAACP) consolidated five separate cases into one for the Supreme Court review. Elementary-aged children were the subjects of each case, which concerned schools for Black children that provided an inferior education with insufficient funding, crowded classrooms, and deficient facilities, compared to those for white children (Stratton 2024) "We conclude that, in the field of public education, the doctrine of 'separate but equal' has no place. Separate educational facilities are inherently unequal."—Chief Justice Earl Warren (National Archives, *Brown v. Board of Education* 1954) This milestone decision drove future changes in laws and culture norms. Future laws like the Civil Right Act of 1964 and the Voting Rights Act of 1965 grew from the *Brown* decision One likely unexpected result of this law was the loss of Black teachers from segregated schools, now integrated. "There were approximately 82,000 African-American teachers across the South at the time of the *Brown* decision. As schools were integrated, those serving African-Americans were closed, and their teachers were fired. Over the next decade, nearly half had been fired" (Eckert 2018, para. 4)

(continued)

Table 2.1 (continued)

Date	Event
1955	**Montgomery Bus Boycott**: On December 1, 1955, Rosa Parks would not give up her seat to a white person in a local Montgomery bus, and as a result, she was arrested. That action began a bus boycott by almost all Black riders lasting more than a year. Carpools were organized; city officials arrested people; houses of noted Black leaders were bombed (Martin Luther King, Jr. and E. D. Nixon) The boycott only ended after a Supreme Court decision on June 5, 1956 (*Browder v. Gayle*): segregation on public buses was unconstitutional. This boycott did not happen without long years of fighting to eliminate bus segregation, led by the Women's Political Council. While Parks was not the first to protest, not give up a seat, and get arrested, her arrest sparked an action that reached the Supreme Court, knocking down one more segregation practice (Stanford MLK R&E Institute n.d.)
1955	**Polio vaccine**: "At its peak incidence in the USA, in 1952, approximately 21,000 cases of paralytic polio (a rate of 13.6 cases per 100,000 population) were recorded" (Britannica 2024b, para. 3) The Salk injectable polio vaccine was administered to young children after being approved April 12, 1955, with "more than 450 million doses" (para. 3) injected

References

Ackerman, E. 2016/2024. A brief history of the microwave oven. *IEEE Spectrum.* September 30. updated January 11. https://spectrum.ieee.org/a-brief-history-of-the-microwave-oven

Alwayn, V. 2004. A brief history of fiber-optic technologies. In *Fiber-Optic Technologies*, Cisco Press. https://www.ciscopress.com/articles/article.asp?p=170740

Britannica, T. Editors of Encyclopaedia 2024a. Levittown. *Encyclopedia Britannica*. September 3. https://www.britannica.com/place/Levittown-New-York

Britannica, T. Editors of Encyclopaedia 2024b. Polio. *Encyclopedia Britannica.* December 16. https://www.britannica.com/science/polio

Chu, E. & Tarazano, D. L. n.d.. A Brief History of Solar Panels. Retrieved from https://www.smithsonianmag.com/sponsored/brief-history-solar-panels-180972006/

Eckert, J. 2018. The Brown decision should have included integration of teachers and administrators. In *Wheaton Faculty Experts on the Impact of Brown v. Board of Education.* April 3. Retrieved from https://www.wheaton.edu/news/recent-news/2018/april/the-impact-of-brown-v-board-of-education/

Gormley, J. 2022. *The Story Behind The Accidental Invention Of Teflon.* July 24. https://www.grunge.com/935674/the-story-behind-the-accidental-invention-of-teflon/

Ingels, M. 1952. Petticoats and Slide Rules. *Western Society of Engineers.*September 4.

Geselowitz, M. N. and Leder, R. S. 2017. *Pacemakers*. Engineering and Technology History Wiki. https://ethw.org/Pacemakers

ISES. 2020. ISES SWC50 - The Century of Solar-Stories and Visions. https://www.swc50.org/celebratory-booklet

Keyes, R. W. 2009. The Long-lived Transistor: The basic unit of the computer revolution has changed in form, material and fabrication, but nothing has come along in half a century to displace it. *American Scientist*, *97*(2), 134–141. http://www.jstor.org/stable/27859300

Lemelson-MIT. n.d.. Charles Ginsburg, Videotape Recorder. https://lemelson.mit.edu/resources/charles-ginsburg

Lloyd, J. 2024. See a timeline of key moments in the history of the storied Rose Bowl game. *NBC Los Angeles.* December 30. https://nbclosangeles.com/news/sports/rose-bowl-football-game-history/3591492/

Marsh, A. 2024. The Incredible Story Behind the First Transistor Radio. IEEE Spectrum. September 30. https://spectrum.ieee.org/transistor-radio-invented

National Public Radio (NPR). n.d. *Mrs. America: Women's Roles in the 1950s*. https://www.pbs.org/wgbh/americanexperience/features/pill-mrs-america-womens-roles-1950s/

National Science Foundation (NSF), Division of Science Resources Statistics. 2008. *Science and Engineering Degrees: 1966–2006*. Detailed Statistical Tables NSF 08-321. Arlington, VA. Available at http://www.nsf.gov/statistics/nsf08321/.

National Inventors Hall of Fame. n.d. Charles P. Ginsburg. https://www.invent.org/inductees/charles-p-ginsburg

National Archives. 1954. *Brown v. Board of Education*. May 17. https://www.archives.gov/milestone-documents/brown-v-board-of-education

National Museum, United States Army. n.d.. ENIAC. https://www.thenmusa.org/armyinnovations/innovationeniaccomputer/

The New York Times (NYT). 1951a. *Women Engineers See Field Widening*. March 11. https://timesmachine.nytimes.com/svc/tmach/v1/refer?pdf=true&res=9B01E4D61538E13BBC4952DFB566838A649EDE

The New York Times (NYT). 1951b. *U.S. Agencies Seek Women Engineers*. March 12. https://timesmachine.nytimes.com/svc/tmach/v1/refer?pdf=true&res=9403E3DF1038E13BBC4A52DFB566838A649EDE

Oncea, J. 2024. Delight your friends with these fiber optics stories. May 31. https://www.photonicsonline.com/doc/delight-your-friends-with-these-fiber-optics-stories-0001

Rafferty, J. P. 2024. Urban sprawl. *Encyclopedia Britannica*. June 25. https://www.britannica.com/topic/urban-sprawl

Sammet, J. E. 1978. The early history of COBOL. COBOL Session, *Association for Computing Machinery*. June 1. https://doi.org/10.1145/800025.1198367

Sheidlower, N. 2020. The controversial history of Levittown, America's first suburb. July 31. Retrieved from https://www.untappedcities.com/the-controversial-history-of-levittown-americas-first-suburb/

Shockley, W. 1954. Transistor physics. *American Scientist*, *42*(1), 41–72. http://www.jstor.org/stable/27826528

Smithsonian. n.d. Commercial aviation in mid-century. https://airandspace.si.edu/explore/stories/commercial-aviation-mid-century

SWE. 2000. *Five Decades of the Society of Women Engineers*. Society of Women Engineers Records (LR001539), [unprocessed], Walter P. Reuther Library, Archives of Labor and Urban Affairs, Wayne State University.

SWE. 2001. *Women Innovators: From Ancient Times to the 21st Century*. SWE Magazine. March/April.

SRI. n.d. Color television. https://www.sri.com/hoi/color-television/

Stanford University, The Martin Luther King, Jr., Research and Education Institute. n.d.. Montgomery Bus Boycott. https://kinginstitute.stanford.edu/montgomery-bus-boycott

Stratton, E. 2024. Brown v. Board: Equal Protection. *U. S. Constitution.net*. August 8. https://www.usconstitution.net/brown-v-board-equal-protection/

Tefal. 2024. Innovation is in our DNA. https://www.tefal.com/about-tefal/our-innovations

The New York Times (NYT). 1952. *Women Could Fill Engineering Jobs*. March 16. https://timesmachine.nytimes.com/svc/tmach/v1/refer?pdf=true&res=9C02E1DC133AE53ABC4E52DFB5668389649EDE

Weingroff, Richard. 2023. Betty, Girl Engineer. https://www.fhwa.dot.gov/highwayhistory/betty_engineer.cfm. Accessed May 20, 2025.

Chapter 3
Mária Telkes, PhD (1900–1995) Achievement Award Recipient 1952

In recognition of her meritorious contributions to the utilization of solar energy

Cecilia D. Craig

3.1 Introduction

Born in Hungary in 1900, Mária Telkes (Fig. 3.1) moved to the US in her early adult years to begin a career in physical chemistry, biophysics, and engineering, with a lifetime interest in solar energy, becoming known as the *Sun Queen*. Dr. Telkes's most significant inventions were a solar distillation device and a solar-powered residential heating system. She also invented other devices that harnessed and utilized energy stored from sunlight. She worked for Westinghouse, Massachusetts Institute of Technology (MIT), and Curtis-Wright, among others, and advised the US Office of Scientific Research and Development during WWII. With 27 patents to her name, working into her 80s, she died in 1995 at age 95.

Among Dr. Telkes's many honors, she was the first recipient of the Society of Women Engineers Achievement Award in 1952, she received the Charles Greeley Abbot Award from the American Solar Energy Society in 1977, she was posthumously inducted into the National Inventors Hall of Fame in 2012, and she was the subject of a Google Doodle on December 12, 2022 (Google 2022).

C. D. Craig (✉)
San Jose, CA, USA
e-mail: ceal@research.druai.com

C. D. Craig et al. (eds.), *Women Engineering Legends 1952-1976: Society of Women Engineers Achievement Award Recipients*,
https://doi.org/10.1007/978-3-032-00224-2_3

Fig. 3.1 Portrait of Mária Telkes 1956 (Courtesy Library of Congress)

3.2 Early Life

Born in 1900 and raised in Budapest, Hungary, Telkes was the daughter of Aladar Telkes and Mária Laban de Telkes (Rafferty 2024). She had a chemistry lab at home as a child (age 11). A "simple school experiment, the melting of sulfur, made [her] intensively curious about chemistry" (ICWES 1964, p 281). She set it up "inside her parents' garden house" (Hernandez 2025), where she initiated one "loud, but harmless explosion" (ICWES 1964, p 281) doing experiments (Hernandez 2024; ICWES 1964.

Telkes studied physical chemistry at the University of Budapest, earning a BA and a PhD in physical chemistry in 1924. A book about alternative energy—*Energy Sources of the Future* by Kornel Zelowitch—that she encountered in her freshman year was the catalyst for her lifelong passion for solar energy (SWE Philadelphia n.d.). A newly minted PhD, Telkes visited a relative working at the Hungarian consulate in Cleveland, Ohio. She was inspired to move to the US in 1925, working for the Cleveland Clinic as a biophysicist (NIHF n.d.; Rafferty 2024). She worked under Dr. George Crile, co-founder of the clinic, for 12 years. Together they "invented a photoelectric mechanism capable of recording brain waves" (Saxon 1996) and conducted research on synthetic cells to "better understand cancer cells" (Hernandez 2024).

In 1929, a paper by Dr. Crile, Dr. Telkes, and Amy Rowland, *The Physical Nature of Death*, shared their experimental findings about what happens when a physical body dies. They found:

1. In animals, plants, and fruits, an electric potential exists during life and disappears at death.
2. The potential is varied by insomnia, anesthetics, poisons, hemorrhage, asphyxia, change in electrolytic solution, adrenalin, injury, heat, and cold.
3. At the moment of clinical death, the potential difference between different organs drops to zero for a few moments; following this, each organ regains its potential for a short time, but finally, the potential of all tissues drops to zero, the respiration of the tissues stops, and molecular disintegration sets in. (Crile, Telkes, and Rowland 1929, p 71).

The electrode designed and built by Dr. Telkes was inserted into an amoeba, and electric potential measurements were captured under different conditions (e.g., radiation, temperature changes) and influences (e.g., adrenalin or oxygen). A figure in the paper showed 50 apple halves in a vertical column as a *battery* with an electrical potential of more than 1 V (Crile et al. 1929).

In 1932 while she was at the Cleveland Clinic, Telkes was interviewed by a noted journalist Arthur Brisbane, who described "Telkes as a 'youthful and beautiful Hungarian highbrow' with a 'deep voice' but [did] not similarly describe her male counterparts' physical characteristics, reflecting the deeply ingrained gender biases of the time" (Hernandez 2024). He did not mention much of their work, focusing on her physical attributes. She fought this double standard throughout her career, though she learned to turn it to her advantage in her later years. By 1934, she was recognized in the press alongside Shirley Temple and other notable women, though she was the only scientist. (Hernandez 2024).

After becoming a US citizen in 1937, Telkes began working as an engineer at Westinghouse, developing "instruments that converted heat into electrical energy" (Rafferty 2024). Two years later, she moved to MIT, where she started her long-term pathway into solar energy research.

3.3 Mid-Career

While working on projects turning heat into electricity at Westinghouse, she heard about an effort at MIT called the *Solar Energy Conversion Project*, a foundation funded by Dr. Godfrey Lowell Cabot (ICWES 1964). She wrote to the project leaders and asked to be interviewed, finally traveling to Cambridge to show her work on a thermopile, which is more efficient than other technologies. Ultimately, she was invited to join the team in 1939, working for Hoyt Hottel, a chemical engineering faculty member. Hottel was never an enthusiastic advocate for solar energy like Telkes was, although he led "MIT's solar team for more than two decades" (Rinde 2020). He developed what research grants asked for instead "approaching solar energy collection as [...an] engineering problem...[not a] societal imperative" (Rinde 2020). Telkes was a visionary, seeing how solar energy could benefit society. However, with the advent of WWII, the MIT team's focus shifted quickly to support the military and its needs.

As an advisor for the US Office of Scientific Research and Development (OSRD), Telkes designed a solution for soldiers downed at sea, who often died from lack of water to drink. Inspired by a jellyfish she had observed, the distillation kit developed by Telkes had a "balloon-like plastic envelope [to hold sea water] that fit a paper-cup size package when empty" (Hernandez 2024). When the seawater evaporated, drinkable water remained. Unfortunately, the WWII order for 100,000 solar stills was not delivered by the summer of 1945, as Hottel dithered over the details (Hernandez 2024). Specific references suggest the device was used on life rafts during the war (Encyclopedia 2025). Regardless, the distillation kit finally became part of survival kits and Telkes' 1966 patent no. 3,415,719 (Fig. 3.2) documents this invention (USPTO 1968). The portable still could generate up to "one quart of fresh water" (Lemelson-MIT n.d.) daily.

3.3.1 Solar House at MIT

Over the following decades, Telkes explored, invented, and produced devices and processes to convert solar energy efficiently and store it for later use. First, using sodium sulfate, known as Glauber's Salt, she led the construction on the MIT grounds of a house using this material to "absorb heat during the daytime and release it when the building cooled at night" (Hernandez 2024). Unfortunately, this approach was unsuccessful in the long term as the material caused container corrosion.

By some accounts, Telkes's assertive personality may have caused friction with the MIT team and with Hottel, potentially contributing to the failed venture and ultimately to her being fired from MIT. She was the only woman on the team, and few welcomed assertive women then. Nonetheless, her niece shared that Telkes could be "unsteerable away from her vision" at times. (Hernandez 2024). Many inventors are known for this singularity of focus, and for Telkes, like male inventors, this focus may have been an essential part of who they were and why they accomplished so many innovations. She was a visionary with a passion for solar energy to help solve the world's energy problems and help developing economies. (Hernandez 2024).

3.3.2 Dover Sun House

Telkes took her passion for solar energy to other partners: Eleanor Raymond, an architect, and Amelia Peabody, a sculptor/philanthropist. Peabody provided funding for the Dover Sun House, Raymond designed it, and again, Telkes used Glauber's Salt as the basis for energy storage that "melts in sunlight, accumulating energy in the process, and releases that energy in the form of heat when it cools and hardens"

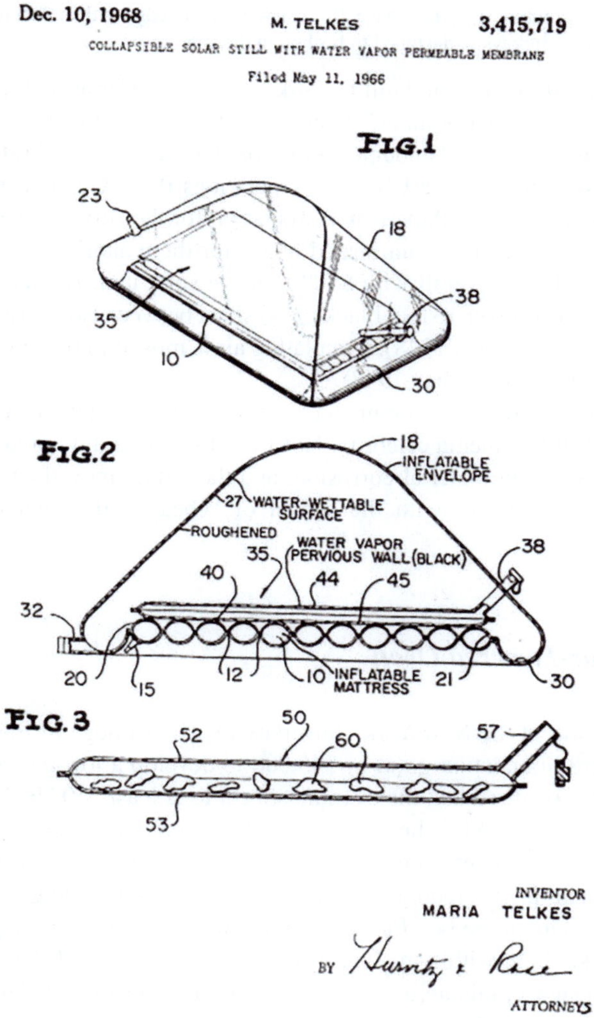

Fig. 3.2 Collapsible solar energy conversion and storage granted to Mária Telkes, US patent no. 3,415,719

(Boyd n.d.). Using Glauber's Salts instead of water for heat storage, the Dover Sun House, built in 1949 (ISES SWC50 2025), differed from prior solar houses.

The shape of the house built in 1948–49 was different from that of the "stately Colonials" (Mayer et al. 2023) in the rural (at that time) area of Dover, MA. Mayer described it in an interview with *Scientific American* (2023):

> Some say the house resembled a wedge of cheese. Imagine buying a triangular wedge of cheddar and placing it on the table with the wide side facing down. That's the shape of this house. The slanted part of the cheese wedge – the back of the house – looked fairly normal, with brown siding and some windows. But the other side of the wedge - where the rind

> would be - was a futuristic space house. It was lined edge to edge with eighteen 10-foot-tall, imposing glass windows… designed to beckon in the sun.

The solar house designed and built by Telkes went "comfortably through two winters heated by solar energy alone" (Daniels 1950). Storage of the sodium sulfate beside the house could hold enough energy for "ten sunless days" (Daniels 1950).

In 1948, as reported in the US Naval Operations Base, Guantanamo Bay, Cuba newspaper, Telkes told the American Association for the Advancement of Science, "she planned to move into a sun-heated house for the winter" estimating a cost of $20,000 for the house overall with $3000 for the solar unit. A heating unit on the roof made of "an 800-square foot black metal sheet between two glass plates [acted] as a heat trap" (The Indian 1948). Circulating air is moved to bins in strategic locations in the house. (The Indian 1948).

By now, Telkes could use the media to the project's advantage. The Dover Sun House made Telkes a media celebrity. However, the house did not last more than a few years, plagued by material corrosion, and the public moved away from being interested in solar power with the arrival of "cheap and abundant petroleum" (Gomez 2023).

3.3.3 Solar-Powered Oven

When Telkes was at the New York University (NYU) College of Engineering, she received a Ford Foundation grant of $45,000 to develop a low-cost solar-powered oven. The grant required that the oven must heat to at least 350 °F (175 °C) and be easy to set up and use. While her oven was not the first of its kind (French scientists were building solar cookers in the 1870s), its low cost and simplicity made it attractive for rural areas in developing countries. It used a heat-absorbing chemical, where the sun's rays entered the box through a window. Four mirrors arranged inside the box reflected sunlight within the box. It stayed hot as long as the sun was shining. The "stove design eliminate[d] the need for solar reflectors" (Magat 1954) and retained heat for about 30 minutes after sunlight disappeared.

> Her stove is a closely insulated box, roughly triangular in shape. Four ordinary mirrors fan out from the tilted face of the stove. At the rear of the stove is a removable drawer through which the food is placed (Magat 1954).

Telkes projected the oven's cost to be $5 in production. She asked a reporter, Sidney Fields, from the *Daily News* in 1964, "Can you imagine what a fantastic boon a solar stove would be…in countries with no fuel or electricity?" (Fields 1964). Several newspapers showed "Countess Stella Andrassy…[who] carried out cooking experiments…on a Manhattan roof" (Ubell 1954) using the oven, baking cakes, steaks, lentils, and rice. This NYU project used staff from its planned introduction locations: For example, Fatolah Sotoodeh, an NYU doctoral student in industrial engineering from Tehran (Magat 1954).

3.3.4 Other Research

In 1958, Dr. Telkes was named the Director of joint programs between NYU and Curtis-Wright Corporation for solar energy research and applications, establishing an NYU Solar Research Laboratory at the Curtis-Wright Princeton division (SWE 1958).

3.4 Later Career

In 1960, Telkes finished a solar energy center in Princeton, NJ, with solar-heated rooms, solar stoves, solar-heated pools, and much more (SWE Philadelphia n.d.). About a decade later, at the University of Delaware, she was involved with another solar-based structure (NIHF n.d.), Solar One, was built in 1971. This house generated electricity and heat from solar photovoltaic cells (Rinde 2020).

Around 1961, she also researched space-proof and sea-proof materials to protect equipment in extreme temperatures at Cryo-Therm. These included ideas to ensure a stable temperature for delicate instruments in shipping containers used for missile and satellite systems (ICWES 1964). Telkes predicted, "Within 10 years, we should have solar traps on roofs to convert sunshine into electricity for our homes and even for a whole community" (Fields 1964).

After using the solar-powered oven developed by Telkes in 1954, Countess Andrassy and Dr. Telkes collaborated on another invention, a cold frame (US-3206892-A in 1965), to stimulate plant propagation and growth.

In 1972, Dr. Telkes moved to the University of Delaware's Institute of Energy Conversion. The Institute aimed to "substitute the clean energy of the sun for at least some of the polluting fuels" (Telkes 1972a). She was also the Adjunct Professor of Energy Conversion at that University. (Telkes 1972b).

Mária Telkes died in 1995, in her native country, Hungary, where she had visited for the first time in 70 years (Saxon 1996).

3.5 Patents

Dr. Telkes was awarded 27 patents in her lifetime. The range of her interests is given in Table 3.1.

Table 3.1 US Patent and Trademark Office (US PTO) Patents by Mária Telkes

#	US PTO number	Patent name	Description	Owner	Date	Assignor (if any)
1	US-2229481-A	Thermoelectric couple	Conversion of thermal energy to electrical energy for thermocouples	Mária Telkes	January 21, 1941	Westinghouse Electric & Mfg. Co.
2	US-2229482-A	Thermoelectric couple	Conversion of thermal energy to electrical energy for thermocouples	Mária Telkes	January 21, 1941	Westinghouse Electric & Mfg. Co.
3	US-2246329-A	Heat absorber	Thermoelectric generators…heat absorber utilized in transmitting heat	Mária Telkes	June 17, 1941	Westinghouse Electric & Mfg. Co.
4	US-2289152-A	Method of assembling thermoelectric generators	Method for assembling thermoelectric generators	Mária Telkes	July 7, 1942	Westinghouse Electric & Mfg. Co.
5	US-2366881-A	Thermoelectric alloys	Thermoelectric alloys for use in thermocouples	Mária Telkes	January 9, 1945	Research Corporation
6	US-2595905-A	Radiant energy heat transfer device	Radiant energy heat transfer devices, including a heat-absorbing cell, contain a heat storage and transfer medium.	Mária Telkes	May 6, 1952	None
7	US-2677243-A	Method and apparatus for the storage of heat	Apparatus for storing heat, utilizing the principle of the heat of fusion, and a process for releasing heat thus stored	Mária Telkes	May 4, 1954	None
8	US-2677367-A	Heat storage unit	Radiant energy heat storage devices… employing cells or bins containing compositions of matter for heat storage	Mária Telkes	May 4, 1954	None
9	US-2677664-A	Composition of matter for the storage of heat	Composition of matter for use in the storage of heat	Mária Telkes	May 4, 1954	None

(continued)

Table 3.1 (continued)

#	US PTO number	Patent name	Description	Owner	Date	Assignor (if any)
10	US-2808494-A	Apparatus for storing and releasing heat	Heating furnaces are space heaters in which heat may be stored	Mária Telkes	October 1, 1957	None
11	US-2856506-A	Method for storing and releasing heat	Method of storing and releasing heat	Mária Telkes	October 14, 1958	None
12	US-2915397-A	Cooking device and method	An article comprising material to be cooked is brought to cooking temperature by directly contacting it with a surface heated to a cooking temperature or above	Mária Telkes	December 1, 1959	None
13	US-2936741-A	Temperature stabilized fluid heater and a composition of matter for the storage of heat therefor	Fluid heater	Mária Telkes	May 17, 1960	None
14	US-2989856-A	Temperature stabilized container and materials therefor	Improved container for maintaining articles packed therein within a specified range of temperatures	Mária Telkes	June 27, 1961	None
15	US-3206892-A	Collapsible cold frame	Adapted to provide passage of light and infrared rays for propagation and stimulation of plant growth	Mária Telkes & Stella Andrassy	September 21, 1965	Dow Chemical
16	US-3248464-A	Method and apparatus for making large-celled material	Improved methods for making panels of improved large celled, substantially transparent materials and equipment for making such panels	Mária Telkes	April 26, 1966	Dow Chemical

(continued)

Table 3.1 (continued)

#	US PTO number	Patent name	Description	Owner	Date	Assignor (if any)
17	US-3270515-A	Dew collecting method and apparatus	Methods of and apparatus for collecting water from the atmosphere	Mária Telkes	September 6, 1966	Melpar, Inc.
18	US-3415719-A	Collapsible solar still with water vapor permeable membrane	Solar still has a water vapor permeable bag for retention of brackish water and for use as the evaporator-absorber unit of the still	Mária Telkes	December 10, 1968	Melpar, Inc.
19	US-3440130-A	Large-celled material	Light transmitting panel having a core of compacted hollow cells of light transmitting material	Mária Telkes	April 22, 1969	Dow Chemical
20	US-3695903-A	Time/ temperature indicators	Device for detection and indication of thawing a frozen object	Máia Telkes & henry Hahn	October 3, 1972	American Std, Inc.
21	US-3986969-A	Thixotropic mixture and making of same	Mixture for storage of heat energy utilizing a heat of fusion material (sodium sulfate decahydrate, borax, and aluminum silicate)	Mária Telkes	October 19, 1976	University of Delaware
21	US-3986969-A	Thixotropic mixture and making of same	Mixture for storage of heat energy utilizing a heat of fusion material (sodium sulfate decahydrate, borax, and aluminum silicate)	Mária Telkes	October 19, 1976	
22	US-4011190-A	Selective black for absorption of solar energy	Improved selective black particles for use in solar receivers	Mária Telkes	March 8, 1977	Ses Inc.

(continued)

Table 3.1 (continued)

#	US PTO number	Patent name	Description	Owner	Date	Assignor (if any)
23	US-4010620-A	Cooling system	A system for using solar energy for cooling a building	Mária Telkes	March 8, 1977	University of Delaware
24	US-4034736-A	Solar heating method and apparatus	Solar air heating, and more particularly to a method and apparatus for heating gases which selectively uses heat from the winter sun and rejects heat from the summer sun	Mária Telkes	July 12, 1977	University of Delaware
25	US-4187189-A	Phase change thermal storage materials with crust forming stabilizers	Relates to materials which utilize solar energy or ambient temperature change for the storage of thermal effects	Mária Telkes	February 5, 1980	American Technological University
26	US-4250866-A	Thermal energy storage to increase furnace efficiency	Relates to air heating furnace systems	Mária Telkes	February 17, 1981	Research Institute for Advanced Technology
27	US-4954278-A	Eutectic composition for coolness storage	Phase change material which melts and freezes at a temperature below 32 degrees F	Calvin D. MacCracken & Mária Telkes	September 4, 1990	Calmac Mfg. Co.

3.6 Awards, Recognition, and Professional Organizations

Dr. Mária Telkes received what was initially named "the SWE Award for Meritorious Contribution of Engineering", *in recognition of her pioneering work in the utilization of solar energy* (SWE 2020). She is shown receiving the award in Fig. 3.3 in Chicago IL in 1952. This award was later renamed the SWE Achievement Award and marked the beginning of SWE's individual awards program (SWE 1952).

Fig. 3.3 Mária Telkes receives the first Society of Women Engineers Achievement Award during the 1952 American Society of Civil Engineers *Centennial of Engineering* in Chicago, Illinois. Pictured *(l to r)* Rodney Chipp, Beatrice Hicks, Mária Telkes, unknown, Dot Merrill, unknown (Photograph courtesy of Society of Women Engineers Records, Walter P. Reuther Library, Wayne State University)

Dr. Telkes also received these honors and awards (Lemelson-MIT n.d.; Rafferty 2024; Telkes 1977; SWE 1977):

- *OSRD Certificate of Merit*, 1945: Solar distiller.
- *Lifetime Achievement Award*, National Academy of Sciences Building Research Advisory Board. 1977: Significant contributions to building science and technology.
- *Charles Greeley Abbot Award,* American Section of the International Solar Energy Society. 1977: In recognition of (her) outstanding contributions to the utilization of solar energy for humankind.
- *National Inventors Hall of Fame*. 2012: Solar energy work.

Telkes was "a member of the American Chemical Society, the Electrochemistry Society, Sigma Xi, and the Business and Professional Women's Club, and a lifetime member of the Society of Women Engineers" (Telkes 1952).

3.7 SWE Involvement

Dr. Telkes was a SWE Life member and active in the Philadelphia Section from 1964 to 1977. “Mária was a lively conversationalist at Section dinner meetings at the Engineers’ Club of Philadelphia, and she frequently expounded on the virtues of solar power” (SWE Philadelphia n.d.). At one point, she hosted local SWE members on a visit to the Dover Sun House.

In 1964, at the First International Conference of Women Engineers and Scientists (ICWES), sponsored by SWE, Dr. Telkes, then Director of Research and Development, Cryo-Therm, participated in a *Symposium on Developing Engineering and Scientific Talent* (ICWES 1964, p 281). She spoke about her life’s work in solar energy at MIT and her later endeavors, linking her work and vision to the conference theme, *Focus on the Future*. Closing with

> It is most appropriate that our untapped potential should develop the untapped energy source for the most significant benefit of the least developed group of humanity. Perhaps there will be some of you who will “Reach for the Sun.”

A member of the Southwest Texas section for a few years, in 1980, she moved to Maryland to work for Monegon Ltd., a design and analysis of energy systems firm after wishing her fellow section members a fond farewell. (Telkes 1980).

3.8 Legacy

Mária Telkes was known as the *Sun Queen* for her research into solar energy and solar-powered devices. She invented, designed, and constructed solar-powered homes, a solar still converting salt water into fresh water, and a low-cost solar oven for rural and developing areas. Dr. Mária Telkes was the recipient of the inaugural SWE Achievement Award in 1952. Recognized for her significant contributions by the Office of Scientific and Research Development, the National Academy of Sciences Building Research Advisory Board, the American Section of the International Solar Energy Society, and the National Inventors Hall of Fame, Telkes left a mark on the world and inspires all with her accomplishments and passion for helping the world by providing renewable resource products and concepts, harnessing the sun’s energy.

Telkes described herself as “blue-eyed, blond-haired, 5’7” and fond of music” (Telkes 1950). This quotation from the SWE Records sums up her life in her own words:

> I have been interested in Chemistry and Physics during my high school days. My aim was to develop new energy sources, especially the use of Solar Energy. I have studied with this aim in mind, and nothing could have stopped me from working in this field. If I have succeeded to some extent, I feel that I enjoy my scientific work more than any other activity (Telkes n.d.).

References

Boyd, Andrew. n.d. No. 2608: Maria Telkes. The Engines of Our Ingenuity, Cullen College of Engineering, University of Houston. Accessed March 7, 2025. https://engines.egr.uh.edu/episode/2608

Crile, George W., Maria Telkes and Amy F. Rowland. 1929. An Experimental Investigation of the Physical Nature of Death. *Proceedings of the American Philosophical Society*, Vol. 68, No. 2, 69-81. Accessed October 9, 2023. https://www.jstor.org/stable/984307

Daniels, Farrington. 1950. Atomic and Solar Energy. *American Scientist*. Vol. 38, No. 4, October 520-548. Accessed https://www.jstor.org/stable/27826335

Encyclopedia of World Biography. 2025. Telkes, Maria. Encyclopedia.com. Last revision February 10, 2025. https://www.encyclopedia.com/history/encyclopedias-almanacs-transcripts-and-maps/telkes-maria

Fields, Sydney. 1964. Harnessing the Sun. *Daily News*. June 19. From *Maria Telkes*, Society of Women Engineers Records (LR001539). Box 193, Folder 9, Walter P. Reuther Library, Archives of Labor and Urban Affairs, Wayne State University.

First International Conference of Women Engineers and Scientists (ICWES). 1964. Contribution of Maria Telkes. Symposium on Developing Engineering and Scientific Talent. Proceedings. Library of Congress. Catalog Card Number 64-24332.

Gomez, Adrian. 2023. PBS Documentary looks at the life of solar power pioneer Maria Telkes. Albuquerque Journal. April 2 [updated June 7, 2023]. Accessed March 7, 2025 https://www.abqjournal.com/news/local/pbs-documentary-looks-at-the-life-of-solar-power-pioneer-m-ria-telkes/article_1617ee72-d84a-540b-8881-fc85f0a49d71.html

Hernandez, Axel Alfaro. 2024. A Solar Life. USPTO Monthly Review, April 2024. Accessed March 5, 2025 https://www.uspto.gov/learning-and-resources/journeys-innovation/historical-stories/solar-life?

ISES SWC50. 2025. The Story of Solar-Stories and Visions. https://www.swc50.org/celebratory-booklet

Lemelson-MIT. n.d. Maria Telkes: Telkes Solar Cooker. Accessed June 16, 2024. https://lemelson.mit.edu/resources/maria-telkes

Magat, Richard. 1954, November 13. Sun Harnessed for Cooking. *Science News Letter.* From *Maria Telkes*, Society of Women Engineers Records (LR001539). Box 193, Folder 9, Walter P. Reuther Library, Archives of Labor and Urban Affairs, Wayne State University.

Mayer, Johanna, Katie Hafner, and The Lost Women of Science Initiative. 2023. This Biophysicist 'Sun Queen' Harnessed Solar Power. Scientific American. November 16. Accessed June 12, 2024 https://www.scientificamerican.com/article/maria-telkes-the-biophysicist-who-harnessed-solar-power/

Google. 2022.Celebrating Maria Telkes. December 12. Accessed June 16, 2024 https://doodles.google/doodle/celebrating-maria-telkes/

National Inventors Hall of Fame (NIHF). n.d. Maria Telkes. Accessed June 12, 2024 https://www.invent.org/inductees/maria-telkes

Rafferty, J.P. 2024. Mária Telkes. Encyclopedia Britannica, December 8, 2024. https://www.britannica.com/biography/Maria-Telkes

Rinde, Meir. 2020. The Sun Queen and the Skeptic: Building the World's First Solar Houses. Distillations. July. Accessed July 5 2021 https://www.sciencehistory.org/distillations/the-sun-queen-and-the-skeptic-building-the-worlds-first-solar-houses)

Saxon, Wolfgang. 1996. Maria Telkes, 95, an Innovator of Varies Uses for Solar Power. New York Times. August 13. https://www.nytimes.com/1996/08/13/us/maria-telkes-95-an-innovator-of-varied-uses-for-solar-power.html?searchResultPosition=1

Society of Women Engineers Philadelphia (SWE Philadelphia). n.d. Notable Members: Maria Telkes. Hall of Fame (H-Z). Accessed March 7, 2025. https://philadelphia.swe.org/hall-of-fame-h-z/

Society of Women Engineers (SWE). 1952. Maria Telkes, Achievement Award, Centennial of Engineering. November. From Society of Women Engineers Records, Walter P. Reuther Library, Archives of Labor and Urban Affairs, Wayne State University. https://reuther.wayne.edu/node/1380

Society of Women Engineers (SWE). 1958. Dr. Maria Telkes Appointed Solar Research Director. *Newsletter*. November. From Society of Women Engineers Records, Box 193, Folder 9, Walter P. Reuther Library, Archives of Labor and Urban Affairs, Wayne State University.

Society of Women Engineers (SWE). 1977. BRAB honors Telkes for significant contributions. *Newsletter*. May/June. From Society of Women Engineers Records, Box 193, Folder 9, Walter P. Reuther Library, Archives of Labor and Urban Affairs, Wayne State University.

Society of Women Engineers (SWE). 2020. Celebrating 70 years of SWE. *All Together*. May 27. https://alltogether.swe.org/2020/05/celebrating-70-years-of-swe/

Telkes, Maria. n.d. Comment. From *Maria Telkes*, Society of Women Engineers Records (LR001539). Box 193, Folder 9, Walter P. Reuther Library, Archives of Labor and Urban Affairs,Wayne State University.

Telkes, Maria. 1950. Biographical Notes. From *Maria Telkes*, Society of Women Engineers Records (LR001539). Box 193, Folder 9, Walter P. Reuther Library, Archives of Labor and Urban Affairs,Wayne State University.

Telkes, Maria. 1952. Current Biography. From *Maria Telkes*, Society of Women Engineers Records (LR001539). Box 193, Folder 9, Walter P. Reuther Library, Archives of Labor and Urban Affairs,Wayne State University.

Telkes, Maria. 1972a. *Letter dated July 11, 1972, to Carolyn Phillips, SWE Secretary*. From *Maria Telkes*. Society of Women Engineers Records (LR001539). Box 193, Folder 9, Walter P. Reuther Library, Archives of Labor and Urban Affairs,Wayne State University.

Telkes, Maria. 1972b. *Letter dated September 1, 1972, to Carolyn Phillips, SWE Secretary*. From *Maria Telkes*. Society of Women Engineers Records (LR001539). Box 193, Folder 9, Walter P. Reuther Library, Archives of Labor and Urban Affairs,Wayne State University.

Telkes, Maria. 1977. *Letter dated August 1, 1977, to Barbara Krohn, Editor, SWE Newsletter*. From *Maria Telkes*. Society of Women Engineers Records (LR001539). Box 193, Folder 9, Walter P. Reuther Library, Archives of Labor and Urban Affairs,Wayne State University.

Telkes, Maria. 1980. *Note on newsletter dated April 14, 1980, to Southwest Section*. From *Maria Telkes*. Society of Women Engineers Records (LR001539). Box 193, Folder 9, Walter P. Reuther Library, Archives of Labor and Urban Affairs,Wayne State University.

The Indian. 1948. Ain't science grand—Drop that coal shovel. *The Indian*. Vol. III, No. 35. October 9. US Naval Operating Base, Guantanamo Bay, Cuba. Accessed from https://jstor.org, dated September 11, 2023.

Ubell, Earl. 1954. $5 Stove Traps Suns' Rays to Bake, Broil, and Cook. *New York Herald Tribune*. March8. From *Maria Telkes*, Society of Women Engineers Records (LR001539). Box 193, Folder 9, Walter P. Reuther Library, Archives of Labor and Urban Affairs,Wayne State University.

US Patent and Trademark Office (US PTO). 1968. Collapsible solar still with water vapor permeable membrane. https://ppubs.uspto.gov/3415719/

Chapter 4
Elsie Gregory MacGill (1905–1980) Achievement Award Recipient 1953

In recognition of her meritorious contributions to aeronautical engineering

Janet L. Williams

4.1 Introduction

Elsie MacGill (Fig. 4.1) was a pioneering force in aeronautical design and manufacturing, and a powerful voice for women's rights throughout her lifetime. Like all pioneers, she marked many "firsts" in her education and career. Her work as chief aeronautical engineer from 1938–1943 at The Canada Car and Foundry Company, Ltd. (Can Car), where she was engaged in the manufacture of Hawker Hurricanes, contributed to the war effort and earned her the moniker "Queen of the Hurricanes." In her later career, as an aeronautical engineering consultant, she continued to perform aircraft design and served on panels and committees aimed at international aircraft safety. MacGill was honored by the Society of Women Engineers as the 1953 Achievement Award recipient. She also pursued equity for women in Canada, serving as one of seven commissioners on the Royal Commission on the Status of Women and tirelessly advocating for the adoption of the recommendations generated by the Commission after publication of its report in December 1970 (Morris 2006).

J. L. Williams (✉)
Albuquerque, NM, USA

C. D. Craig et al. (eds.), *Women Engineering Legends 1952-1976: Society of Women Engineers Achievement Award Recipients*,
https://doi.org/10.1007/978-3-032-00224-2_4

Fig. 4.1 Elsie MacGill (Library and Archive Canada, Public domain)

4.2 Early Life

Elizabeth "Elsie" MacGill had a remarkable home environment for the early 1900s. Her father was an immigration attorney, and her mother, Helen MacGill, was a college-degreed journalist who advocated for women's rights in British Columbia. She was the first woman to graduate with a degree from Trinity College (now the University of Toronto), and she went on to earn an MA degree in 1890. When Elsie MacGill was 12, her mother was appointed judge of the Juvenile Court of Vancouver, making her the first woman jurist in British Columbia (Black 2008). As a result, Elsie MacGill grew up in an environment where social justice, feminism, and equal education were persistent themes. Both her mother and grandmother were suffragists and influenced MacGill in this regard (Sissons 2007).

MacGill and her sister Helen enjoyed an enriched home-school educational setting, enhanced during preschool years by studies in art, swimming, and other skills. She attended Lord Roberts Elementary School and King George Secondary School. The family had moved from their beloved home when her father's law business fell victim to racial unrest, and her mother bought a cabin outside of town as a refuge, while her father spent most of his time in town. As this older cabin required periodic repairs, MacGill acquired a reputation as "Miss Fix-it," attending to various electrical and mechanical tasks around the house (Bourgeois-Doyle 2008).

When she was 16 and her sister 18 they both enrolled at the newly established University of British Columbia (UBC). Due to her interest in science and her penchant for home repairs, Elsie declared she would major in engineering. However, after one term, the Dean of Applied Sciences insisted she could not study engineering at UBC. She spent 2 years in the general education curriculum before embarking on the next chapter of her education (Bourgeois-Doyle 2008).

MacGill heard that she could study electrical engineering at the University of Toronto. She applied and matriculated in 1923. When she graduated in 1927 (Fig. 4.2), she became the first woman in Canada to obtain a BS in Electrical Engineering. She was one of just two women among the 98 members of the Class of 1927 pictured in Applied Sciences and Engineering in the yearbook. The other woman took a degree in architecture (Internet Archive 1927).

4.3 Early Career

After graduation, MacGill found a job in the burgeoning motor car industry in Pontiac, Michigan, and moved there in the summer of 1927. She was first drawn to aeronautical engineering during this period of her first paid employment. This field of engineering had only recently been introduced at the University of Toronto as a standalone discipline. Fortunately for MacGill, one of the institutions with a pioneering program in aeronautical engineering was the University of Michigan in Ann Arbor, not far from Pontiac. Enrolling with a Fellowship in the Master of Science Program in 1927, she graduated in 1929 as the first woman to receive a master's degree in Aeronautical Engineering in North America (Canada: The Story of Us 2017).

However, before embarking on her career, she had to endure a major setback in her life. Just before her final exams in 1929, she awoke and was unable to feel her

Fig. 4.2 Elsie MacGill graduation photo from the University of Toronto, 1927. (Courtesy of University of Toronto Archives and Robert Lansdale Photography)

legs. She was diagnosed at the University hospital with "acute infectious myelitis." Though the word polio was not used, she suffered exactly the kind of debilitation that a polio victim would experience. She was told she would likely never walk again. Consistent with the determination she had shown to date, MacGill did not accept that prognosis. She moved back to Vancouver with her family and began the lengthy process of rehabilitation and recovery. It was more than a year before she began to recover movement in her lower extremities. As she became more mobile with the help of a wheelchair, she began reading journals and jotting down design ideas. She wrote articles about aeronautics, engineering, and other technical topics. MacGill developed a talent for writing about technical topics in terms accessible to the public and generated articles for mainstream publications such as *Vanity Fair* and *Chatelaine* as well as technical journals. Slowly, MacGill began to regain sensation in her legs, and by 1931 had taught herself to walk with two canes. Finally, 3 years after receiving her master's degree, she felt ready to resume her career, but felt she needed additional education (Bourgeois-Doyle 2008).

4.4 Career Pivot

By 1932, the Massachusetts Institute of Technology (MIT) had become an American powerhouse of aeronautical engineering. With financial help from her parents, she moved to Boston and took up post-graduate studies with the idea of obtaining a doctorate, but her depleted finances stymied this goal. In 1934, she dropped out and gratefully accepted a position at Fairchild Aircraft's new plant in Longueuil, Quebec, offered through the connections of one of her professors. The new plant began manufacturing the American-designed Fairchild 71 and proceeded to development and manufacture of several Canadian designs. MacGill and her team designed the first Canadian-made all-metal Fairchild Super 71 and 71P, performing stress analysis and wind tunnel testing to solve difficult design issues. These aircraft were not commercial successes but taught the budding aeronautical engineer valuable lessons.

In 1937, she heard of an opening at Can Car in Fort William, Ontario (now Thunder Bay) through her friends Helen and Bill Soulsby. He had taken the plant manager job, and the company needed additional creative talent as they entered the aircraft market. Shortly thereafter, in May 1938, she accepted a position as Chief Aeronautical Engineer (Fig. 4.3). Just before her move, she received the thrilling news that she had been accepted as the first woman member of the Engineering Institute of Canada (Bourgeois-Doyle 2008).

MacGill settled in at Can Car and worked on several aircraft projects—with mixed commercial results—but she impressed the management with her expertise. Her first big opportunity to design an aircraft from the ground up came when the Mexican government asked for two-seat trainers that would function in high-altitude conditions. She had the freedom to incorporate many ideas from her portfolio to meet the exacting specifications for the trainer, dubbed the Maple Leaf Trainer II (Fig. 4.4). It was a technical success, though not a commercial success due to

Fig. 4.3 Elsie MacGill, chief aeronautical engineer (Courtesy of Ingenium; 16035)

competition in the military aircraft market. This project made MacGill the first woman engineer to design and build an aircraft (Sissons 2007).

4.5 World War II

In 1938, with the threat of war looming, the British began quietly seeking ways to manufacture more Hawker Hurricanes, their fighter plane of choice. In December, they awarded a contract to Can Car for production of 40 planes (Fig. 4.5).

As Chief Aeronautical Engineer, MacGill's job was to use the 5000-plus aircraft drawings provided to design all the tooling and equipment needed to manufacture and assemble the Hurricanes, as quickly as possible (Mackie 2020). On September 9, 1939, Britain declared war on Germany, making speed imperative. The Battle of Britain, with its onslaught of German aircraft wreaking havoc over England, highlighted the demand for increased production capacity for fighter planes like the

Fig. 4.4 Elsie MacGill and Bill Soules with Maple Leaf Trainer II (Courtesy of Ingenium; 0497)

Hawker Hurricane. Over 1400 Hurricanes were rolled off the assembly line at Can Car over the next 5 years. Britain's ability to replace planes lost in combat more quickly than the enemy was a decisive factor in this conflict. Though Canada's contribution in the total number of planes used in WWII was relatively small, its ability to provide Canadian pilots in Canadian-built planes was symbolic of Canada's pride in their role in the war effort, and 1400 Hurricanes were no small contribution. MacGill's role in this accomplishment earned her the nickname "Queen of the Hurricanes." She was featured in many wartime articles, including a *True Life* comic book (Fig. 4.6) that featured stories of heroic wartime efforts (True Comics #8 1942).

Her efforts included designing a winterized version of the Hawker, building the first de-icer on a fast fighter plane and a design for adding skis that would allow it to land on snow and ice—a necessity for the cold conditions at the Eastern front (Baguely 2024). She and her staff also worked on designs to modify the Hurricane for catapult take-off from ships at sea (as used on aircraft carriers) (Bourgeois-Doyle 2008).

During this period, MacGill published technical papers that described her concepts for streamlined production, separation of component manufacture and assembly, and standardization of materials and components. These concepts allowed more rapid production of the Hurricanes and ensured the interchangeability of parts on location in Europe (Valour Canada 2024). Her ideas captured the interest of her

Fig. 4.5 Hawker Hurricanes at Can Car, Sept 1941 (Military Aviation Museum)

Fig. 4.6 True Life Comics January 1942 story of Èlsie MacGill (Canada Aviation and Space Museum)

colleagues, who saw long-term applicability in the aviation industry. In 1940, the Engineering Institute of Canada awarded her the Gzowski Medal, their highest award, in recognition of her novel ideas to improve production in aeronautical engineering.

In May 1942, as Hurricane production was winding down, Can Car won a contract to build American Curtiss SB2C Helldivers for the United States Navy. MacGill oversaw design of the retooling for production of this effort. She and plant manager Bill Soulsby, now widowed, struggled to incorporate numerous design changes from Curtiss (ordered by the Navy). Things did not proceed smoothly, delaying the production of the aircraft as promised. There were also whispers that MacGill and Soulsby were in a relationship only a few months after his wife had died. Before the year ended, they either left of their own accord or were dismissed from Can Car. Undaunted, the two married on June 4, 1943, near Chicago, IL, with her sister and brother-in-law in attendance. After a brief honeymoon, they moved to Toronto, where MacGill began her own engineering consulting firm, another first for a woman engineer (Bourgeois-Doyle 2020). Soulsby became plant manager at nearby Victory Aircraft Limited (Bourgeois-Doyle 2008).

4.6 Later Career and Advocacy

In the 1940s, MacGill became aware of technical issues in commercial aviation through her contacts in government. In 1946, she was appointed as a Technical Advisor for the International Civil Aviation Organization (ICAO), affiliated with the United Nations, working on the aircraft structures subcommittee. This appointment presented her many opportunities to keep up with the latest developments in aviation and aircraft design. In her consulting firm, government work was abundant throughout the war. When it tapered off in 1945, her stature grew in the ICAO, and she added speaking engagements on aviation to her business activities. She was also appointed chair of the Stress Analysis Committee under the ICAO, making her the first woman to chair a United Nations Committee (Bourgeois-Doyle 2020; Sissons 2007). As an independent female businesswoman in the 1950s, she affiliated with women's business and professional clubs, moving beyond engineering-related organizations. This marked the beginning of her increased interest in political and social issues (Bourgeois-Doyle 2008).

MacGill suffered a broken leg from a fall at home in 1953, forcing her back into a wheelchair for an extended convalescence. She took advantage of her forced immobility by writing a biography about her mother, Helen Gregory MacGill, who had retired from her position as juvenile court judge in 1945, and had died in 1947. The book kindled a heightened interest in the struggles for social justice that had dominated her mother's life (Bourgeois-Doyle 2008, 2020). She became involved in the Toronto Business and Professional Women's clubs, ascending to serve at the provincial level of the organization in Ontario. She was a member of the Soroptimist Club, which named her Soroptimist of the Month in their November 1953 publication of The American Soroptimist (Soroptimist of the Month 1952). She was elected national president of the Canadian Federation of Business and Professional Women's Clubs in 1962, a position she held for 2 years. These years in service organizations propelled her into a second career of advocacy for women's rights and other social issues.

4.7 Advocacy

After President John F. Kennedy launched a Commission on the Status of Women in 1961, women in Canada started lobbying their prime minister and other officials for a similar body to be established in Canada to examine discrimination against women and gaps in women's rights. Eventually, in 1967, Prime Minister Lester Pearson's government established the Royal Commission on the Status of Women in Canada. Ottawa journalist and broadcaster Florence Bird was named chair of the commission. Two men and four women were appointed to serve under Ms. Bird. One of the new commissioners was Elsie MacGill (Fig. 4.7).

The Commissioners investigated and reported on the status of women in Canada through public hearings and collection of personal testimonies. At the conclusion of their activities, they provided recommendations to ensure equality for all women in Canada. They published their report in 1970, with 167 recommendations on topics ranging from birth control to childcare to maternity leave, pay equity, and family law reform (Morris 2006). MacGill was not content to merely finish the report. She

e010858739

Fig. 4.7 Royal Commission on the Status of Women in Canada; Elsie MacGill seated on the far left. (W. H. "Bill" Olson, Dominion Wide Photographs Limited, Library and Archives Canada)

began to actively promote and publicize the findings across the country, through speaking engagements with women's groups and clubs, including those of which she remained a member. She spread the word about women's right to be in the workplace and to receive equal pay to their male counterparts. Her activism after the report was released helped establish paid maternity leave for new mothers in Canada. It also led to equity in marriage laws, in matters such as the waiting period for divorce, the right to keep one's maiden name, and the ability for a married woman to obtain her own passport (Championing Women in Canada 2023).

4.8 Honors and Legacy

The list of awards and honors received by Elsie Gregory MacGill is long and distinguished. Her technical accomplishments have been honored many times, from the Gzowski Medal in 1940 to her honorary doctorates from the University of Toronto (Fig. 4.8), conferred in 1973, and the University of Windsor in 1976. If the SWE Achievement Award in 1953 opened the door to numerous technical recognitions, her service on the Royal Commission of the Status of Women in Canada brought her many more honors for her service to women in Canada. MacGill was bestowed the Order of Canada in 1971 and the Queen's Silver Jubilee Medal in 1977 (SWE Records n.d.).

In 1975, the Ninety-Nines, an exclusive club of licensed women pilots formed in 1929, awarded her the Amelia Earhart Medal, named after one of its most famous

University of Toronto Archives. 2008-58-2MS

Fig. 4.8 MacGill receiving an Honorary Degree at the University of Toronto in 1973 (Courtesy of University of Toronto Archives and Robert Lansdale Photography)

founders. Multiple Canadian schools are named after MacGill, and in 2016 a street in Montreal was named for her (Gyulai 2016). She is featured in numerous air, space, and flight museums across Canada, and she continues to accumulate honors in more recent years. In 2019, Canada Post included her image in a series on Canadians in Flight (Fig. 4.9). The Royal Canadian Mint issued a circulating $1C coin in her honor in 2023 and published an illustrated biography as part of the release (Honouring Elsie MacGill 2023; SWE Records n.d.).

Articles and blogs document her innovations and lasting contributions to the aerospace industry. These accolades are partly attributable to her litany of firsts as a woman engineer, from her education to her technical and business accomplishments. They are also attributable to her grit, determination, and perseverance in overcoming obstacles to success for pioneering women in technical fields as well as people with disabilities. Her ability to turn challenges and setbacks into opportunities to move forward is often noted and admired among authors who have written about MacGill. Finally, her contributions to aeronautical engineering design and production and to global aviation safety have certainly contributed to her continued fame and public admiration.

A University of Toronto newsletter opinion article from 2016 summed up the national ethos and consummate admiration for MacGill elegantly:

> Elsie became a symbol of the economic changes in Canada during World War II... Her contribution to the war effort made her a Canadian war hero. She contributed to the freedoms we take for granted today…. Elsie defined Canada and Canadian values through her actions and accomplishments. She was a pioneer for women in engineering and business, a war hero and a role model (Amon et al. 2016).

Fig. 4.9 Stamp issued in 2019 as part of "Canadians in Flight" series (©Canada Post Corporation, 2019. Reproduced with permission)

References

Amon, Christina; Wells, Mary; and Woodhouse, Kim. 2016. Why the next Canadian woman on a banknote should be Elsie MacGill. *University of Toronto News*, December 7. https://www.utoronto.ca/news/why-next-canadian-woman-banknote-should-be-elsie-macgill

Baguely, Richard. 2024. Queen of the Hurricanes. *Hackaday Blog,* December 27. https://hackaday.com/2019/05/07/the-queen-of-the-hurricanes/

Black, Naomi. 2008. "Helen Gregory MacGill." *The Canadian Encyclopedia.* Historica Canada. Article published April 01; Last Edited October 28, 2020. https://www.thecanadianencyclopedia.ca/en/article/helen-gregory-macgill

Bourgeois-Doyle, Richard I. 2008. *Her Daughter the Engineer: The Life of Elsie Gregory MacGill.* National Research Council of Canada Press. ISBN 978-0-660-19813-2

Bourgeois-Doyle, Dick. 2020. The Rise of Elsie MacGill. *University of Toronto Engineering News,* October 1. https://news.engineering.utoronto.ca/the-rise-of-elsie-macgill/

Canada: The Story of Us. 2017. Canadian Elsie MacGill was the first female aeronautical engineer in the world. *Canada: the Story of Us* blog series, The Canadian Broadcasting Company, May 5. https://www.cbc.ca/2017/canadathestoryofus/canadian-elsie-macgill-was-the-first-female-aeronautical-engineer-in-the-world-1.4099967

Championing Women in Canada. 2023. Championing Women in Canada: 4 Ways Elsie MacGill Moved the Dial. *Royal Canadian Mint.* August 23. https://www.mint.ca/en/blog/2023-08-elsie-macgill-championing-women-in-canada

Gyulai, Linda. 2016. Elsie MacGill an 'icon' of why Montreal became an aerospace centre. *Montreal Gazette*. September 19. https://www.montrealgazette.com/news/article273415.html

Honouring Elsie MacGill. 2023. *Royal Canadian Mint.* https://www.mint.ca/en/discover/canadian-commemorative-circulation-program/2023/elsie-macgill?rcmeid=con_2308_ElsieMacGill_LandingPagefromBlog_EN

Internet Archive. 1927. Students' Administrative Council of the University of Toronto. *The Torontonensis,* (Yearbook for the University of Toronto), pp. 88-96.

Mackie, John. 2020. Elsie MacGill. *The Canadian Museum of Flight.* http://www.canadianflight.org/content/elsie-macgill

Morris, Cerise. 2006. Royal Commission on the Status of Women in Canada. *The Canadian Encyclopedia.* Historica Canada. Article published February 07; Last Edited March 22, 2016. https://www.thecanadianencyclopedia.ca/en/article/royal-commission-on-the-status-of-women-in-canada

Sissons, Crystal. 2007. Elsie MacGill. *The Canadian Encyclopedia.* Historica Canada. Article published December 18; Last Edited August 04, 2023. https://www.thecanadianencyclopedia.ca/en/article/elizabeth-muriel-gregory-macgill

https://archive.org/details/torontonensis27univ/page/96/mode/2up

Society of Women Engineers (SWE) Records (LR001539). n.d. *Elsie MacGill.* Box 193, Folder 9, Walter P. Reuther Library, Archives of Labor and Urban Affairs, Wayne State University.

Soroptimist of the Month. 1952. Soroptimist of the Month—High Flying Designer. *The American Soroptimist*, November. (courtesy of SWE Archives)

True Comics #8. 1942, pp. 17-21. January. https://comicbookplus.com/?dlid=17045

Valour Canada. 2024. Elsie MacGill: Queen of the Hurricanes. Valour Canada Military History Library. https://valourcanada.ca/military-history-library/elsie-macgill-queen-of-the-hurricanes/. accessed December 2024.

Chapter 5
Edith Clarke (1883–1959) Achievement Award Recipient 1954

In recognition of her many original contributions to stability theory and circuit analysis

Jill S. Tietjen

5.1 Introduction

A trailblazing electrical engineer, Edith Clarke's textbook on circuit analysis of AC power systems was the standard college textbook for years for power engineers. Clarke (Fig. 5.1) was the first woman to graduate with an MS degree in electrical engineering from MIT, the first professionally employed female electrical engineer in the USA, the first full-time female professor of electrical engineering at any college in the USA, and one of the first three women to become a fellow of what is today the Institute of Electrical and Electronics Engineers (IEEE). Clarke was inducted into the National Inventors Hall of Fame for her graphical calculator, which was used for decades by engineers planning transmission lines and is the basis for the algorithms in computer programs today. Her patent for the "Clarke Calculator" was one of the three patents that she was awarded.

J. S. Tietjen (✉)
Technically Speaking, Inc., Greenwood Village, CO, USA

C. D. Craig et al. (eds.), *Women Engineering Legends 1952-1976: Society of Women Engineers Achievement Award Recipients*,
https://doi.org/10.1007/978-3-032-00224-2_5

Fig. 5.1 Edith Clarke. (Courtesy of Society of Women Engineers Records, Walter P. Reuther Library, Wayne State University)

5.2 Life

Edith Clarke was born on a farm near Ellicott City in Howard County, Maryland, the daughter of a lawyer-gentleman farmer. The sixth of nine children, three of whom died young, and the fifth of seven daughters, she grew up on a 400-acre farm. Not a child prodigy, she probably had what today would be categorized as a learning disability. She did not read well in the early years of her schooling though she progressed by having family members read to her. She loved math and playing games (Goff 1946; Ogilvie & Harvey 2000; Sicherman et al. 1980).

Her father died when she was 7 years old and her mother when she was 12. A maternal uncle became her guardian, and her eldest sister oversaw the younger children. After completing her secondary education at age 16 from Briarley Hall, a boarding school for girls, she returned home. Not expected to work, Clarke enjoyed tennis, swimming, skating, and target practice. She was also very interested in foreign countries and their people. She loved all kinds of games, particularly card games. Clarke especially liked the games of Duplicate Whist (a predecessor to the game of bridge) and Auction. However, she realized her life could not continuously consist of card games. If she could have work, however, as interesting as Whist, then she thought she could enjoy that type of work (Goff 1946; Ogilvie & Harvey 2000; Sicherman et al. 1980).

When she turned 18, Clarke became of age and decided to use her inheritance to attend college. Her relatives were aghast at her decision to spend the principal from the inheritance. But, she remembered a conversation with her mother who had expressed admiration for a lawyer who had used his inheritance to finance his education. She was admitted to Vassar College. Clarke majored in mathematics and astronomy, became involved in Kappa Kappa Gamma, theatricals, athletics, and other activities, and graduated Phi Beta Kappa with an AB in 1908 with honors (Goff 1946; Howes 1939; Ogilvie & Harvey 2000; Sicherman et al. 1980).

Like many women of her time, Clarke became a teacher. She taught math and science for 3 years, first in San Francisco, then at Marshall College in West Virginia. "Teaching is not at all like a game of Duplicate Whist! I therefore turned to something else." After suffering a serious illness in 1911, she later told a reporter, "Thinking that I was going to die, I just decided to do what I wanted to do. Study engineering." Now healthy, she enrolled as a civil engineering undergraduate (sophomore) at the University of Wisconsin and remained there for the 1911–1912 academic year. In 1912, she took a summer job at the American Telephone & Telegraph Company (AT&T) as a computer assistant for Dr. George A. Campbell. The mathematical equations she was solving were so interesting that she stayed for 6 years. These calculations helped AT&T build transcontinental telephone lines from the East Coast to the West Coast (Brittain 2009; Clarke n.d.; Goff 1946; Howes 1935; National Inventors Hall of Fame 2018; Ogilvie & Harvey 2000; Rosengren 2023; Sicherman et al. 1980).

During the last of those 6 years (1915–1918), she supervised the women at AT&T who did computations for research engineers in the Transmission and Protection Engineering Department. She simultaneously studied radio at Hunter College and electrical engineering at Columbia University at night. The World War I outbreak led her to do her bit toward victory which she thought required a focus on her education. That decision was facilitated by a fellowship that she received from Vassar College. Thus, she enrolled at the Massachusetts Institute of Technology (MIT) in 1918 and received her master's degree in electrical engineering in 1919; she was the first woman to be awarded that degree from MIT. Her thesis was titled "Behavior of a lumpy artificial transmission line as the frequency is indefinitely increased" (Clarke n.d.; Goff 1946; Hobbs 2004; Ingels 1952; Ingram 2022; Ogilvie & Harvey 2000; Rosengren 2023; Sicherman et al. 1980).

Upon graduation, she wanted to work for either General Electric (GE) or Westinghouse. But World War I was over, and even with her stellar credentials, no one would hire her as an engineer because of her gender—they had no openings for a woman engineer! She later said, "There is no demand for women engineers, as such, as there are for women doctors; but there's always a demand for anyone who can do a good piece of work." From 1919 to 1921, she again worked as a "computer" at GE, directing women computers who were calculating the mechanical stresses in turbines for the turbine engineering department. She also served as president of the General Electric Women's Club in 1921 (Clarke n.d.; Goff 1946; Howes 1935; Ingram 2022; Proffitt 1999; Rosengren 2023; Sicherman et al. 1980).

But Clarke wanted to be an electrical engineer! Since that was not the job she was offered and she wanted to travel the world, she left GE in 1921 to teach physics at the Constantinople Women's College in Turkey. She satisfied some of her travel bug with visits to many European countries and Egypt. A year later, GE did offer her a job as an electrical engineer in the central station engineering department. When she accepted this job at age 39 (!), she became the first professionally employed female electrical engineer in the USA (Brittain 2009; Clarke n.d.; Goff 1946; Sicherman et al. 1980).

Clarke's area of specialty was electric power systems and problems related to their operation. The GE engineers needed to know how currents were distributed among the transmission lines and what happened to the systems in case of faults, including lightning strikes. Clarke prepared charts that were calculating devices putting into graphical format ways to solve problems without resorting to solving the equations underlying the charts themselves. Her charts would save much valuable calculating time during World War II (Goff 1946; Sicherman et al. 1980).

Clarke received a patent in 1925 (No. 1,552,113) (Fig. 5.2) for her "graphical calculator"—a method of considering the impacts of energy storage and dissipation in long electrical transmission lines. It greatly simplified the calculations that needed to be done. Her calculator does graphically by means of movable vectors in a smaller fraction of the time, the work previously done through formulas. The "Clarke Calculator" simplified the processes used for calculations associated with planning transmission lines for 40–55 years—until the advent of mainframe computers and associated customized software for electric utility planning engineers. That customized software incorporated algorithms to perform the calculations via computer that had formerly been done by hand (generally using a slide rule) or by using Clarke's graphical calculator (Brittain 2009; National Inventors Hall of Fame 2025).

Clarke made innovations in long-distance power transmission and the development of the theory of symmetrical components and circuit analysis. Symmetrical components are a mathematical means by which engineers can study and solve problems of power system losses and performance of electrical equipment. Her work was critical for the growth and development of the US electric grid in the 1920s and 1930s. In 1930, Clarke became the head of the engineers at GE studying power system stability. In 1933, Clarke moved to GE's Analytical Division where she performed apparatus and system analysis (Clarke n.d.; Sicherman et al. 1980).

Clarke wrote the textbook *Circuit Analysis of AC Power Systems, Symmetrical and Related Components* (1943) (Fig. 5.3) with a second volume in 1950. The book grew from her notes for lectures that she had given to the Central Station Engineering Department engineers at GE. In its two volumes, this textbook was used to educate all power system engineers for many years (Brittain 2009; Davis 2024; Goff 1946; Proffitt 1999; Sicherman et al. 1980).

Clarke authored or co-authored 19 technical papers during her employment at GE and later, while at the University of Texas (1947–1956), reflecting her status as an authority on the topics of hyperbolic functions, equivalent circuits, and graphical analysis within electric power systems (Brittain 2009; Clarke n.d.; Hobbs 2004; National Inventors Hall of Fame 2025). "Simplified Transmission Line Calculations," which appeared in the *General Electric Review* in May 1926, provided charts for transmission line calculations (Brittain 2009). Clarke (Fig. 5.4) enjoyed developing labor-saving charts from which required values could be read directly. She also designed hydroelectric dams in the Western USA (Brittain 2009; National Inventors Hall of Fame 2025).

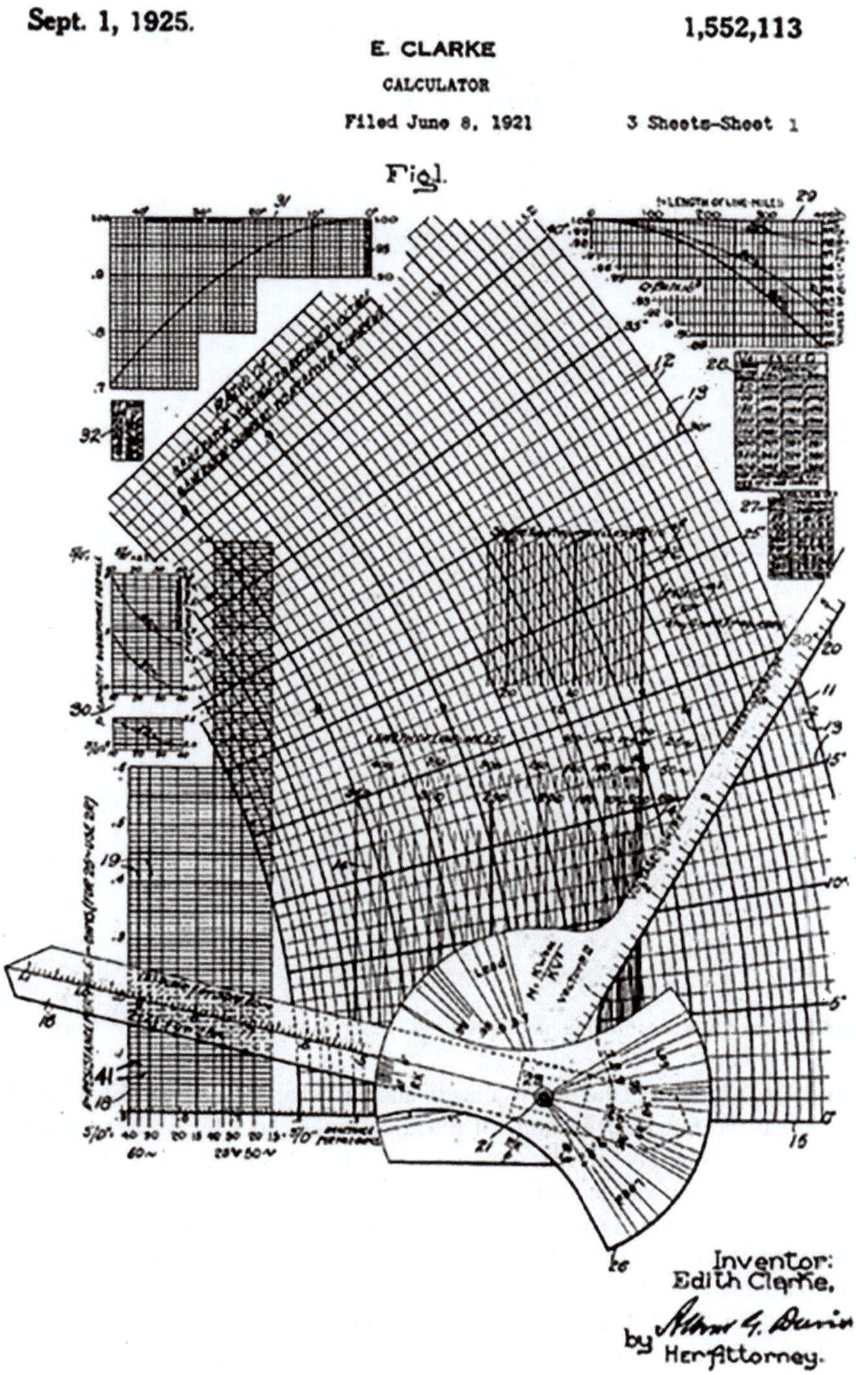

Fig. 5.2 Edith Clarke's Patent Number 1,552,113 calculator

TECHNICAL INFORMATION CENTER
STONE & WEBSTER ENGINEERING CORP.
P. O. BOX 5406
DENVER, COLORADO 80217

CIRCUIT ANALYSIS
OF A-C POWER
SYSTEMS

VOLUME I
Symmetrical and Related Components

By
EDITH CLARKE
CENTRAL STATION ENGINEERING DEPARTMENT
GENERAL ELECTRIC COMPANY

One of a Series Written in the Interest of
the General Electric Advanced
Engineering Program

LIBRARY COPY

NEW YORK • JOHN WILEY & SONS, INC.
LONDON • CHAPMAN & HALL, LIMITED

Fig. 5.3 Signed title page from Edith Clarke's 1943 textbook. (Courtesy of Jill S. Tietjen)

In 1926, Clarke was the first woman to address what is today the IEEE—at the time it was the American Institute of Electrical Engineers (AIEE). Her topic was "Steady-State Stability in Transmission Systems" (Texas Engineer 2023). The paper's publication was another first—the first woman to have her work appear in the *AIEE Transactions* (a journal) (Davis 2024). In 1927, Clarke's second patent was issued. Patent No. 1641, 737 "relates to electrical power transmission lines and more particularly to the transmission of electrical power at high voltage over long distances." In 1932, Clarke's paper before the AIEE, "Three-Phase Multiple Conductor Circuits," was named the best paper of the year in the northeastern district. This paper examined the use of multiple conductor transmission lines, intending to increase the capacity of the power lines. Clarke designed the turbine system

Fig. 5.4 Edith Clarke in her office. (Courtesy—MIT Museum)

Fig. 5.5 Edith Clarke in the lab. (Courtesy—MIT Museum)

for the Hoover Dam during the 1930s. Her system was subsequently installed at other hydroelectric power plants in the western USA (Brittain 2009; Davis 2024; Sicherman et al. 1980).

Clarke's 1941 AIEE paper, "Stability Limitations of Long-Distance A-C Power Transmission Systems," containing an analysis of stability issues with long transmission lines, was judged the best AIEE paper of the year. In 1944, her third patent was issued. Patent No. 2,339,052 "relates to electric circuits and more particularly to improvements in open-delta-connected transformer systems." In 1948, Clarke (Fig. 5.5) was named one of the first three women fellows of IEEE. She had previously become the first female full voting member of IEEE. Clarke was one of the few women licensed professional engineers in New York State (Brittain 2009; Ingram 2022; National Inventors Hall of Fame 2025; Sicherman et al. 1980).

After her retirement from GE in 1945, Clarke bought a farm in Howard County, Maryland. But, she was enticed back into the workforce by the University of Texas. In early 1947, she joined the faculty as an associate professor of electrical engineering, becoming the first full-time woman professor of electrical engineering in the USA. She was promoted to full professor later that year. She served on numerous committees, provided special assistance to foreign students, and served as a graduate advisor. She encouraged graduate students and assistant professors to publish early in their careers. After a decade at the University of Texas, she retired in 1956 and lived on her farm for the last years of her life. Clarke had always loved the out-of-doors and played tennis and skied throughout her life, and she loved to camp. She also played contract bridge and loved puzzles (Brittain 2009; Clarke n.d.; Howes 1935, 1939; National Inventors Hall of Fame 2025; Ogilvie & Harvey 2000; Sicherman et al. 1980; Texas Engineer 2023).

5.3 Awards and Honors

Clarke received Tau Beta Pi's Women's Badge—#95 (at a time when Tau Beta Pi did not admit women to membership), a distinction that she was awarded for accomplishments in the field. In 1954, Clarke received the Society of Women Engineers' Achievement Award "in recognition of her many original contributions to stability theory and circuit analysis." She was listed in *Who's Who in Engineering* and *Who's Who in New York*. She was profiled before her death in *Women Can be Engineers* (Goff) (Clarke n.d.; Engineering and Technology History Wiki n.d.; Tau Beta Pi 2024).

Clarke received several honors posthumously. In 2003, Clarke was inducted into the Maryland Women's Hall of Fame. In 2015, she was inducted into the National Inventors Hall of Fame for her invention of the graphical calculator. In 2016, the University of Texas at Austin established the Edith Clarke Woman of Excellence Award to honor outstanding female faculty (Rosengren 2023).

5.4 Legacy

In 1940, on Vassar College's 75th anniversary, Clarke gave a speech titled "College Women in Professions: The Experience of an Engineer." She wrote to every female member of AIEE and asked them about their experiences working as engineers and finding jobs. Clarke said,

> In times of business expansion, the best men in any graduating class have a wide choice of openings. With electrical engineering at its present state, there is no job ready and waiting for a woman when she gets her degree…She can, with perseverance, find an opening (Rosengren 2023).

Clarke believed that engineering would open up as a field for women, as she said, "There is a future for women in engineering and someday the only limitation will be their own lack of ability, as we are fast approaching an age in which men and women will be measured by their worth as individuals" (Texas Engineer 2024). As James Brittain wrote, "Her [Edith Clarke's] achievements provided an inspiring example for the next generation of women with aspirations to seek a career in electrical engineering" (Brittain 2009).

Tau Beta Pi, the engineering honor society, has named one of its Donor Recognition Clubs after her. She was Women's Badge #95 and she is cited as the inventor of the graphic calculator.

Clarke is credited with helping the electric grid grow, helping that grid remain stable and reliable, and laying the foundation for the smart grid—the grid of the future. She was the first to use an analyzer to obtain data about power networks—a critical step toward understanding, estimating, and projecting electron flows in the electric grid. The US Department of Energy calls her efforts "the first steps toward smart grid technology. She could be called the Smart Grid's 'Founding Mother'" (Rosengren 2023).

References

Brittain, James E. 2009. From Computer to Electrical Engineer: The Remarkable Career of Edith Clarke. In *Women in Engineering: Pioneers and Trailblazers*, ed. Margaret E. Layne, P.E., 141-154. Reston, Virginia: American Society of Civil Engineers Press.

Clarke, Edith. Curriculum Vitae. n.d. From Society of Women Engineers Records (LR001539), Box 187, Folder 34, Walter P. Reuther Library, Archives of Labor and Urban Affairs, Wayne State University.

Davis, Amanda. 2024. "Edith Clarke: Architect of Modern Power Distribution – The first U.S. woman EE designed the Hoover Dam's turbine system." IEEE Spectrum, July 10.. https://spectrum.ieee.org/edith-clarke-modern-power-distribution.

Engineering and Technology History Wiki. n.d. SWE Awards. Accessed December 15, 2024. https://ethw.org/SWE_Awards.

Goff, Alice C. 1946. *Women Can Be Engineers.* Ann Arbor, Michigan: Edwards Brothers, Inc.

Hobbs, Amy. 2004. Archives of Maryland (Biographical Series): "Edith Clarke (1883-1959)" MSA SC 3520-14065. https://msa.maryland.gov/megafile/msa/speccol/sc3500/sc3520/014000/014065/html/14065bio.html.

Howes, Durward, ed. 1935. *American Women: The Official Who's Who Among the Women of the Nation 1935-36*. Los Angeles, California: Richard Blank Publishing Company.

Howes, Durward, ed. 1939. *American Women: The Official Who's Who Among the Women of the Nation 1939-40*. Los Angeles, California: American Publications, Inc.

Ingels, Margaret. 1952. Petticoats and Slide Rules. *Western Society of Engineers*. September 4.

Ingram, Elizabeth. 2022. "Meet Edith Clarke, the first female electrical engineer in the U.S." *Hydro Power*, March 14. https://www.hydroreview.com/business-finance/jobs-and-people/meet-edith-clarke-the-first-female-electrical-engineer-in-the-u-s/.

National Inventors Hall of Fame. 2025. Edith Clarke. Accessed November 30, 2018. https://www.invent.org/inductees/edith-clarke.

Ogilvie, Marilyn and Joy Harvey, eds. 2000. *The Biographical Dictionary of Women in Science: Pioneering Lives from Ancient Times to the Mid-20th Century.* New York, New York: Routledge.

Proffitt, Pamela, ed. 1999. *Notable Women Scientists*. Detroit, Michigan: The Gale Group.

Rosengren, Paul Lief. 2023. Famous Women Engineers in History – Book 2: Edith Clarke, Trailblazer in Electrical Engineering. *IEEE USA E-Books,* June 30.

Sicherman, Barbara, and Carol Hurd Green with Ilene Kantrov and Harriette Walker. 1980. *Notable American Women: The Modern Period.* Cambridge, Massachusetts: The Belknap Press of Harvard University Press.

Tau Beta Pi. 2024. Tau Beta Pi Women's Badge. Accessed December 15, 2024. https://www.tbp.org/about/Dist/WomensBadge.cfm.

Texas Engineer. 2023. Edith Clarke: A Woman of Many Firsts." Accessed December 9, 2024. https://magazine.engr.utexas.edu/2023/edith-clarke.

Chapter 6
Margaret H. Hutchinson Rousseau, ScD (1910–2000) Achievement Award Recipient 1955

In recognition of her significant contributions to chemical process and production plant design

Holly J. Teig

6.1 Introduction

Dr. Margaret Hutchinson Rousseau was a pioneer in chemical production plant design, performing an impactful role in the fledgling chemical industry which expanded rapidly during World War II. She rose to the challenge of designing production plants for strategic resources needed in the war effort—penicillin, synthetic rubber, and aviation fuel—contributing to the betterment of disease treatment and providing valuable technical expertise during the global conflict.

Throughout her career, Dr. Hutchinson (Fig. 6.1) designed many types of chemical and petrochemical plants used across a variety of industries. Her specialization in the research and development of fractionation methods resulted in the development of groundbreaking processes and a "ripple tray" patent.

Dr. Hutchinson was recognized by the American Institute of Chemical Engineers (AIChE) for her contributions to the chemical processes industry with the prestigious Founders Award (AIChE.org 2008). She was honored by the Society of Women Engineers as the 1955 Achievement Award recipient.

H. J. Teig (✉)
Peoria, IL, USA

C. D. Craig et al. (eds.), *Women Engineering Legends 1952-1976: Society of Women Engineers Achievement Award Recipients*,
https://doi.org/10.1007/978-3-032-00224-2_6

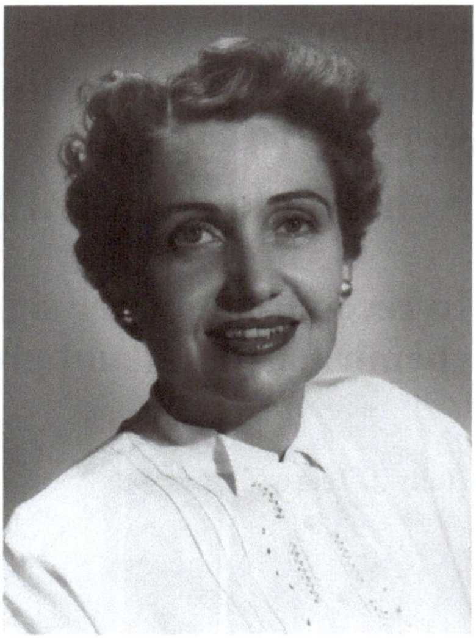

Fig. 6.1 Margaret Hutchinson Rousseau, Portrait (1955). (Photograph courtesy of Society of Women Engineers Records, Walter P. Reuther Library, Wayne State University)

6.2 Early Life

As a girl growing up in the oil fields of Houston, Texas, Hutchinson enjoyed music and ballet. Her scientific interests began in high school chemistry class, where "a talented chemistry teacher opened the windows of my mind to an exciting new world" (SWE Records n.d.). The fact that her father was a petroleum engineer may have factored into her decision to study chemistry.

Despite lack of encouragement from family, teachers, and others, Hutchinson decided to enter the field of engineering. Her high school chemistry teacher had lamented that he "had not been able to attend MIT to fulfill his ambition," which clearly planted a seed in Hutchinson's mind about where to pursue her studies (SWE Records n.d.).

Hutchinson enrolled at Rice Institute in Houston as a chemistry major, switching to chemical engineering a year later. She received her Bachelor of Science degree in chemical engineering in 1932. In retrospect she stated, "I went on to Rice Institute where I finally realized it was the engineering phase of chemistry which had its appeal for me." She pursued graduate studies at Rice, receiving her graduate degree as a chemical engineer from Rice in 1933 (SWE Records n.d.).

Hutchinson was admitted to MIT and went on to receive a Doctor of Science degree in chemical engineering from MIT in 1937, the first woman to earn a doctorate in the subject at that institution (Mit-amita.org 2016). Her thesis topic *Liquid film in gas absorption* was published that same year in *Industrial and Engineering Chemistry*, volume 29 (Madhavan 2015).

"There was no field work for women in engineering, except library and laboratory work. She continued with graduate work, determined that she would prepare herself to design oil refineries and petrochemical plants," reflected W.F. Ryan,

Engineering Manager and Vice President of Stone and Webster Corporation, where she would later be employed (SWE Records n.d.).

6.3 Early Career

Dr. Hutchinson found work immediately upon graduation with E.B. Badger & Sons company, who were engaged in engineering for the petroleum and petrochemical industries. Hired as a process engineer, Dr. Hutchinson developed process calculations and estimates on projects for topping and vacuum units in oil refineries and sulfur dioxide extraction units. She correlated data on existing designs of bubble caps, tunnel caps, and similar equipment leading to the development of a radically improved design (SWE Records n.d.).

In 1939, she married William C. Rousseau, a co-worker at E.B. Badger & Sons. They met in a bubble tower, the chemical engineer's favored equipment for the separation of materials by boiling, while they were developing separate laboratory experimental work for the improvement of bubble tower design (Findagrave.com n.d.).

Dr. Hutchinson earned a Professional Engineering license in 1945, registered in Massachusetts, and joined the National Society of Professional Engineers.

6.4 World War II

During World War II, Dr. Hutchinson was assigned to various defense projects, performing impactful work in connection with the war effort. She designed butadiene plants for the United States government's synthetic rubber plants (Hatch 2006; SWE Records n.d.). Synthetic rubber was in high demand in the US and for the troops in Europe and the Pacific theater due to the disruption in the supply of natural rubber from the tropics due to the war in the Pacific.

Dr. Hutchinson performed process design work on the production of oxygen with Arthur D. Little, Inc., a chemical engineering consulting company (SWE Records n.d.). That firm had been given the contract to develop oxygen masks for pilots of high-altitude unpressurized US war planes (Hatch 2006; Little n.d.).

Dr. Hutchinson was also engaged on the design of refining alkylation units and gas plants, which enabled the development of high-octane gasoline for aviation fuel (Madhavan 2015; SWE Records n.d.).

In 1941, the United Kingdom sought help from US scientists in solving the problem of scaling up the production of penicillin. Lab techniques had produced only limited quantities of the valuable drug, with varying degrees of shelf stability (American Chemical Society 2008). A significant breakthrough in production methods was needed to supply the war effort for the Allied forces and deliver the life-saving medicine to the operating theaters. Twenty-one chemical and pharmaceutical companies joined the race with Pfizer, Merck, Squibb, and Abbott leading the efforts (American Chemical Society 2008).

The focus of production shifted rapidly to deep-tank fermentation—a sterile process to grow large quantities of oxygen-dependent microorganisms. Two companies already had deep-tank fermenters; therefore, they raced ahead of others. Commercial Solvents Corporation (CSC) was using this technique for the production of acetone and butanol, using a plant designed by Dr. Hutchinson. Pfizer was using deep-tank fermentation for production of citric acid from lemons and limes (Pitt 2023). Dr. Hutchinson (Fig. 6.2) is credited as contributing to Pfizer's effort to scale-up and design the first deep-tank penicillin fermentation commercial plant (AIChE.org n.d.; IChemE 2025; Madhavan 2015).

Pfizer's project went on to use deep-tank fermentation to produce the vast majority of penicillin used for soldiers' medical needs during the war (American Chemical Society 2008). By June 1944, sufficient quantities of penicillin had been produced to allow each soldier to carry it in the D-Day invasion. The estimate of how many soldiers' lives were saved by penicillin in WWII is 15%. Compare that to the

Fig. 6.2 Margaret Hutchinson looking through a microscope certifying penicillin samples at the Food and Drug Administration (Photograph courtesy of National Archives)

American Civil War, where 3/5 of the Union troops and 2/3 of the Confederate troops died from infectious bacterial disease (Harris Sittig 2020).

As David Wilson wrote in his book *In Search of Penicillin*, "It is the biggest single failing of the myth about penicillin that it ignores the technological breakthrough of deep fermentation, a breakthrough that was every bit as vital to the successful development of penicillin as any of the more dramatic laboratory work" (American Chemical Society 2008).

6.5 Post-war Career

After supporting the war effort, Dr. Hutchinson went back to her previous career interests, focusing on chemical plant design and research into manufacturing process methods. She developed process designs for methanol synthesis and formaldehyde manufacture. She also conducted projects that correlated vapor liquid equilibrium data for low temperature hydrocarbons and developed design data on absolute alcohol and glacial acetic acid. Now working as a chemical engineering consultant, she handled all the process design work for an ethyl oxide and an ethylene glycol plant, built for Jefferson Chemical Company, a joint venture of Texaco Inc. and American Cyanamid (SWE Records n.d.).

Dr. Hutchinson's expertise and professional reputation was acknowledged by AIChE in 1945, when she was admitted as the first woman "active" member of the society (AIChE.org 2008; Tietjen 2017). She authored multiple papers that were presented at AIChE conferences including: *Liquid Capacity of Bubble Plate Caps* and *Aerated Flow Principle Applied to Sieve Plates* (SWE Records n.d.).

Colleagues in the industry took notice of her publications. "I have used these two papers, particularly the second, in the course of my work and consider them a major contribution to the subject," noted Carol R. Bauer of Foster Wheeler Corporation and fellow SWE member (SWE Records n.d.).

As her career advanced, Dr. Hutchinson continued to expand on her technical accomplishments. She directed research and development of a unique fractionating tray (Fig. 6.3) of her own design, patented in 1956, in the US and ten other countries (US Patent Office 1956).

A company merger with Stone & Webster Engineering Corporation of Boston progressed Dr. Hutchinson's consulting work. She continued her specialization in research and development relating to fractionation, but was also handling obtuse problems in fractionating column design, sieve trays, and abnormal operating requirements for the clients of the firm. She studied and correlated pilot plant data on the manufacture of ethylene by the chlorohydrin process (SWE Records n.d.).

Dr. Hutchinson (Fig. 6.4) was active with the Society of Women Engineers (SWE) throughout her career, joining as a member in 1953. She was elected and served on SWE's Board of Trustees for many years, with oversight of the headquarters and scholarship endowment funds (SWE Records n.d.). She also served as general chairman for the 1961 National SWE Convention.

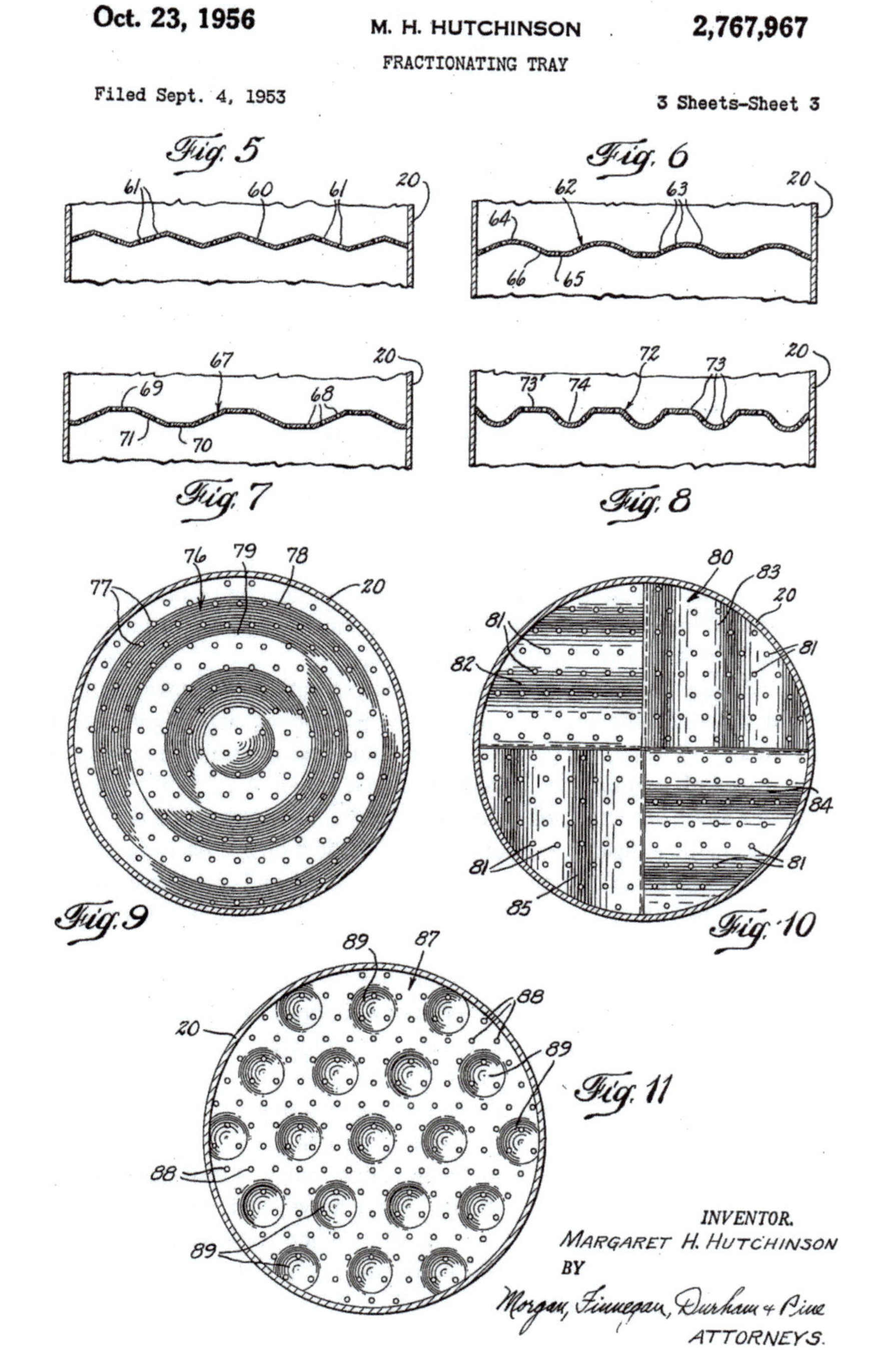

Fig. 6.3 Fractionating Tray patent drawing, Margaret Hutchinson Rousseau, inventor, Patent Number 2767967-A

Fig. 6.4 Margaret Hutchinson Rousseau, Portrait (1961). (Photograph courtesy of Society of Women Engineers Records, Walter P. Reuther Library, Wayne State University)

Dr. Hutchinson attended the first International Conference of Women Engineers and Scientists in June 1964, held in New York City. She and other past SWE Achievement Award recipients inspired students at a luncheon held in their honor. Her advice to aspiring students about engineering was that students "must be in love with the subject and willing to work hard" (SWE Records n.d.).

Following her retirement in 1961, Dr. Hutchinson turned her attention to supporting music and cultural programs. She became an overseer with the Boston symphony (MIT.edu 2016) where she and her husband William endowed the principal tuba musician's chair in perpetuity.

6.6 Awards and Recognition

In 1955, Dr. Hutchinson received the Achievement Award of the Society of Women Engineers (SWE Records n.d.; MIT.edu 2016; Olsen 1994) for her meritorious contributions to the field of chemical engineering (Fig. 6.5). In a letter supporting her nomination, W.F. Ryan, Engineering Manager and Vice President of Stone and Webster Corporation wrote,

> Dr. Hutchinson is unique. We are well aware of the large number of women in engineering who are doing highly important work, but probably none is doing work of the same complexity and requiring the highly specialized attainments required for Dr. Hutchinson's work. We feel that your organization would be hard put to find a more deserving woman engineer for your annual award (SWE Records n.d.).

Fig. 6.5 Margaret Hutchinson (left) accepts the Achievement Award from President Katharine Stinson, 1955 National Convention. (Photograph courtesy of Society of Women Engineers Records, Walter P. Reuther Library, Wayne State University)

Newspaper articles interviewing Dr. Hutchinson about her SWE Achievement Award tried to make her more relatable to their readers. In 1955, it was a radical idea to have a woman engineer with a doctorate degree and a young son working outside the home. The headline in the *Los Angeles Examiner* was *Mother and Housewife Reaches Top Despite Disheartening Setbacks,* while the *Boston Globe* reported *She Hates to Cook, Now She's Top Engineer,* with the subtitle of *Hubby Likes to Cook and He's an Engineer, too* (Findagrave.com n.d.). Dr. Hutchinson was listed in *Who's Who in American Women* that same year (SWE Records n.d.).

In 1983, Dr. Hutchinson was honored with the prestigious Founders Award of the AIChE (AIChE.org 2008), recognizing her outstanding contributions and advancements to the field of chemical engineering.

6.7 Legacy

Dr. Hutchinson broke an educational glass ceiling as the first woman to earn a Doctorate of Science degree in chemical engineering from MIT in 1937, 30 years after MIT established the program (Mit-amita.org 2016). Only 6 years earlier, Yunhe ("Ruth") Feng had earned her doctoral degree in chemical engineering at

Ohio State, the first woman to be granted a PhD in this field in the US (The Ohio State University 2020). These trailblazers paved the way for many woman chemists and chemical engineers to follow in their footsteps at MIT, Ohio State, Pennsylvania State College, and other higher educational institutions.

Dr. Hutchinson became the first woman "active" member of the AIChE in 1945 (AIChE.org 2008; Ellis 2024; Tietjen 2017), a testament to her standing and reputation in the industry. During chemical engineering's first half-century, the profession was dominated by men (Ellis 2024). W.F. Ryan reflected on this accomplishment:

> The membership grade of 'active' was very prestigious, given only to engineers who had actually been the principal engineer in charge of the design of an important and successful installation. This grade of membership could not be attained by teaching, working in research, or by functioning as a consulting engineer. Dr. Hutchinson was the only woman elected to this grade of membership at the time (Ellis 2024).

As most women engineers were not given opportunities to work on job sites, very few met the criteria established by the AIChE governing body. As recently as the early 1970s, only 1 in every 200 AIChE members was a woman (Ellis 2024).

Dr. Hutchinson was also the first woman recipient of the prestigious Founders Award of the AIChE in 1983 (AIChE.org 2008), recognizing her outstanding contributions and advancements to the field of chemical engineering.

In 2017, AIChE established the Margaret Hutchinson Rousseau Pioneer Award for Lifetime Achievement by a Woman Chemical Engineer (AIChE.org 2008). The award honors a woman member of AIChE who has made significant contributions to chemical engineering research or practice. The honoree's accomplishments also include a component of service, mentorship, or leadership in helping to raise the visibility of women engineers and paving the way for women to have a greater impact in chemical engineering (AIChE.org n.d.).

Within her marriage, Dr. Hutchinson was ahead of her time in deciding to continue to use her maiden name throughout her professional career and in publications. In 1947, the family welcomed a son, also named William. She and husband William organized household duties similar to a modern dual engineering career household. In a news article from 1956, she noted that "I hate to cook" so her husband managed the kitchen while she interacted with her son's teachers and served on the local school PTA group. The couple, working for competing chemical firms at the time, even had a rule to "not talk shop" at home (Findagrave.com n.d.).

References

AIChE.org. 2008. Milestones. American Institute of Chemical Engineers. April. AIChE Centennial Celebrations: "Milestones". Accessed November 2, 2016.

AIChE.org. n.d. Margaret Hutchinson Rousseau Pioneer Award for Lifetime Achievement by a Woman Chemical Engineer. American Institute of Chemical Engineers. https://www.aiche.org/community/awards/margaret-hutchinson-rousseau-pioneer-award-lifetime-achievement-woman-chemical-engineer. Accessed December 12, 2024.

American Chemical Society. 2008. Penicillin Production through Deep-tank Fermentation. June 12. https://www.acs.org/education/whatischemistry/landmarks/penicillin.html. Accessed 12 December 2024.
Ellis, Gordon. 2024. History of Women in Chemical Engineering: Margaret Hutchinson Rousseau AIChE. March 12. Accessed November 6, 2024.
Findagrave.com. n.d. Memorial of Margaret Hutchinson Rousseau. https://www.findagrave.com/memorial/. Accessed 11 December 2024.
Harris Sittig, Linda. 2020. Margaret Hutchinson Rousseau: Bringing Penicillin to the Masses. December 2. https://strongwomeninhistory.com/2020/11/02/margaret-hutchinson-rousseau-bringing-penicillin-to-the-masses/. Accessed 12 December 2024.
Hatch, Sybil E. 2006. *Changing our world : true stories of women engineers* (2nd ed.). Reston: ASCE Press. p. 179. ISBN 9780784408414. Retrieved 20 November 2016.
IChemE. 2025. Ten chemical engineers that shaped our world (Day 356). May 18. https://icheme-blog.org/2015/05/18/ten-chemical-engineers-that-shaped-our-world-day-356/comment-page-2/. Accessed December 12, 2024.
Little, Arthur D. n.d. Timeline. https://www.adlittle.com/en/timeline. Accessed 12 December 2024.
Madhavan, Guruprasad. 2015. *Think Like an Engineer*. Oneworld Publications. pp. 83–85, 91–93. August 20. ISBN 9781780746371. Retrieved 20 November 2016.
Mit-amita.org. 2016. MIT Women's Association Archived. March 3. https://www.mit-amita.org/esr/prewar.html. Accessed December 12, 2024.
MIT.edu. 2016. In Memorium: Margaret H. Rousseau'37, First Woman to Receive Doctorate From MIT ChemE. Accessed November 6, 2024.
Olsen, Kirstin. 1994.*Chronology of women's history* (1st ed.). Westport, Conn.: Greenwood Press. p. 283. ISBN 9780313288036. Retrieved November 20, 2016.
Pitt, Martin. 2023. Brewing Up a Storm: The Origins of Biotechnology Part 2. April 27. https://www.thechemicalengineer.com/features/ta/.Accessed 12 December 2024.
Society of Women Engineers (SWE) records (LR001539) n.d. *Margaret Hutchinson.* Box 189 Folder 40, Walter P. Reuther Library, Archives of Labor and Urban Affairs, Wayne State University. Accessed December 12, 2024.
The Ohio State University. 2020. Nations first female PhD in engineering was an Ohio State chemical engineering alumna. April 10. https://cbe.osu.edu/news/2020/04/nations-first-female-phd-engineering-was-ohio-state-chemical-engineering-alumna. Accessed December 11, 2024.
Tietjen, Jill S. 2017. *Engineering Women: Re-visioning Women's Scientific Achievements and Impacts*. Springer. p. 17. ISBN 978-3-319-40800-2. Retrieved 20 November 2016.
US Patent Office. 1956. Patent US-2767967-A Fractionating tray issued October 23, 1956. Patent Public Search Basic USPTO. Accessed November 6, 2024.

Chapter 7
Elise F. Harmon (1909–1985) Achievement Award Recipient 1956

In recognition of her significant contributions to the area of component and circuit miniaturization

Janet L. Williams

7.1 Introduction

Elise Harmon (Fig. 7.1) was a chemist and biologist by education, and a science teacher immediately after graduation from college. She evolved to work as a chemist, a government researcher and investigator, and a research and development engineer in the manufacture of printed circuit boards. After transitioning to electronics in the 1950s, she invented and patented several new technologies in the emerging printed circuit board industry that allowed efficient printing of circuit elements on plastic substrates.

Earlier in her engineering career, she worked for the government, conducting critical research on the performance of electrical generators in aircraft and worked with suppliers to provide longer-life carbon brushes for aircraft generators that would perform reliably at high altitude, critical to the war effort.

Keeping abreast of the rapidly evolving technologies in production and miniaturization in the 1960s, she presented seminars and published papers at conferences in printed circuit technology and manufacture. She founded her own consulting

J. L. Williams (✉)
Albuquerque, NM, USA

C. D. Craig et al. (eds.), *Women Engineering Legends 1952-1976: Society of Women Engineers Achievement Award Recipients*,
https://doi.org/10.1007/978-3-032-00224-2_7

Fig. 7.1 Portrait of Elise F. Harmon. (Courtesy of Society of Women Engineers Records, Walter P. Reuther Library, Wayne State University)

company in 1970 and consulted with domestic and international clients in printed circuit production and circuit miniaturization into the 1980s. Harmon was honored by the Society of Women Engineers as the 1956 Achievement Award recipient (SWE Records n.d.).

7.2 Early Life and Education

Elise Harmon was born in the small town of Mt. Enterprise, Texas and graduated from high school in Marshall, Texas in 1927. A capable student, she attended North Texas State College (now University of North Texas) in Denton TX, earning a Bachelor of Science in chemistry in 1931 (North Texas State 1931). During her time at North Texas State, in 1930, she was elected president of the W.N. Masters Chemical Society (Chemistry Society Elects Officers 1930), named after the professor who had founded the chemistry department (Down the Corridor 2011). Her first teaching experience occurred in 1929–1930, during her time at North Texas State, where she taught freshman classes in chemistry (SWE Records n.d.).

When her mother died in 1931 of cancer, her brother Ham had just graduated from high school, and the youngest sibling, Georgianna (Ann) Harmon was 12. Harmon, who had just finished her undergraduate degree, and her aunt helped raise the younger siblings (Elise Harmon n.d.).

7.3 Early Career

Harmon began teaching chemistry, first at Rochester High School in Rochester, Texas from 1931 to 1933 and then at Brownsville (Texas) Junior College from 1933 to 1936. Harmon was one of the first teachers to introduce science fairs in high schools. She served as the head of the Science Department at Iraan High School, an experimental school in the Texas oil fields supported by the University of Texas, from 1936 to 1939. During this period, she also taught practical and applied chemistry at night to employees of five oil companies. In 1938, she received an MS in Biology from the University of Texas, Austin (SWE Records n.d.).

In the 1930s, Harmon began to consult part time for various companies, in addition to teaching. From 1933 to 1936, she performed investigations for the Davies Fruit Company, analyzing sugar and acid contents of citrus juices and exploring the use of sulfate as a preservative. She subsequently worked part time for Pan American Airways in Brownsville, Texas for 2 years investigating carburetor de-icing products. She then worked for the Standard Oil Company from 1938 to 1941, analyzing various properties of crude oil. These consulting jobs began a shift in her interests and presaged a career change (SWE Records n.d.).

7.4 Career Pivot

Harmon gave up teaching about 1940. After a decade, she no longer found it satisfying (Ogilvie and Harvey 2000). She sought other employment and worked for a very short period for the War Department, Ordnance in St. Louis as a lab technician before landing a position at the National Bureau of Standards, Heat and Power Division, in Washington, DC, US, in 1942 as a chemist and materials engineer. After she left teaching and entered the engineering world, she knew she had found her passion. Years later, in a SWE Newsletter published just after she received the Achievement Award, she said, "I think this was the turning point of my life…I got the feel of engineering and worked with engineering instruments and blueprints. I was crazy about it" (SWE Records n.d.).

In this new milieu, she proved herself to be a competent and productive researcher who developed innovative solutions to problems assigned to her. She worked on insulating materials for cables and performed temperature surveys on aircraft electrical equipment and investigations on grease and lubricant for high-speed bearings. Then she was given full responsibility to investigate an issue with the performance of carbon brushes used in aircraft motors and generators.[1] The brushes in the aircraft generators she worked on were found to deteriorate into powder when operated at altitudes above 10,000 ft. She began to investigate the problem, setting up test

[1] In motors and generators, the brushes transfer the electricity generated from the static parts to the rotating parts.

equipment, conducting tests, and analyzing data. She worked with manufacturers to introduce various organic and inorganic solid and liquid lubricants to protect the brushes and prolong their life. Her efforts resulted in manufacturers producing carbon brushes with extended lives that performed well at high-altitude conditions. Allowing planes to fly above 15,000 ft. was an important contribution to the war effort and resulted in one of nine patents Harmon earned (Ogilvie and Harvey 2000; SWE Records n.d.; There is no prejudice 1967).

Harmon sought additional education to bolster her knowledge base while working for the National Bureau of Standards. She took graduate courses in engineering at George Washington University and additional courses in polymer chemistry, mathematics, and mechanical engineering at the University of Maryland, College Park. These courses reinforced her entry into the world of printed circuit design (SWE Records n.d.).

Harmon was loaned to the Aircraft and Electrical Division of the Naval Research Laboratory from 1945 to 1948 to work on the development of testing procedures to obtain temperature characteristics of aircraft electrical power equipment. She also investigated the effects of fungus, mold, and oxidation on electrical equipment (SWE Records n.d.).

In 1948, Harmon transferred to the Ordnance Division of the National Bureau of Standards, where she worked on engineering development of proximity fuses, guided missiles, and other Army ordnance devices. It was in this position that she was first introduced to printed circuit boards, a very new technology that was not released by the Army to the public until 1950. Printed circuits allowed the electronics in proximity fuses to be miniaturized for use in mines, bombs, and artillery shells (Jinghongyi PCB (HK) Co. n.d.). Harmon conducted detailed investigations for the Army on process methods, machinery, economics, and efficiencies of manufacturing techniques in coordination with other government agencies and suppliers.

In 1951, Harmon's career pivoted again, from government work to private industry, when she was hired by the Aerovox Corporation in New Bedford, Massachusetts as Director of Research and Engineering, reporting to the Vice President. In a recorded panel discussion at the 1968 SWE National Convention, she acknowledged her employer as a private industry that "had the nerve…to hire a woman as director of research and engineering of new products." She mentioned it was the first time they'd ever hired a woman in management at that company (SWE Records n.d.).

In her new role, Harmon researched, developed, and tested three new electronic products, put them through pilot processes, and scaled them for full production. She also developed the production equipment needed to manufacture the products. Her most noted accomplishment was the development of a new method of producing printed circuit boards without chemical etching. Known as the hot die stamp method, it created printed circuits in which silver conductors were infused on thermoplastics and thermosetting materials (Stanley 1993). She was co-awarded a patent (Fig. 7.2) for this method in 1953 (Harmon and Franklin 1951).

Another patent awarded in 1958 was for a silk screen stretcher (Fig. 7.3) to stretch fabric screens for use in printing electrical circuits (SWE records n.d.).

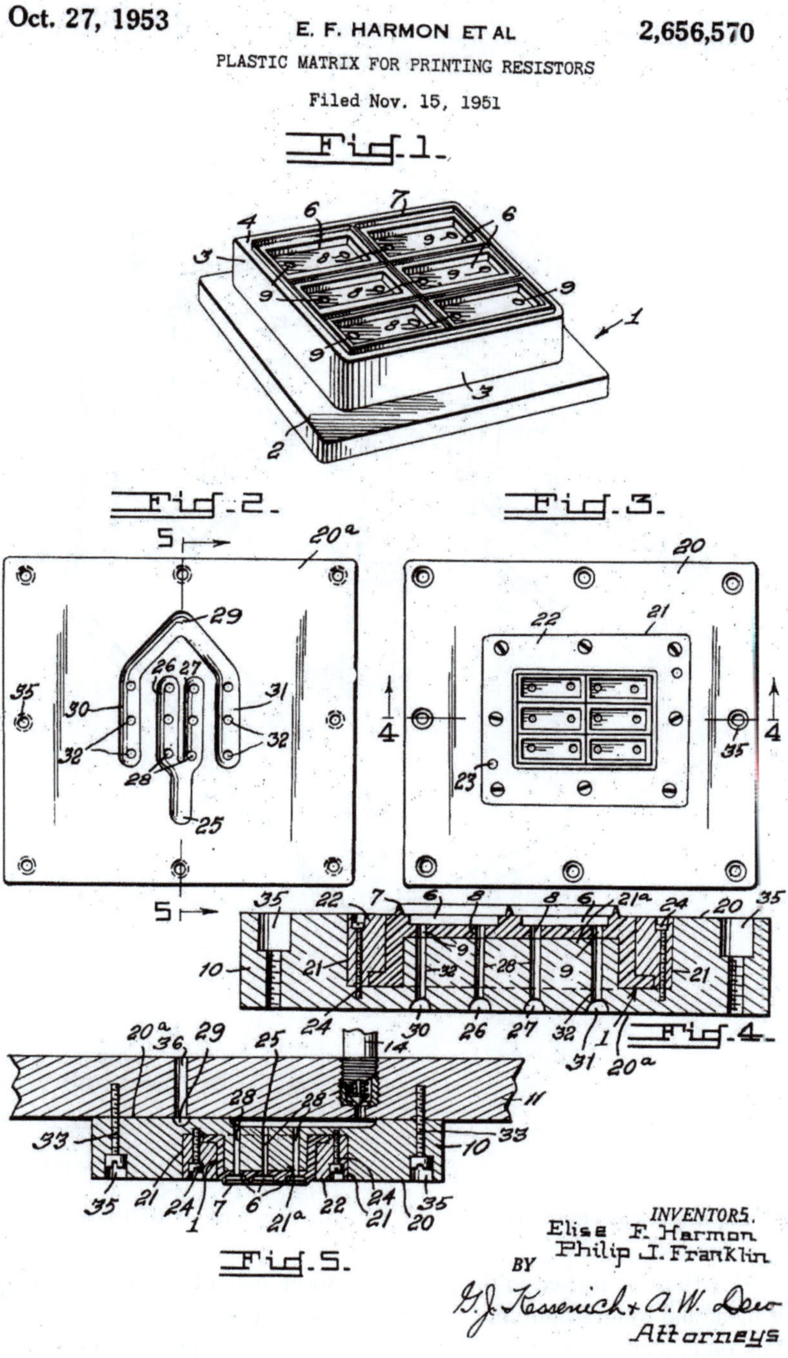

Fig. 7.2 Patent granted to Elise Harmon and Philip Franklin for plastic matrix for use in printing resistors, circuits and the like on plastic, ceramic, glass, or similar non-porous materials. US Patent No. 2656570

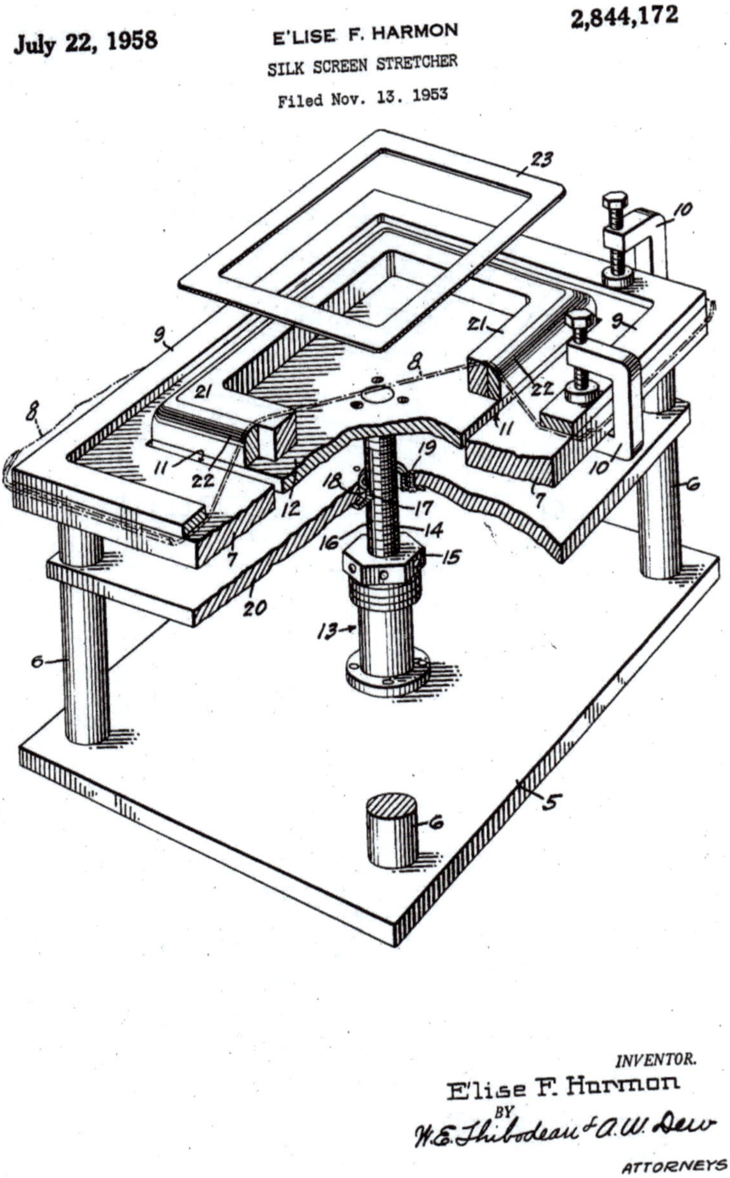

Fig. 7.3 Patent granted to Elise Harmon in 1958 for silk screen stretcher. US Patent No. 2844172

In 1957, Harmon moved to the American Bosch Arma Corporation, Computer Department and Manufacturing, Garden City, New York, in charge of the company's microminiaturization research and development programs. She was the liaison between the Engineering, Manufacturing, and Computer Departments for Arma's Inertial Guidance System for the Atlas ICBM rocket. Her focus switched to studying the state of the art in microminiaturization applied to space vehicles and telemetry, including component manufacturing and packaging and studying new approaches for production (SWE Records n.d.).

Harmon's next career move took her to Autonetics, a Division of Rockwell Corporation in Anaheim, CA in 1962. As senior staff engineer, she was responsible for technologies related to fabrication of multi-layer circuit boards for the computers in the Minuteman. This involved testing fabrication and packaging technologies, establishing standards and specifications, development and operation of a prototype facility, and training of personnel (SWE Records n.d.).

Through the 60s, she became an expert in printed board circuitry and microminiaturization. Recognized as a subject matter expert, she often spoke at seminars and conferences and published numerous technical papers. The SWE Archives contain a letter to North American Rockwell specifically asking for Elise Harmon's participation in organizing and chairing sessions for the 1971 National Electronics Packaging and Production Conference (NEPCON). Conference materials show that she personally chaired several technical seminars as well as organizing the program itself (SWE Records n.d.).

7.5 Later Career

In 1970, Harmon started her own consulting firm, Harmon Associates. She parlayed her expertise gathered through the last two decades into a business with an impressive list of domestic and international clients in the printed circuit industry and companies using printed circuits in their products. Her 1973 resume lists domestic clients with familiar names such as Magnavox, Raytheon, and Honeywell, to whom she provided analysis and advice in various aspects of printed circuit production. Outside the US, she consulted with Nippon Aviatronics, assisting with problems in solder plating and flow, and with officials in the Republic of China in Taiwan to "discuss the introduction of printed circuit fabrication in Taiwan" (SWE Records n.d.).

7.6 Honors

Receiving the SWE Achievement Award in 1956 opened the door to additional honors for Harmon. As a panelist at the 1968 SWE conference, she stated that 1952–1957 were her best years, not only due to receiving the Achievement Award, but also because she discovered SWE as an organization (SWE Records n.d.).

In addition to SWE, she joined several technical organizations, including the Society of Plastics Engineers, American Chemical Society, American Electroplating Society, and the Institute of Printed Circuits (IPC). In 1968, she received IPC's President's Award, given to IPC members who have exhibited ongoing leadership in

IPC and have made significant contributions of their time and talent to the association and the electronics interconnect industry. She also received a merit award in 1973 from the Institute for the Advancement of Engineering in Los Angeles (IPC 2025; L.A. Institute. 1973; SWE Records n.d.).

7.7 Legacy

Elise Harmon was a genuine pioneer for women in engineering, serving in responsible technical roles and forging paths for the women who followed her in the printed circuit industry. Her position as the first woman manager at Aerovox, as the Director of Research and Development, certainly broke a glass ceiling long before the phrase was coined by Marilyn Loden in the 1970s. She was intrepid in her early career, stepping into roles in which she had little direct experience, but using her creativity, problem-solving skills, diverse education and perseverance to succeed. As her career and expertise in printed circuit manufacturing matured, she exuded confidence and denied the existence of barriers that may have hindered her career. On the contrary, Harmon never admitted any kinds of issues based on her gender. In one interview in the early 1970s, she asserted, “Once you’ve proved yourself capable, there’s no barrier” (SWE Records n.d.). Another article reports, “Although Miss Harmon works in a predominantly man’s world, she adamantly denies any obvious or subtle prejudice on their part” (There is no prejudice 1967).

Harmon was described in a Fullerton, California newspaper article published after she started her consulting firm as “direct and imaginative” and “impatient of trivialities, preferring to get to the heart of the matter without delay” (SWE Records n.d.). These are qualities that must have served her well as she navigated the male-dominated field of printed circuit design and consulted with men in many industries using printed circuit technology during her years of consulting.

All her interviews and newspaper articles portray her as down-to-earth, practical, and matter-of-fact about her accomplishments. In an article in the SWE Newsletter after receiving the 1956 Achievement Award, she admitted that getting the award was a “complete surprise.” However, Harmon did employ humor on occasion. In a 1968 panel discussion with several past Achievement Award recipients, she joked, “If you look back to 1956…maybe there weren’t too many of us to choose from” (SWE Records n.d.).

Harmon felt that her eclectic educational background in science and engineering was a positive factor in her career success. In a SWE newsletter from 1956 she observed: “I don’t consider myself an exceptional chemist, biologist, or engineer. But the combination of all three has given me the background I’ve needed when attacking and solving problems that have confronted me in my work” (SWE Records n.d.).

References

Chemistry Society Elects Officers. 1930. *Denton Record-Chronicle*, June 20, pg. 3. Denton, TX.

Down the Corridor. 2011. Down the Corridor: Chemistry Centennial. *North Texan,* March 16.

Elise Harmon. n.d. *Scholarly Community Encyclopedia.* https://encyclopedia.pub/entry/39516. Accessed February, 2025

Harmon, E.F., Franklin, Philip J. 1951. Plastic Matrix for Printing Resistors US Patent 2,656,570. Filed November 15, 1951. Issued October 27, 1953.

IPC. 2025. IPC President's Award. https://www.ipc.org/ipc-presidents-award. Accessed February 1, 2025.

Jinghongyi PCB (HK) Co. n.d. The historical timeline and outlook of printed circuit boards. *Jinghongyi PCB (HK) Co. Limited blog, News and Information.* https://www.pcbelec.com/the-history-of-pcb-evolution.html. Accessed February 2025.

L.A. Institute. 1973. 2 OC Engineers Awarded Honors by L.A. Institute. *The Anaheim Bulletin,* February 13. Anaheim, CA.

North Texas State 1931. Commencement Program for North Texas State Teachers College, 1931. August 25, pg 2. https://digital.library.unt.edu/ark:/67531/metadc174776/m1/2/?q=harmon.

Ogilvie, Marilyn, and Harvey, Joy. 2000. *The Biographical Dictionary of Women in Science*, Vol 1. Routledge. pgs 557-558. Accessed January 2025.

Society of Women Engineers (SWE) Records (LR001539). n.d. *Elise Harmon.* Box 189, Folder 4, Walter P. Reuther Library, Archives of Labor and Urban Affairs, Wayne State University.

Stanley, Autumn.1993. *Mothers and Daughters of Invention: Notes for a Revised History of Technology*. Rutgers University Press.

There is no prejudice. 1967. *The Sydney Morning Herald,* December 28, Page 12. Sydney, Australia.

Part II
1957–1961

Chapter 8
Setting the Stage: 1957–1961 The Space Age Launches

Holly J. Teig and Janet L. Williams

8.1 Introduction

The space age began in 1957 with the launch of the *Sputnik I* satellite by the Soviet Union. So many technologies developed and refined during this time—lasers, solar cells, integrated circuits, atomic power, polymers, color television—would end up being used in some form to support America's space program. Engineers and scientists were determined to accept President John F. Kennedy's challenge to put a man on the moon. The pace of scientific and technological development accelerated, led by the aerospace industry and the newly created National Aeronautics and Space Administration (NASA 2016).

The availability of the birth control pill in 1960 ushered in a new era for family planning, and more women entered the workforce. This labor trend was driven by a desire for a higher standard of living, the easing of household tasks through the use of modern appliances, and changes in traditional attitudes toward women's work outside the home. However, working women still encountered social and legislative barriers to full acceptance in professional roles, which would not be addressed for another decade.

H. J. Teig (✉)
Peoria, IL, USA

J. L. Williams
Albuquerque, NM, USA

C. D. Craig et al. (eds.), *Women Engineering Legends 1952-1976: Society of Women Engineers Achievement Award Recipients*,
https://doi.org/10.1007/978-3-032-00224-2_8

The world's first commercial atomic electric power plant began operating in 1957 at Shippingport, Pennsylvania.

Rebecca H. Sparling received the Achievement Award for her contributions to high temperature metallurgy and nondestructive testing of metals, including preparing the first book on malleable iron structures to be written in the USA in 20 years.

Mabel MacFerran Rockwell received the Achievement Award for her significant contributions to the field of electrical control systems, including electrical control systems for the Polaris missile launcher.

1957

1958

The Civil Rights Act of 1957 gave women the right to serve on federal juries, but a few states still did not allow women to serve on state and local juries until the late 1960s.

In 1958, DuPont chemist Joseph Shivers invented fibers that could expand up to five times their length. Marketed as Lycra, it revolutionized fashion.

Dr. Rune Elmqvist and Ake Senning implanted the first internal cardiac pacemaker in Arne Larsson in Sweden in 1958.

Désirée le Beau, PhD, received the Achievement Award for her significant contributions to the field of rubber reclamation, including a patented process for producing reclaimed rubber in particulate form.

1959

American Airlines offered the first commercial US jet service in a Boeing 707-123 Astro Jet with a flight from New York to Los Angeles in January 1959, marking the beginning of the "Jet Age"

STATUS OF WOMEN IN 1960

- The participation of women in the workplace was 38%. They worked primarily as teachers, nurses, and secretaries.
- About 7% of practicing physicians and 3% of lawyers were women; women engineers still hovered near 1%.
- Most Ivy League schools did not admit women.
- Women's ability to obtain credit in their own name was still limited.

Esther M. Conwell, PhD, received the Achievement Award for her contributions to solid-state physics research; her research in electrical properties of semiconductors contributed to improved transistors.

Theodore Maiman operated the first laser in 1960 at the Hughes Research Laboratory in California, shining a high-powered flash lamp on a ruby rod with silver-coated surfaces. By bringing together optics and electronics, lasers opened vast new fields of science and technology.

1960

- Searle obtained approval from the FDA to sell the drug Enovid as a contraceptive, making it the first birth control pill. It still was illegal to use for birth control in many states and could only be prescribed for married women.
- Two out of 100 US senators and 15 out of 435 representatives in the US Congress were women –3.5% of Congress.

In 1961, after testing more than 600 compounds, a British pharmacologist named Stewart Adams and two doctors filed a patent for an alternative drug to aspirin for pain and inflammation, later to be called ibuprofen.

Laurel van der Wal received the Achievement Award in recognition of her significant contributions to the developing field of space biology.

1961, Soviet Yuri Gagarin became the first person to orbit Earth. In May, President John F. Kennedy announced to a joint session of Congress the quest to put a man on the moon before the end of the next decade, sparking the US launch of the Apollo program.

1961

President Kennedy created the Commission on the Status of Women by Executive Order 10980 in late 1961. He appointed Eleanor Roosevelt its chair.

8.2 State of Technology

8.2.1 In Space

Satellites The space age began when the Soviet Union launched *Sputnik I*, the first man-made satellite, in October 1957. At 23 inches in diameter, it was about the size of a beach ball and weighed nearly 200 pounds. *Sputnik I* caught the world's attention. The US was still scrambling to complete its own artificial satellite program *Vanguard*, which carried a smaller sized payload of cameras and instruments. A month later, before the US program had even proven itself, the Soviets launched *Sputnik II*, carrying a much heavier payload, including a dog named Laika. This panicked the American public, fearful that the Russians would soon launch ballistic missiles toward American soil. The US Department of Defense doubled down on its development efforts by funding a second, concurrent satellite program named *Explorer*. The *Vanguard I* satellite was launched in January 1958, featuring the first solar cells in space. The space age ushered in new political, military, technological, and scientific developments and led to the creation of NASA in 1958 (Garber 2007).

Humans in Space Yuri Gagarin of the Soviet Union was the first human to journey into outer space in April 1961. Gagarin spent 108 min completing one orbit at an altitude of 187 miles above Earth. Three weeks later, the US launched Alan Shepard into space aboard the *Freedom 7* capsule atop a Mercury-Redstone rocket. Shepard's journey was a brief 15-min flight to evaluate key systems flying at an altitude of 116 miles above Earth. While weightless in space, Shepard took control of the spacecraft, manipulating the pitch, yaw, and roll axes, testing vehicle control (NASA 2016). After Shepard's successful return to Earth, capsule project engineer Walt Kapryan commented, "We felt we were on the threshold of something very important. We were on the threshold of opening a new frontier" (NASA 2016). In May, President John F. Kennedy announced to a joint session of Congress the following technological challenge:

> I believe that this nation should commit itself to achieving the goal, before this decade is out, of landing a man on the moon and returning him safely to the Earth (JFK Library 1961).

NASA engineers accepted this challenge, knowing full well that many more technological problems would need to be solved in order to achieve this bold goal. The space race was on! (NASA 2016).

8.2.2 Airlines

Airline Travel In 1959, American Airlines launched the first scheduled transcontinental US passenger service with the Boeing 707, from Los Angeles to New York, ushering in the Jet Age (American n.d.). They also instituted the first electronic reservation system *SABRE* in 1961, replacing the time-intensive process (90 min on average) manual operators had been performing to book each airline ticket for passengers (Sabre 2016). Other airline carriers quickly followed suit, realizing that this automation efficiency would transform the airline industry.

8.2.3 Materials

Polymers In 1951, J. Paul Hogan and Robert L. Banks of Phillips Petroleum Company discovered a white, solid polymer while investigating the conversion of propylene and ethylene, released in refining, into gasoline components. They immediately filed for patent rights, naming the substance *polypropylene.* Continued process development yielded a second material, *high-density polyethylene (HDPE)*, which is far stiffer and more heat-resistant than anything on the market. Within 6 years, HDPE became the preferred material for plastic baby bottles (it could withstand the high sterilization temperatures), shatter-proof food containers, and hula hoops. A new industry was born. Today, billions of pounds of polypropylene and HDPE are used to make everything from milk jugs to furniture to industrial piping (ACS 1999).

Elastic Fiber In 1958, DuPont chemist Dr. Joseph Shivers invented LYCRA fibers that could expand up to five times their length and be spun into fine filaments, yet return to their original shape. Softer, stronger, and more durable than conventional elastic thread, the fashion industry put LYCRA to use in undergarments, pantyhose, and swimwear. LYCRA debuted in athletic gear sewn as ski suits worn by the French men's alpine ski team at the 1968 Winter Olympic Games (LYCRA 2025).

8.2.4 Computer Hardware and Software

Integrated Circuits In 1958, Jack Kilby at Texas Instruments came up with the idea of the integrated circuit (a microchip) using gold flying-wire connections on a germanium substrate. An integrated circuit consists of a group of resistors, capacitors, and transistors connected on a solid surface to perform electronic functions (Computer History Museum 2025a). Robert Noyce at Fairchild Semiconductor improved Kilby's design, changing the substrate to silicon and the interconnecting lines to aluminum, which enabled practical manufacture (Computer History Museum 2025b). This construct remains the basic design for all microchips today—albeit they continue to shrink in size while expanding in function. Nancy D. Fitzroy, Achievement Award 1972, invented a thermal chip to measure temperature in integrated circuits, along with special thermal analysis techniques for integrated circuit chips used in the semiconductor industry (ETHW n.d.).

8.2.5 Business and Communications

Photographic Copies In 1959, the Haloid Xerox Company introduced the 914 xerographic copier, which could produce seven copies per minute. The process to make photographic copies onto plain, uncoated paper had been known for some time, but this was its first commercial application (Britannica 2025b). While Model 914 was cost-effective and had high speed for business usage, it tended to catch fire when it overheated. Xerox thoughtfully included a small fire extinguisher with each copier (Hemmungs Wirtén 2004).

Typewriters IBM forever changed how typewriters were designed when it introduced the Selectric electric typewriter in 1961. Featuring a spinning, bobbing typeball instead of individual type keys, it solved the imperfections of the moving carriage and nest of type bars that had plagued writers and secretaries for 90 years. The Selectric, costing $495, could print 186 words a minute while its rotating camshaft allowed print characters to strike as quickly as 20 milliseconds apart (IBM n.d.). IBM was inundated with orders, eventually selling 13 million units before electronic word processors took over the market.

8.2.6 In the Home

Television The first color TV units rolled off the production line in 1954, priced at $1000 (about the same price as an automobile), far too expensive for most households (Fisher et al. 2025). Consumers were persuaded to buy color sets by the rapid increase of color programming available by the mid-1960s, including *Walt Disney's Wonderful World of Color* which aired from 1961 to 1969 (McGinn Spring 2001). Early live color broadcasts had to be captured to be preserved. Ampex introduced a color videotape recorder in 1958, which NBC used to tape *An Evening with Fred Astaire*, the oldest surviving live network color videotape (Britannica 2016). Television has had a considerable influence on society since its inception, having gone from a novelty to the dominant mass media communication tool for news and entertainment (Walsh 2010).

8.2.7 Medical

Medical Devices In 1958, engineer Earl Bakken produced the first wearable external pacemaker (University of Minnesota Medical School 2018). That same year in Sweden, 43-year-old Arne Larsson received the first implantable cardiac pacemaker, thanks to pioneering surgery performed by Drs. Rune Elmqvist and Ake Senning. Larsson lived a long and active life, going through 22 pacemakers before passing away at the age of 86 from an unrelated illness (St. Jude Medical 2008).

8.2.8 *Energy*

Lasers The US National Academy of Engineering chose the combination of lasers and fiber optics as one of the 20 most important engineering developments of the twentieth century. In 1960, Theodore H. Maiman, a physicist at Hughes Research Labs in California constructed the first laser, shining a high-powered pulsed lamp on a ruby rod with silver-coated surfaces. The laser's ability to generate an intense, very narrow beam of single-wavelength light quickly found applications in science, technology, and medicine (Townes 2003). There are at least ten Nobel Prize winners whose work was made possible by lasers. Esther Conwell PhD, (Achievement Award 1960), used miniature lasers in her research on semiconductors and Elsa Garmire PhD, (Achievement Award 1994), performed her doctoral research on a ruby laser (ETHW n.d.). The ruby laser was used to send a light beam to the Moon in 1969, where it hit a reflector placed on the Moon's surface by the Apollo astronauts. The round-trip travel time of the pulse provided a very accurate measurement of the distance to the Moon. Developed by Robert Hall and his associates at General Electric in 1962, semiconductor lasers were small and inexpensive, able to be used in supermarket bar-code readers, in optical-fiber communications, and in laser pointers (Townes 2003).

Nuclear Power The first commercial atomic electric power plant in the US began operating in 1957 at Shippingport, PA (US Department of Energy n.d.). Atomic energy expanded rapidly after 1967, peaking at 100,000 MW of power generation capacity nationwide by 1990 (EIA 2023).

8.3 State of Women

From 1950 to 1960, the number of women working outside the home grew by 5%. While most worked in traditional female jobs such as teachers, secretaries, and nurses, the number of women in professional roles was creeping upward. Job expansion in professional fields during this decade included roles in finance, real estate, insurance, and professional services (Peterson and Willard Wirtz 1962). Women engineers were still less than 1% in their profession (Marczoch 1964).

There was a great influx of married women joining the workforce—from 7.7 million in 1950 to 12.4 million in 1960—which amounted to four-fifths the total gain over the time period. When the children got older, women were rejoining the workforce. This trend was driven by a desire for a higher standard of living, the easing of household tasks through use of modern appliances, and changes in traditional attitudes toward women's work outside the home. The age of working women was also on the rise, with 35% of working women aged 45–64 years in the 1960 census data (Peterson and Willard Wirtz 1962).

Women faced many barriers in the workforce: a labyrinth of antiquated laws governing their every action, overt discrimination, and pay differentials were commonplace. In Michigan, a woman could not work as a bartender in a city with a large population as *the profession of bartending could lead to the potential for moral and social problems for women* (Justia 1948). However, it was allowed that women could be waitresses to serve the liquor in those same establishments (without any concern for their moral or social problems). As part of the 1959 California Fair Employment and Housing Act, the law specified: *It shall be an unlawful employment practice for an employer to refuse to permit an employee to wear pants on account of the sex of the employee.* The accompanying discrimination complaint form even had an option to check "denied the right to wear pants" (FindLaw 2023).

Census data records show that women workers received median annual earnings of $2230 in 1959, which was less than half as much as the $4595 average for men. Non-white women fared worse, with median average earnings of $909. While some of the disparity reflects that many women worked part time, it also exemplifies the pay inequities of the time (Peterson and Willard Wirtz 1962).

One of the largest hurdles for working women was unplanned pregnancy, which took time away from their jobs or forced them to quit. In 1960, the first hormonal birth control pill was approved by the FDA for widespread use, distributed by Searle. The push for development of a birth control pill occurred after the Great Depression, when having smaller families meant more resources for each child, a strategy to combat poverty. Population control advocates Margaret Sanger and Gregory Pincus led the charge. Sanger had long believed that family planning was essential to giving women more rights and improving their lives.[1] Katharine McCormick contributed the majority of funding for development from her personal inheritance with the remainder provided by the Planned Parenthood Federation of America (Sanger was a co-founder). Both Sanger and McCormick felt that if a new contraceptive method were created, then it should be controlled by women since they are the ones who get pregnant and bear the responsibility for child rearing. New insights into endocrinology and physiology of the female reproductive system led to a pill comprising synthetic estrogen and progestin. At long last, most married women in the workplace had a much more reliable means of family planning (Buttar and Seward 2009). By 1965, in *Griswold v. Connecticut,* the Supreme Court ruled that remaining state bans on birth control violated marital privacy, overturning lingering access limits to married couples. In 1972, *Eisenstadt v Baird* extended birth control access to unmarried people (Justia 1965, 1972).

Women inventors of the period tinkered with the products they encountered in their lives. Evelyn Berezin received a patent in 1959 for a data processing system. African American Jessie Pope patented a thermostatically controlled curling iron in 1958 and went on to establish a company to manufacture it. In 1960, African American I.O. Carter received a patent for a nursery chair (SWE Magazine 2001).

[1] Sanger's mother had experienced 18 pregnancies with 11 live births, all while debilitated with tuberculosis. She died at age 49.

8.4 State of SWE[2]

The SWE organization faced growing pains during its early years. Catherine Eiden assumed a full plate of issues upon her election as SWE President in 1958. These included financial solvency questions, bylaws refurbishment, organizational leadership sprawl, and an Internal Revenue Service ruling against the Society which required rewording of the 1952 Articles of Incorporation (Kindya 1990; SWE Records 2000).

In 1959, the existing Board of Directors structure was dissolved, and a smaller Executive Committee was put in place. It consisted of four officers (President, Vice President, Secretary, and Treasurer) plus five additional voting members. The Council of Section Representatives, operating as the legislative branch, met once a year to implement the recommendations of the Executive Committee (Kindya 1990; SWE Records 2000).

In 1958, the Lillian Moller Gilbreth Scholarship was awarded for the first time, furthering SWE's mission to support women pursuing their technical education. The great lady herself had contributed $500 toward the endowment. Life membership was a new option to support SWE that same year, with the donations flowing into the headquarters endowment fund. The first life member was Maryly Van Leer Peck, who came from a family of all engineers (Kindya 1990). Membership statistics compiled in 1960 reported SWE having 560 members, including 34 student members. Corporate memberships were incorporated into the bylaws in 1961 (Kindya 1990; SWE Records 2000).

Maryly Van Leer Peck had been busy overhauling the Student Affairs committee, envisioning a pipeline of future SWE members upon graduation. She initiated a mailing of letters to colleges requesting women engineering students' enrollment figures. Results showed that there were 779 women engineering students in the 93 schools that responded. That number was greater than the entire membership of SWE! Yet, most students had never heard of SWE. Work began on a student section manual to encourage groups of students to form sections in their universities. In 1958, there were only two strong student sections: Drexel and Purdue, whose roots pre-dated SWE as a national organization (Kindya 1990; SWE Records 2000).

Women engineering students during this time faced many obstacles to completing their coursework, many of which were environmental. Most Ivy League schools did not yet admit women to their degree programs. On campus, the only other women were secretaries and librarians. The faculty and students were all male.

> I could not attend Harvard because I was a woman. I attended Radcliffe College and was given a diploma from them. I had to live in Radcliffe Quad, almost a mile from classes. There were only three women physics students, and I don't think there were any women in engineering. I also could not use the Harvard undergraduate library, because women were considered a distraction. Many classes were held in buildings that had no women's restrooms. We were considered not as smart as the men, because of our lack of preparation in high school. I loved it anyway… With the help of Nobel Prize winner Edward Purcell

[2] This section includes personal recollections about the time period and SWE from a survey of SWE Fellows and Late Career/Retiree Affinity group members conducted in February 2025.

> [Physics 1952], I got into both Harvard and MIT graduate schools. I chose MIT and worked for the Nobel Prize winner Charles Townes [Physics, 1964].
>
> —Elsa Garmire, SWE Fellow and Achievement Award recipient 1994 (personal correspondence 2025).

Women engineers in the workforce during this era were the trailblazers, navigating a time of great technology change. Practicing engineers and academics got in on the ground floor of the newly burgeoning space program, computer industry, nuclear energy program, consumer electronics industry, and medical device field. Their contributions and diverse perspectives reverberate through technology to this day. These pioneers were also on the front line of establishing what a professional workplace culture looked like with their male colleagues and customers. This sometimes meant scolding a colleague when they voiced inappropriate comments or correcting a customer who asked why you were not at home minding the children. Traditionally, women engineers with young children relied on relatives or paid help for childcare and housework. Public child daycare was not commonly available until the late 1960s.

SWE was changing and modernizing, along with the rest of society. Construction of an Engineering Center was being planned in Manhattan, across from the United Nations. It would have offices for all the engineering societies and SWE was delighted to be included. In 1961, SWE moved to its new headquarters and finally had a seat at the table alongside the other engineering societies (Kindya 1990; SWE Records 2000).

8.5 Laws and Culture

The Space Race *Sputnik I*'s launch in 1957 initiated the space age. Engineers and scientists were determined to accept President Kennedy's challenge to put a man on the moon, pushing technologies forward at a rapid pace.

Societal Change The availability of birth control pills in 1960 gave women more control over family planning. Legislative changes began opening up women to juries and politics.

A Young, Vibrant President with 'Conscious' Constituents President John F. Kennedy swept into office in 1961 on a wave of young voter support, representing completely different demographics than those of previous Democratic candidate victories. These constituents had a passion for social change and social justice, with the belief that reasoned argument was a powerful non-violent tool.

> Younger Americans at the start of the 1960s were better educated than their parents and more apt to have careers requiring professional accreditation. They were postponing marriage and childbearing. They had the time, the financial stability, the credentials and the self-confidence to become activists challenging conventional authority. They were referred to [in politics] as the *conscience constituency* (Isserman and Kazin 2004).

Summary Table 8.1 provides the examples of these changes.

Table 8.1 Culture and legal highlights for 1957–1961

Dates	Events
1957	**Civil Rights Act**: The Civil Rights Act of 1957, enacted under President Dwight D. Eisenhower, was the first significant measure to address African American civil rights since 1875. It established the US Commission on Civil Rights, created a civil rights division in the US Justice Department, and authorized the US Attorney General to seek federal court injunctions to protect the voting rights of African Americans. It also gave women the right to serve on federal juries, but a few states still did not allow women to serve on state and local juries until the late 1960s (US House of Representatives 1957)
1961	**Man on the Moon Challenge**: President John F. Kennedy announced to a joint session of Congress the following technological challenge: "I believe that this nation should commit itself to achieving the goal, before this decade is out, of landing a man on the moon and returning him safely to the Earth (JFK Library 1961)"
1961	**Peace Corps:** In his inaugural address, President John F. Kennedy stirred up the idealistic appeal to defend freedom abroad, through opportunities such as the Peace Corps. The Peace Corps assisted other countries in their advancement through education, agriculture, health, trade, technology, environmental, and community development. The Peace Corps grew from 900 to nearly 16,000 volunteers serving in a peak of 52 countries in 1966 (Britannica 2025a)

References

American. n.d. History of American Airlines. AA.com. https://www.aa.com/i18n/customer-service/about-us/history-of-american-airlines.jsp. Accessed February 21, 2025.

ACS. 1999. Discovery of Polypropylene and the Development of a New High Density Polyethylene. American Chemical Society. November 12. https://www.acs.org/education/whatischemistry/landmarks/polypropylene.html. Accessed February 20, 2025.

Britannica, T. Editors of Encyclopaedia Britannica. 2016. Videotape. *Encyclopedia Britannica.* November 18. https://www.britannica.com/technology/videotape. Accessed March 11, 2025.

Britannica, T. Editors of Encyclopaedia Britannica. 2025a. Peace Corps. *Encyclopedia Britannica.* January 18. https://www.britannica.com/topic/Peace-Corps. Accessed February 26, 2025.

Britannica, T. Editors of Encyclopaedia. 2025b. Xerox. *Encyclopedia Britannica.* February 15. https://www.britannica.com/money/Xerox-Corporation. Accessed February 21, 2025.

Buttar, Aliya; Sheraden Seward. 2009. Enovid: The First Hormonal Birth Control Pill. January 20. *Embryo Project Encyclopedia.* ISSN: 1940-5030. https://embryo.asu.edu/pages/enovid-first-hormonal-birth-control-pill. Accessed March 7, 2025.

Computer History Museum. 2025a. 1958: All Semiconductor "Solid Circuit" is Demonstrated. https://www.computerhistory.org/siliconengine/all-semiconductor-solid-circuit-is-demonstrated/. Accessed March 11, 2025.

Computer History Museum. 2025b. 1959: Practical Monolithic Integrated Circuit Concept Patented. https://www.computerhistory.org/siliconengine/practical-monolithic-integrated-circuit-concept-patented/. Accessed March 11, 2025.

EIA. 2023. US Energy Information Administration. August 24. https://www.eia.gov/energyexplained/nuclear/us-nuclear-industry.php. Accessed March 11, 2025.

Engineering and Technology History Wiki (ETHW) n.d. SWE Awards. https://ethw.org/SWE_Awards. Accessed March 22, 2025.

FindLaw. 2023. California Code, Government Code - GOV 12947.5. January 1. https://codes.findlaw.com/ca/government-code/gov-sect-12947-5.html?/. Accessed March 7, 2025.

Fisher, D.E., Fisher, M.J., Fink, D.G., Noll, A.M. 2025. Television. January 21. *Encyclopedia Britannica*, https://www.britannica.com/technology/television-technology, accessed February 20, 2025.

Garber, Steve. 2007. Sputnik and The Dawn of the Space Age. NASA. October 10. https://www.nasa.gov/history/sputnik/index.html. Accessed February 20. 2025.

Hemmungs Wirtén, Eva. 2004. *No Trespassing: Authorship, Intellectual Property Rights, and the Boundaries of Globalization*. University of Toronto Press. pp. 61–64. ISBN 9780802086082.

IBM. n.d. The IBM Selectric. https://www.ibm.com/history/selectric. Accessed February 21, 2025.

Isserman, Maurice and Kazin, Michael. 2004. *America Divided: The Civil War of th*e 1960s. Second Edition. Oxford University Press. ISBN 0-19-516046-0

John F. Kennedy (JFK) Library. 1961. Historic Speeches. May 25. https://www.jfklibrary.org/learn/about-jfk/historic-speeches/address-to-joint-session-of-congress-may-25-1961. Accessed February 17, 2025.

Justia. 1948. Goesaert v. Cleary, 335 US 464. https://supreme.justia.com/cases/federal/us/335/464/. Accessed March 7, 2025.

Justia. 1965. Griswold v. Connecticut. 381 U.S. 479. https://supreme.justia.com/cases/federal/us/381/479/. Accessed March 7, 2025.

Justia. 1972. Eisenstadt v Baird. 405 U.S. 438. https://supreme.justia.com/cases/federal/us/405/438/. Accessed March 7, 2025.

Kindya, Marta. 1990. Four Decades of the Society of Women Engineers. Society of Women Engineers Records (LR001539). Box 180, Folder 9, Walter P. Reuther Library, Archives of Labor and Urban Affairs, Wayne State University. Accessed January 23, 2025.

LYCRA. 2025. See how LYCRA fiber has shaped the world. The LYCRA Company, https://www.lycra.com/en/lycra-about-lycra-fiber/see-how-lycra-fiber-has-shaped-world. Accessed February 20, 2025.

Marczoch, Sigrid. 1964. Why Don't American Women Stay in Engineering? August. PDF document. Accessed March 11, 2025.

McGinn Spring, Kathleen. 2001. The Story of Color Television. November 14. https://web.archive.org/web/20131004220132/http://161.58.97.168/200111/11114c01.html. Accessed March 11, 2025.

NASA. 2016. Shepard's Mercury Flight Was First Step on the Long Journey to Mars. May 2. https://www.nasa.gov/humans-in-space/shepards-mercury-flight-was-first-step-on-the-long-journey-to-mars/. Accessed February 20, 2025.

Peterson, Esther; W. Willard Wirtz. 1962. Women Workers in 1960. US Department of Labor Women's Bureau Bulletin 284. https://fraser.stlouisfed.org/files/docs/publications/women/b0284_dolwb_1962.pdf. Accessed March 7, 2025.

Sabre. 2016. Sabre History. August 16. https://www.sabre.com/files/Sabre-History.pdf. accessed February 21, 2025.

Society of Women Engineers (SWE) Records. 2000. SWE 50 Years 1950-2000. Society of Women Engineers Records (LR001539). Box 180, Folder 9, Walter P. Reuther Library, Archives of Labor and Urban Affairs, Wayne State University. Accessed January 10, 2025.

St. Jude Medical. 2008. St. Jude Medical Celebrates 50th Anniversary of the World's First Implantable Pacemaker. St. Jude Medical Diagnostic and Interventional Cardiology. October 7. https://www.dicardiology.com/content/st-jude-medical-celebrates-50th-anniversary-worlds-first-implantable-pacemaker. Accessed March 12, 2025.

SWE Magazine. 2001. Women Innovators from Ancient Times to the 21st Century. March/April. Society of Women Engineers Records (LR001539). Box 180, Folder 9, Walter P. Reuther Library, Archives of Labor and Urban Affairs, Wayne State University. Accessed March 7, 2025.

Townes, Charles H. 2003. The First Laser. https://press.uchicago.edu/Misc/Chicago/284158_townes.html. *A Century of Nature: Twenty-One Discoveries that Changed Science and the World.* ISBN: 0-226-28413-1. Accessed March 11, 2025.

University of Minnesota Medical School. 2018. Remembering Earl Bakken. October 22. https://med.umn.edu/surgery/news/remembering-earl-bakken. Accessed March 12, 2025.

US Department of Energy. n.d. Shippingport Atomic Power Station becomes Operational. https://www.energy.gov/management/december-23-1957-shippingport#:~:text=The%20Shippingport%20Atomic%20Power%20Station,becomes%20operational%20at%20Shippingport%2C%20Pennsylvania. Accessed March 11, 2025.

US House of Representatives. 1957. The Civil Rights Act of 1957. June 18. https://history.house.gov/Historical-Highlights/1951-2000/The-Civil-Rights-Act-of-1957. Accessed March 11, 2025.

Walsh, Kenneth. 2010. The 1960s: A Decade of Promise and Heartbreak. March 9. https://www.usnews.com/news/articles/2010/03/09/the-1960s-a-decade-of-promise-and-heartbreak. Accessed February 20, 2025.

Chapter 9
Rebecca H. Sparling (1910–1996) Achievement Award Recipient 1957

In recognition of meritorious contributions to high-temperature metallurgy and nondestructive testing of metals

Debra A. Kimberling

9.1 Introduction

Rebecca Sparling was an expert in metallurgy and nondestructive testing. She pioneered the application of immersed ultrasonic inspection to gas turbine parts and was an industry leader in the development of liquid dye penetrant surface inspection methods commonly used in the aerospace industry. She wrote the comprehensive 1944 revision of the *American Malleable Iron* Handbook, which became an industry standard and textbook.

Sparling was a leader in the aerospace industry when there were few women in engineering. She was Chief Metallurgist at Northrop Aircraft from 1944 to 1951, and then a Staff/Design Specialist at Convair/Pomona from 1951 to 1968 representing her company on agencies including the National Academy of Sciences, Materials Advisory Board, Aircraft Research and Testing Committees, and the materials and structures engineering section of the Aircraft Industries Association. She published more than 30 articles in technical and professional journals.

After Sparling received the Society of Women Engineers' (SWE) Achievement Award in 1957, she became an environmental activist. Upon retirement in 1968, Sparling joined the San Bernardino County Air Pollution Appeals Board. She represented the American Association of University Women (AAUW) before the State Energy Commission.

D. A. Kimberling (✉)
San Diego, CA, USA

C. D. Craig et al. (eds.), *Women Engineering Legends 1952-1976: Society of Women Engineers Achievement Award Recipients*,
https://doi.org/10.1007/978-3-032-00224-2_9

Sparling graduated with an MS degree from Vanderbilt University, in Physical Chemistry, although her dream had been to study engineering. Her early career was adversely impacted by the Great Depression, but she gained practical metallurgical experience through a series of manufacturing, foundry, and automobile plants as a metallurgist and as a consultant from 1931 to 1943.

Rebecca Sparling (nee Hall) married Edwin K. Smith in 1935, having one son Douglas born in 1938. She later married Joseph E. Sparling in 1948 (Oakes 2020).

9.2 Early Life and Education

Rebecca Hall Sparling (Fig. 9.1) was born in Memphis, Tennessee in 1910 to Kate Sampson and Robert Meredith Hall. The youngest of ten children, she was raised in Tennessee. None of her family or acquaintances were in technical or scientific fields although both parents were college graduates. Her father, who held a law degree, was a business industrialist. Sparling's dream was to make great discoveries. Thus after a year at Hollins, a women's college in Virginia, she enrolled at Vanderbilt University. Her family was not exactly pleased about the decision stating "it wasn't the proper thing for Southern ladies to do" (ICWES 1964; Munday 1955; Norris 1977; Oakes 2020).

Needing to fund her education herself, Sparling received an undergraduate scholarship working as a laboratory instructor, then she received a graduate fellowship as she earned her master's degree teaching laboratory. She also worked at the *Nashville Banner*, a daily newspaper, reporting news and writing publicity. She was a member of the Delta Delta Delta sorority (Munday 1955; SWE Records n.d.).

Fig. 9.1 1957 Society of Women Engineers Achievement Award Recipient Rebecca Sparling at work. (Photograph courtesy of Society of Women Engineers Records, Walter P. Reuther Library, Archives of Labor and Urban Affairs, Wayne State University)

Enjoying the practical side of scientific work, Sparling first became interested in engineering while she was a sophomore in college (SWE Records n.d.). A guest speaker addressed "The Romance of Cast Iron" and the role it plays in civilization (Nordyke 1965). She was fascinated with how the metallurgist attributed metals to world events.

> Specifically, he connected the fall of the Roman Empire with the lead pipes that carried its water, theorizing the soft metal adversely affected Roman minds. And he conjectured that the rise of subsequent civilizations was due in part to replacing lead with iron. I really got caught up in it (Norris 1977).

Because she was not able to accompany an all-male group on a required 2-week summer field trip to the mountains, Sparling did not meet the requirements for an engineering degree. Rather she went towards a technical degree in Chemistry as a way to enter the field of engineering through a back door. She earned her 1930 BA cum laude in Physical Chemistry, and a 1931 MS in Physical Chemistry (Norris 1977; SWE Records n.d.).

9.3 Early Career/Great Depression

Sparling gained practical metallurgical experience through a series of manufacturing, foundry, and automobile plants as a metallurgist and as a consultant from 1931 to 1943. Her first jobs were in foundries, dealing primarily with the production of cast metal parts. She worked on gray and malleable iron, steel castings, and special metals for specific service requirements such as high-temperature, corrosive chemicals, or abrasive wear (ICWES 1964).

She left her first job at the American Cast Iron Pipe Company in Birmingham, Alabama (1931–1932) after 1 year when it became apparent that she wasn't going to advance in her career. She decided it was futile to balk against stereotyping that "probably would have never allowed me to be anything other than an executive secretary" (Norris 1977).

Her next job was at the Lakeside Malleable Castings Company in Racine, Wisconsin (1933–1934). She investigated causes of foundry scrap and developed special alloyed malleable irons to meet customer applications. From Wisconsin, she found there were few job opportunities for engineers but she was able to take advantage of other opportunities that were available—in technical writing and technical translation. She interpreted metallurgical progress into everyday words in technical and non-technical briefings for William H. Baldwin Company/International Nickel Company in New York City (1934–1935).

> When I got to Chicago in the [19]30s, I found that there weren't many jobs for engineers. But there was quite a bit of technical writing to be done. And luckily I fell right into it. The same way with technical translation. I never was interested in languages, but my mother, who was quite brilliant, insisted that we learn them (Sparling 1957).

Sparling opened her own consulting office in Detroit from 1936 to 1941, specializing in brake drums and cylinder improvements (Munday 1955). She investigated failures, especially of high alloyed iron and steel, for several foundries and automobile firms (SWE Records n.d.).

> Her reception in the Windy City was a chilly one; most firms weren't even willing to interview her ... One foundry superintendent she met greeted her with this blunt dismissal: "You're not going to work here. I've got a daughter your age and she doesn't have enough sense to come in out of the rain." To which Sparling replied: "I know a man your age who's in an insane asylum, but I don't think you're insane." She got the job (Norris 1977).

With the advent of World War II in 1942, Sparling answered an advertisement for women engineers to go to Washington, DC. However, she couldn't get a professional rating, "with my 11 years of engineering experience they wanted me to work in an administrative or clerical position." After 4 months, she eventually found a job in the Metallurgical and Testing Division of the Naval Gun Factory staff, part of the US Navy Bureau of Ordnance. Sparling conducted testing and troubleshooting for such items as failed anti-aircraft guns and worn gun barrels (Norris 1977).

Commonly then, she and the other women were paid less than the men who were doing the same jobs. She felt more opposition in the Washington job than in any other, chalking it up to the strain of those early war days (Norris 1977).

> No one I think would ever feel that the war was a fortunate bit of timing, but we all know that there was a great deal of technical progress made at that time. … I'm quite aware that I had many interesting problems which would never have come to me if the automotive plants hadn't been so busy even before we entered the war and making ordnance for the allies (Sparling 1957).

After Washington, Sparling moved back to Detroit to consult for a number of clients and begin work for the Malleable Founder's Society.

9.4 Professional Affiliations

In 1943, the Malleable Founders' Society contracted with Sparling to update the handbook *American Malleable Iron*. It was the first comprehensive update in 20 years. She wrote the entire manuscript with the sole exception of two chapters on casting design and pattern making. The handbook was a standard reference in the castings industry since its 1944 publication and was used as a college textbook (Oakes 2020; SWE Records n.d.).

Throughout her career, Sparling served on joint committees and panels concerned with properties and testing of metals in several industries. She was the Acting Technical Secretary for the American Foundrymen Association in Chicago where she translated foreign papers and edited papers for the 1934 Fifth International Foundry Congress and was an officer of the Detroit Section for the American Institute of Metallurgical Engineers in 1937. She was active in the Society for

Nondestructive Testing, American Society for Metals, and Society of Automotive Engineers throughout most of her career.

Sparling credits her engagement in technical societies to her first manager and addresses some of the culture of the era, being one of the only women engineers in the field, and the support and mentors she had along the way (Sparling 1957):

> I had a very lucky break in my first job too. I went down to the American Cast Iron Pipe Company in Birmingham 1931 Metallurgy Department. The man who was then head of it, ... was ... very active in technical societies…And he encouraged…his employees…often [taking us] to the different technical meetings, introducing us… After I had left his company…, he was nice enough to send me data from his files so that I could write one of my first technical articles…, I have found a great deal of pleasure and satisfaction in working in the different technical organizations just as he started me out many years ago.

Sparling experienced challenges in these trips but persevered, sometimes with the help of others.

> Going to a meeting, the girls at the registration desk would say, "well, now, here's your ticket for the ladies fashion show and tomorrow we take you on a tour of the homes." And I'd say "no, I want to go to technical meetings." And I always won out but sometimes you had to be very firm with them. But last month I went to one and the girl greeted me with joy and said, "Oh, I remember you Miss Sparling. You're not a lady, you're an Engineer." I'd like to be both.
>
> I know when I was in New York, oh 1934 and 35, we had a very big chapter there of the American Society for Metals. There must have been 300 or 400 people at the meetings.
>
> [Sometimes] I was hesitant about going. I wanted to hear the talks, but I…hated to go by myself…Dr. Frances Clark…chief metallurgist of Western Union [would] take me under her wing and introduce me to people. And Bill Isamar…[ASM] executive secretary [did] the same thing…Dr. Gilbreth doesn't remember it, but she helped me a great deal about that time. We had lunch together one day when I was getting a little discouraged and it's just one of the things she did without even thinking about. But it made a big impression on me.

In 1958, Sparling was a SWE Los Angeles Section officer and served at the national level on the 1977 Scholarship Overview Committee, as 1958 Scholarship Committee Chair, and 1967 Award Committee Chair (SWE Los Angeles Section 1983). She chaired the Gilbreth Scholarship committee, the committee that set up the scholarship rules and selected the first recipient (Sparling 1968).

9.5 Later Career/Post World War II

Sparling moved to Southern California in 1944, in part to improve the health of her then 6-year old son. It was the ideal time to enter the aircraft industry. She describes her unique interviews as a materials engineer as follows (Sparling 1957):

> …They didn't know what a metallurgical engineer was. And they'd say, "Are you a metal worker, A or B?" And when I admitted that I couldn't hammer out a piece of sheet metal no matter how hard I tried, why my interview would be at an end.
>
> So after two or three plants, I went to Northrop Aircraft. The interviewer looked me over, rather critically and said, "Well, we have a chief research engineer who is hiring a lot of strange characters. And I'll send you to see him." And he hired me.

> We had a peculiar interview, because he wouldn't tell me what the company was doing, or what the product was, or what type of job he was considering me for, as this was all classified. He just asked me one question after another. And finally he called in the project engineer and said, "Bill, I think we've got the man you want to handle that metallurgy." Bill looked around the room and I was the only other person there. He said, "Ha, ha, you don't mean her." But he did. I was the man who happened to be there right then, and that was the beginning of seven very happy years I spent at Northrop.

From 1944 to 1950, Sparling worked at Turbodyne Corporation, a subsidiary of Northrop Aircraft Inc. (later called Northrop Grumman). The Hawthorne, California facility was working on a 10,000 shaft horsepower turbo propeller engine. Sparling was the Chief Metallurgist and the head of the Materials and Process group. She recommended materials, design specifications, and supervised the metallurgical laboratory and nondestructive tests conducting *post-mortem*s as needed. By setting the allowable stress as a dynamic variable, Sparling could monitor the material's performance output based on input variables such as time, temperature, and load for high-temperature metals (SWE Records n.d.).

Sparling established industry standards for non-destructive test methods, a way for evaluating engineering functionality without destroying the part. Under her watch, immersed ultrasonic inspection, or "immersed scanning" was first employed on Turbodyne turbine wheels. She co-authored the first paper on immersed ultrasonic inspection, in the 1949 publication *The Iron Age* (Smith & Erdman 1949; SWE Records n.d.). She also developed visible, liquid dye penetrant inspection methods that ultimately became the industry standard for surface inspections.

Later, Sparling became the Structures Staff Engineer reporting directly to the Chief of Structures of Research and Development with both technical and administrative responsibility for the Air Force *Cast Wing* project. Sparling directed the design of the largest casting ever undertaken, a 16-foot one-piece magnesium airplane wing, down to ¼″ thickness and 4-foot width. She proved that a commercial foundry could achieve the critical structural specifications and tight dimensional tolerances (Munday 1955; SWE Records n.d.).

The Flying Wing, an aeroplane with no fuselage or tailplane, was a passion of the founder of Northrop Aircraft, John Knudsen Northrop. Sparling thought very highly of Northrop, the man, and the workplace culture he instilled from the top down (Sparling 1957).

> Dr. [sic] John K Northrop is another person whom I've met who had, I think, quite an influence on my thinking. [He was an] aeronautical designer, [and] president of Northrop when I was there. [He] has freedom from prejudice...probably typical of the aircraft industry, I hope so anyway.
>
> I know I was surprised when I first walked into an aircraft engineering plant and saw there all kinds of people working and engineering. Men and women, young and old. All races, creeds, nations...We had quite a few people who were paraplegic. We had one man who used to come with his seeing eye dog...the only thing that mattered was the ability to do the job...I think all of us hoped that someday it'll be that way everywhere.

In 1951, Sparling (Fig. 9.2) began work for Convair, a division of General Dynamics Corporation, as a Materials Design Specialist. The Convair/Pomona plant was just opening, giving Sparling the opportunity to work at a facility starting

Fig. 9.2 Rebecca Sparling works with materials, circa 1960–1969. (Photograph courtesy of Society of Women Engineers Records, Walter P. Reuther Library, Wayne State University)

from the ground floor (Sparling 1957). The Convair-operated facility of the Naval Industrial Reserve Ordnance Plant was engaged in the design and manufacture of guided missiles and other special ordnance weapons (Sparling 1959).

Sparling directed materials research, and later became Staff Assistant to the Chief of Process Control and Reliability. She investigated new uses for materials and processes in guided missiles, electronic components and spacecraft. Her work included the use and development of radioactive and innovative test techniques, and long-term forecasting of materials requirements for missiles and other special ordnance weapons for the US Navy. She worked with agencies such as the National Academy of Sciences, Materials Advisory Board, and the Department of Defense as well as technical groups, technical societies, and materials' suppliers. She represented Convair/Pomona on the aircraft Research and Testing Committee, the materials and structures engineering section of the Aircraft Industries Association (SWE Records n.d.).

9.6 Retirement/Advocacy/Environmental Activism

While at Convair, Sparling supported K-12 community outreach initiatives to encourage more young women into the field of engineering. During her 1957 SWE Achievement Award acceptance speech, she noted (Sparling 1957) the work at Convair to bring more young people into engineering:

> …a team of us [visited] junior high schools [talking to] boys and girls in the seventh and eighth grades because if they go into the ninth grade and do not take their mathematics, we find it is very difficult later on to get them to go back. When they get to college, if they want to go into science and engineering they aren't prepared…my schedule was so heavy that I

thought maybe somebody else would go instead of me. But the training director said, well, we can tell them that we have women engineers, but there's nothing like a live working model, so that's what I am.

Sparling moved to working half-time at Convair in 1964. She served on educational boards and advisory committees including East Los Angeles College (1962) and California State Board of Education on Technical and Semi Professional Education of Women (1963–1964). Sparling addressed the significance of having received the SWE Achievement Award, during a panel discussion with her fellow award recipients at the 1968 SWE National Convention (Sparling 1968):

I believe winning the award gives you a lot of publicity, and people hear of you and ask you to do things that would never have occurred otherwise. And I have particularly enjoyed things like being on [these state and local boards and] giving … talks to young people. … I'm sure I would never have been asked to do it if it hadn't been for the award. It's not only been a pleasure, but I think it is a sort of responsibility, too, because you don't like to get up somewhere—more or less representing the Society of Women Engineers—unless you feel that you can do a good job. And I've tried very hard to do that.

Sparling officially retired in 1968. Thereafter, she devoted most of her time to the environment, working primarily on air pollution from 1968 to 1974. She led the successful fight to include strict emissions regulations on a proposed coal-fired power plant by Southern California Edison slated for the Mojave Desert of San Bernardino County (Sparling 1972; Sparling 1981; Tierney 1980).

Sparling was a 1969 charter member of the Desert Environmental Conservation Association, serving as chair of the Air Pollution Control Scientific Committee and president of the Conserve Our Air & Water Committee. In 1973 she joined the San Bernardino County's Air Pollution Appeals Board (SWE Records n.d.).

Sparling testified before government boards, served on committees, and participated on panels, radio, and TV on topics concerning environment and energy addressing anti-smog regulations and actions by the Southern California Association of Governments. In 1974, she became deeply involved in the use and sources of energy. From 1975 to 1978, she represented California's 31,000 members of the AAUW before the State Energy Commission. Later, she joined the Foreign Policy Association, emphasizing the interdependence of strategic materials (Sparling 1981).

9.7 Honors and Legacy

Rebecca Sparling achieved her dream of making great discoveries. She pioneered the application of immersed ultrasonic inspection for gas turbine parts and was one of the developers of liquid dye penetrant surface inspection (Sparling 1959). Her achievements are best summarized in these recognitions: The 1957 Society of Women Engineers Achievement Award (Fig. 9.3) *in recognition of meritorious contributions to high temperature metallurgy and nondestructive testing of metals* (Sparling 1972; SWE Records n.d.); and her 1972 election as Fellow of the American Society of Metals (now ASM International) *for effective contributions in moving*

Fig. 9.3 Rebecca Sparling (left) with Edith Clarke at the March 1957 SWE National Convention. (Photograph courtesy of Society of Women Engineers Records, Walter P. Reuther Library, Wayne State University)

new developments in materials and processes from research into practice. She has contributed much to the profession promoting a better understanding of the role of the metallurgist to both industry and society (Sparling 1972; SWE Records n.d.). Sparling was selected as part of the second class of ASM Fellows; she was the second woman to become a Fellow (Personal correspondence ASM International 2025). Also, Sparling was elected to the College of Fellows by the Institute for the Advancement of Engineering in 1975 *for continuing distinguished contributions to the engineering profession and the public it serves* (LA Times 1979).

Sparling advanced the field of engineering throughout her career. She developed standards and special materials for unique applications. She led the way for brittle, high-strength materials, military standards for corrosion prevention and deterioration control in electronic components and assemblies. She was a prolific technical writer and was extremely active in several professional and technical engineering societies (SWE Records n.d.):

- Sparling wrote the 1944 edition of *American Malleable Iron* Handbook used as an industry wide standard and college textbook.

- Sparling became a registered mechanical engineer in 1949 in the state of California. She likely was the first woman Professional Engineer in San Bernardino County (Nordyke 1965).
- Sparling published over 30 articles in technical and professional journals (Sparling 1959).
- She served on several technical, professional and government agencies, boards, and committees and delivered more than 70 technical presentations.

Sparling was a respected engineering leader. She used humor, persistence, and common sense to navigate the difficult career options presented during the Great Depression and WWII. In her late career, she led the burgeoning aerospace industry as Chief Metallurgist and Staff Materials Design Specialist at Northrop and Convair, respectively, representing her companies and gaining allies along the way. Even in retirement, Sparling used her engineering expertise to advance environmental activism in the community. Sparling was recognized with the Convair/Pomona Management Club Achievement Award in 1960 and has been recognized in several *Who's Who*, including *Who's Who in Engineering*, *Leaders of American Science* and *Dictionary of International Biography* (Nordyke 1965; SWE Records n.d.). She testified before several government boards addressing anti-smog regulations and energy.

Sparling reflected on her career during her remarks at the first International Conference on Women Engineers and Scientists held in 1964 (ICWES 1964):

> The longer I study materials and work with them, the more interested I become in the subject of materials engineering. Every day brings new inventions, new discoveries in processing and treatment, new opportunities to learn. Increasing my knowledge and using it to make better parts for the missiles which defend our country is my life work—stimulating, demanding, and, I feel, very worthwhile. I wish I could live to be 150 and keep studying and working all the time.

Sparling was an advocate advancing the field of engineering as a viable career for women. She joined education boards at a time when there were almost no women in engineering. She was a community volunteer and represented organizations such as SWE and the AAUW through talks and before the California State Energy Commission. Sparling received the 1965 Engineering Merit Award from the Los Angeles Council of Engineers and Scientists (LACES) *in appreciation of her efforts in stimulating wider recognition of women as a national resource for engineering capabilities, and for her leadership in encouraging and supporting great use of women in the engineering profession* (Nordyke 1965). She was the first woman to be so honored by LACES (Daily Press 1972). In 1978, she received the Engineering Merit Award from the Orange County Engineering Council (SWE Records n.d.). In 1984, she was elected Fellow by the Society of Women Engineers (SWE Records n.d.).

The Rebecca Sparling estate donated significantly to the Society of Women Engineers in 1998, 2 years after her death (Oakes 2020).

References

Daily Press. 1972. Rebecca Sparling honored in Ohio. *Daily Press*. October 18. Page 7. Victorville, California.

Los Angeles Times. 1979. Institute for the Advancement of Engineering Inc. *Los Angeles Times*. February 22. Page 58.

Munday, Marian. 1955. Local Women To Wield Gavel Over Engineers. *The Progress Bulletin*. March 1. Pomona, California.

Nordyke, Priscilla. 1965. Lady Engineer Honored. *Daily Press*. March 8. Page 5. Victorville, California.

Norris, Maria. 1977. Southern Belle Enters Male Domain. *The Register*. April 27. Page 67. Santa Anna, California.

Oakes, Elizabeth. 2020. Encyclopedia of World Scientists, Updated Edition. Infobase Holdings, Inc. July 1.

Proceedings of the First International Conference of Women Engineers and Scientists (ICWES). 1964. Focus for the Future Developing Engineering and Scientific Talent. From the University of Illinois Urbana-Champaign Repository. https://uihistories.library.illinois.edu/REPOSITORYCACHE/156/d5ne15BSVY0hV4ogbAmN6p592koYodSl607Z2in6km8a-HI1y4S15vb3R58qwP34ra4tJ4nx1TJgHk55V7f17K3l8FSw6XNK5mjz39ERv2o9_3409.pdf

Smith, Rebecca H.; Erdman, Donald C. 1949. Immersed Ultrasonic Inspection. *The Iron Age* Volume 164. August 4.

Sparling, Rebecca. 1957. Transcript of March 16 Acceptance Speech, Awards Presentation at the SWE National Convention, Houston. Society of Women Engineers (SWE) Records (LR001539), *Rebecca Sparling*. Box 192, Folder 48, Walter P. Reuther Library, Archives of Labor and Union Affairs, Wayne State University.

Sparling, Rebecca. 1959. Biographical Sketch of January 21. Society of Women Engineers (SWE) Records (LR001539). *Rebecca Sparling*. Box 192, Folder 48, Walter P. Reuther Library, Archives of Labor and Union Affairs, Wayne State University.

Sparling, Rebecca. 1968. Audio excerpt from the 1968 SWE National Convention Achievement Award recipient panel. From Society of Women Engineers (SWE) Stories from the Archives Podcast on February 7, 2020. https://alltogether.swe.org/2020/03/tale-from-the-archive-early-achievement-award-recipient-offer-advice/

Sparling, Rebecca. 1972. Letter dated November 14, to Winnie (Winifred D. White), SWE Headquarters. Society of Women Engineers (SWE) Records (LR001539). *Rebecca Sparling*. Box 192, Folder 48, Walter P. Reuther Library, Archives of Labor and Union Affairs, Wayne State University.

Sparling, Rebecca. 1981. Letter dated February 28, to Barbara Krohn, Editor, US Woman Engineer. Society of Women Engineers (SWE) Records (LR001539). *Rebecca Sparling*. Box 192, Folder 48, Walter P. Reuther Library, Archives of Labor and Union Affairs, Wayne State University.

Society of Women Engineers (SWE) Records (LR001539). n.d. *Rebecca Sparling*. Box 192, Folder 48, Walter P. Reuther Library, Archives of Labor and Union Affairs, Wayne State University

SWE Los Angeles Section, 1983. Letter of Nomination for Fellow Grade dated October 31, to SWE Headquarters. From Society of Women Engineers (SWE) Records (LR001539). *Rebecca Sparling*. Box 192, Folder 48, Walter P. Reuther Library, Archives of Labor and Union Affairs, Wayne State University.

Tierney, Jean. 1980. Proposed coal-fired power plant 'cleanest'. *Daily Press*. March 26. Page 1. Victorville, California.

Chapter 10
Mabel MacFerran Rockwell (1902–1979) Achievement Award Recipient 1958

In recognition of her significant contributions to the field of electrical control systems

Vicki S. Johnson

10.1 Introduction

Mabel MacFerran Rockwell was an electrical engineer who was recognized by the Society of Women Engineers (SWE) for substantial contributions in the field of electrical control systems in 1958. Rockwell also significantly contributed to high-voltage power generation and distribution, aeronautical manufacturing research and development, and electrical system integration within facilities. Born in 1902, Rockwell completed a master's degree in electrical engineering in 1926 and started work at Southern California Edison Company. In her initial positions, she was responsible for the design and fielding of high-voltage power generation equipment and transmission lines from the Hoover Dam. During a break to have and raise her baby daughter, Rockwell attended the Cal-Aero Technical Institute, majoring in aeronautical engineering. She then returned to work at Lockheed Aircraft Company; she quickly worked her way from Plant Electrical Engineer to Engineer-in-Charge of Production Research, where she and her team developed processes for spot welding and deep draw metal forming. Her efforts led to spot welding being allowed on aircraft primary structure and vast improvements in production quality and rate which contributed significantly to Lockheed's ability to supply aircraft for World War II. Rockwell's later career involved working for the US Navy and Westinghouse on control systems for naval vehicles, including the control system for the submarine-launched Polaris missile. Rockwell was always working, even as a consultant, when she was at home to raise her daughter.

V. S. Johnson (✉)
Fountain Hills, AZ, USA

C. D. Craig et al. (eds.), *Women Engineering Legends 1952-1976: Society of Women Engineers Achievement Award Recipients*,
https://doi.org/10.1007/978-3-032-00224-2_10

10.2 Early Life and Education

Mabel MacFerran (or spelled Macferran as Mabel herself used) was born in 1902, in Philadelphia, Pennsylvania to Quaker parents, Edgar O. and Mabel Alexander MacFerran. Her father was an electrical engineer. Her mother graduated with honors in mathematics from Swarthmore College in 1898 (Rockwell 1958).

Rockwell did her initial education at Friends School in Germantown, Pennsylvania. In debating potential careers, she told the press when she received the SWE Achievement Award that she was torn between being a lawyer (because she loved to argue with people) and being an engineer (The Pittsburgh Press 1958).

Her first year of college was spent at Bryn Mawr College where she majored in mathematics. Rockwell then transferred to the Massachusetts Institute of Technology (MIT) where she completed a Bachelor of Science in science, teaching, and mathematics in 1925 (Rockwell 1956; Tietjen et al. 2023).

Rockwell received an Elwell Fellowship to attend Stanford University to pursue a Master of Science (MS) degree in Electrical Engineering under the tutelage of Professor Harris J. Ryan in the Million Volt Laboratory (Electrical Engineering 1935-07; Rockwell 1956). She moved to California, where she remained for the rest of her life. In 1926, Rockwell received her MS degree. At Stanford, Rockwell was inducted into Sigma Xi, the Scientific Research Honor Society (Rockwell 1956; Sigma n.d.).

During the early part of her career, Rockwell took evening classes in Business Law from Southwestern University in Los Angeles. Rockwell spent 1938 taking day classes in aeronautical engineering at Cal-Aero Technical Institute which opened a path to the aviation industry for her (Rockwell 1956).

10.3 Career

Rockwell's early career focused on high-voltage power transmission and equipment at the Hoover Dam to generate high-voltage electricity. The second part focused on aeronautical research at Lockheed Aircraft Corporation. The final part focused primarily on electrical control systems for the Polaris submarine-launched missile and on electrical features of various classified underwater propulsion and guidance systems along with a return to work on power transmission and utilization. Interspersed in each of these timeframes were periods of childcare leave and consulting. Rockwell was determined to stay employed and worked hard to decide what each employer really needed so she could give it to them. More than once she was hired to do one thing and, after arriving at work and seeing the needs, wound up doing something else. Her approach to all her jobs would be recognized today as Six Sigma Black Belt-type work with a drive to methodically test and understand resulting in documented processes to standardize work. Rockwell was not afraid of report writing.

10.3.1 Early Career

Rockwell's first job beginning in 1926 was with Southern California Edison Company as an electrical "testman" where she worked with meters, relays, and special tests. Later, she worked to minimize system shutdowns, made the first static and dynamic stability limit studies of the Edison Big Creek transmission system, made the first short-circuit studies using symmetrical components, and helped in field tests of switches and generators. While at Edison, she conducted the first design studies of the Hoover Dam transmission system. In addition, she and three co-workers patented the "Serjdetour" telephone protector, which prevents high voltage from damaging the telephone (Fig. 10.1) (Electrical Engineering 1935-07; Rockwell 1956; US Patent Office 1932).

In 1931, Rockwell joined the Metropolitan Water District of Southern California where she was an Assistant Electrical Engineer. Rockwell was entirely responsible for the basic technical design of an extensive 230,000-volt (230 kV) power system for bringing power from the Hoover Dam to the large pumping plants along the Colorado River Aqueduct, which the Metropolitan Water District was responsible for building. She made all the stability and "pulling-into-step" studies, conducted circuit and relay calculations, and determined the most economic voltage and size conductors. Rockwell wrote the specifications covering all characteristics of the generators, large pump motors, transformers, and substations. She also designed the 66-kV transmission line required for the project (Rockwell 1956).

Rockwell wrote and published papers, especially during her early career in power generation and distribution. Her first paper, "Parallel Operation of Transformers Whose Ratios of Transformation Are Unequal" received the 1929 Initial Paper Prize for the Pacific District of AIEE (American Institute of Electrical Engineers). The paper "Power Limits of 220 Kv Transmission Lines" written with A.A. Kroneberg was awarded the 1934 AIEE Pacific District prize for best paper (Electrical Engineering 1935-09).

In 1935, MacFerran married Edward W. Rockwell. Edward was also an electrical engineer (Goff 1946). The Rockwells had a daughter Margaret Alice in 1936. Rockwell left the Metropolitan Water District in March 1936 to care for Alice (Rockwell 1958).

10.3.2 Mid-Career

In December 1938, Rockwell went to work for Lockheed Aircraft Corporation in Burbank, California, as the Plant Electrical Engineer. She was responsible for the design and installation of all electrical equipment including machinery, controls, lighting, power supply, substations, and so on, in the plant (Rockwell 1956). The spot welding equipment which facilitated mass production was a challenge for the Plant Electrical Engineer. Spot welding for the automotive industry was

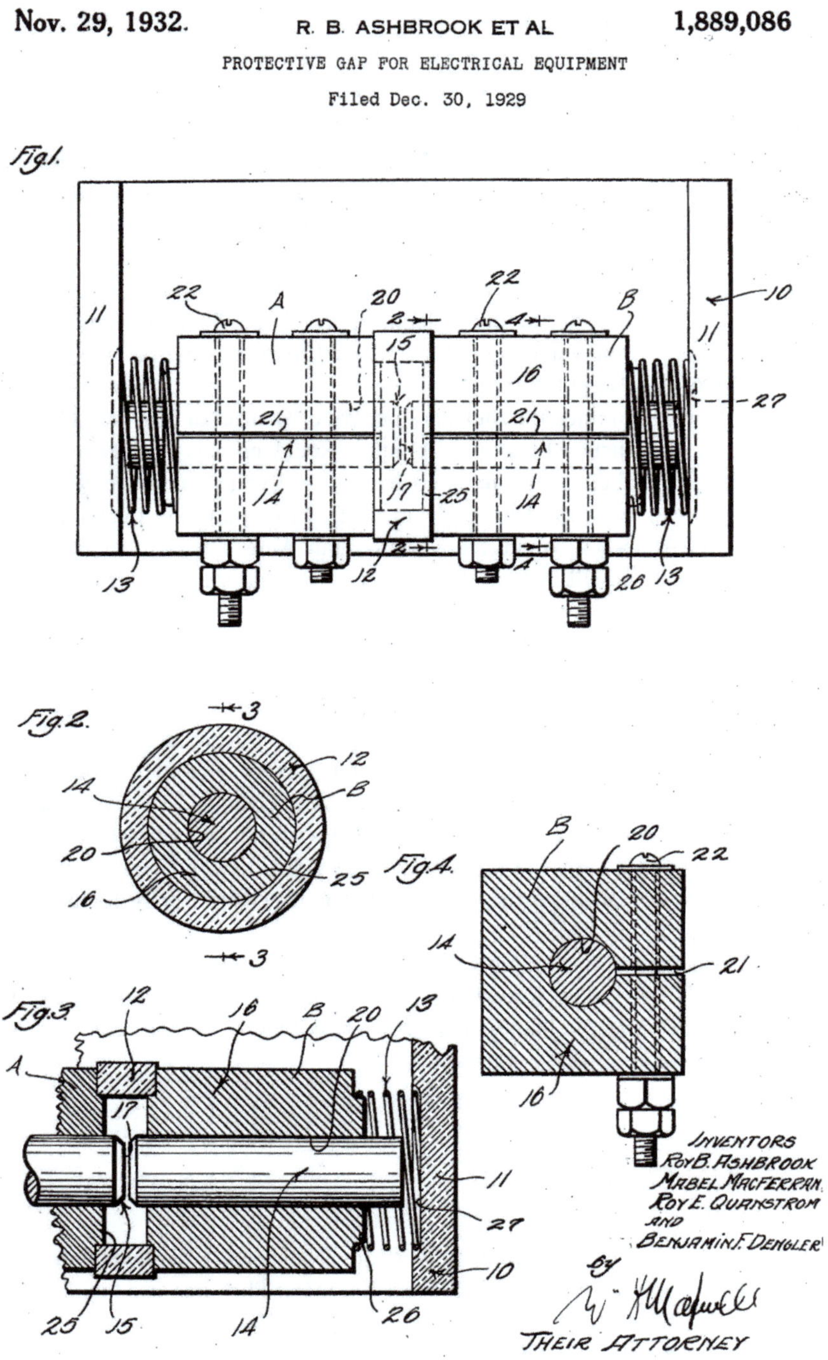

Fig. 10.1 "Serjdetour" telephone protector patent figure. Patent US-1889086A Protective Gap for Electrical Equipment

advanced—steel used for automobile parts has high electrical resistance and low heat conductivity, allowing good spot welds to be made with moderate current. Spot welding for the aircraft industry lagged because of the aluminum alloys used in aircraft parts. Aluminum alloys have low electrical resistance and high heat conductivity, meaning heavy currents and pressures are required (Goff 1946).

The high-power needs and the requirement that the voltage of the spot welder terminals must not drop more than 10% when demand was imposed meant that a very large power supply system was required. A high-voltage line had to be run to the Lockheed plant, and a large capacity substation built. Transformers were required to drop the incoming voltage from 2300 volts to 480 volts. Rockwell was responsible for making this happen (Goff 1946).

There were two spot weld machines that were initially used for secondary aircraft structural parts. The strength of the spot welds and the quality of those parts were not good. Rockwell started a research program to establish the correct calibrations and settings for the machines along with the appropriate procedures. She was quickly promoted to Engineer in Charge—Production Research. Her team turned out more than 100 technical reports for which she was responsible (Rockwell 1956). The spot welded joints were so improved in strength and consistency that the Civil Aeronautics Authority gave its approval for extension of using spot welding to some of the primary aircraft structure—especially skin-stiffener sections (Goff 1946).

With the ability to spot weld primary aircraft structure, new spot weld equipment was required. The new equipment was extremely expensive, and the electrical requirements exceeded the capacity of the local electrical system. The solution was the use of "stored energy" spot weld machines, which were charged up slowly with magnetic energy through three-phase rectifiers and then discharged quickly to make the weld. Rockwell led the effort to determine the correct settings and processes of these new machines (Goff 1946).

Rockwell next investigated roll spot weld machines, which have the advantage of greater speed and consistency than fixed-electrode spot welders. In 1941, the roll welders at Lockheed could turn out 100–140 spots per minute. With work properly designed and scheduled, they could average 30,000 spots in an eight-hour shift. Roll spot weld machines were most efficient on long straight runs like skin panels where numerous stiffeners are attached (Goff 1946).

Even though quality and consistency had been improved, there were still problems with spot welding that prevented it from running smoothly. Initially, Rockwell thought the problem was the complicated electrical control system. Methods were developed to accurately measure current and pressure during the weld as part of the quality process. After investigation, two main problems were discovered: inconsistent pressure due to undersized air lines to the welders and improper surface preparation. The inconsistent pressure problem was solved by the installation of a large air receiver near the welders. The surface preparation included removing the oxide film on the surface of the aluminum alloys by etching with dilute acid and cleaning the outside of the sheets to prevent sticking of the electrodes with consequent burn spots (Goff 1946).

Rockwell led studies to prove that spot welds were at least as good as rivets in fatigue strength under small torsional vibrations representative of aircraft in flight. When parts were designed for spot welding, work was properly scheduled, and roll spot welding was maximized, costs were reduced to a low level and production was sped up significantly—critical for supporting aircraft production for World War II (Goff 1946).

Rockwell was also engaged in research at Lockheed on the forming of sheet metal parts. The old way was "rule of thumb" methods for deep draw forming, often dependent on having an experienced person with the resulting part needing a significant amount of hand work. The effect of forming on material properties was not known. The team developed standards for characteristics of parts best formed by single-action and double-action draw machines. They also found that coarse-grained material was superior to fine-grain material for deep drawing (Goff 1946).

While at Lockheed, Rockwell published papers along with many magazine articles sharing discoveries on both spot welding and deep draw forming. Rockwell described developments in aircraft spot welding in an article in the magazine *Aviation*. Papers on deep draw forming were presented at meetings of the American Welding Society as well as published in the *Welding Journal*. Rockwell co-authored a series of articles in *Metal Progress* on deep draw forming (Goff 1946).

Rockwell left Lockheed Aircraft Company in August 1944 because she was "needed at home" due to a growing daughter. To help with finances, she engaged in various part-time consulting enterprises. She documented and standardized production techniques, wrote technical articles, and organized courses in aeronautical production training, among other things (Rockwell 1956). Rumor had it that Rockwell preferred to work all night; it may have been because her daughter was sleeping then. There were also stories that Rockwell did not like to cook and ate out in all-night restaurants rather than cooking (Engineering and Technology History Wiki - Mabel MacFerran Rockwell n.d.-b).

10.3.3 Late Career

In December 1946, Rockwell returned to full-time employment as a General Engineer at the US Naval Ordnance Test Station, Underwater Ordnance Division, in Pasadena, California. She was initially hired as a Physicist to do electrical engineering work on the design and procurement of equipment for a new underwater missile launcher. As happened so many times, Rockwell recognized that the Underwater Ordnance Division had a massive amount of classified research report material piled up. She was converted to a Technical Editor where she was responsible for working with the authors, helping them express their results clearly and get their papers completed. Rockwell supervised the final publication in some cases. Once the backlog was eliminated, Rockwell converted to General Engineer and became involved in the development work on electrical features of various classified underwater propulsion and guidance systems. Rockwell left the US Naval Ordnance Test Station in July 1951 to join the US Bureau of Reclamation (Rockwell 1956).

At the US Bureau of Reclamation, San Joaquin Valley District, Rockwell was Chief of the Power Branch. She was responsible for all studies on the design of proposed new hydroelectric power projects in the area. She was also responsible for the design of transmission and distribution systems to bring power from Central Valley Project (CVP) dams (Shasta and Folsom) to irrigation districts and other public agencies, which were entitled under the Reclamation Law of 1902. Rockwell made comparative rate and economic studies along with serving as an engineering consultant for CVP customers regarding problems of electrical service. She was also responsible for the crews who read and tested meters at points of delivery to customers. Rockwell became unemployed in June 1953 when her job was abolished due to a reduction in the scope of the US Bureau of Reclamation in public power projects (Rockwell 1956).

From October 1953 to June 1954, Rockwell undertook a special assignment at the US Naval Shipyard at Mare Island. The special assignment included conducting research and development on special instrumentation for submarine guidance. She completed the test and redesign program on an electro-mechanical device. Her complete report and recommendations were followed in the production design of the instrument (Rockwell 1956).

Rockwell returned to self-employed consultant work in June 1954. She prepared designs and cost estimates for electric power transmission and distribution systems for irrigation districts and other agencies in the San Joaquin Valley. These agencies were exploring the possibilities of decreasing their electric power costs by furnishing their own distribution systems and purchasing power at wholesale from utilities or from the CVP. Studies and reports were made for Westlands Water District, Orange Cove Irrigation District, and others (Rockwell 1956).

In December 1954, Rockwell was employed by the United States Air Force, McClellan Air Force Base, Installations Division as an Electrical Engineer. She was responsible for performing analytical design studies, making plans, and preparing specifications for the construction of new electrical facilities on the base. She served as project engineer during construction of two general building projects, including structural and mechanical phases (Rockwell 1956).

At some point, Rockwell transferred to Beale Air Force Base as the Chief Electrical Engineer. After only a few months there, she received a "very attractive offer" from Westinghouse (she commented, "Electrical engineers are in clover right now!"). By March 1958 she was working for Westinghouse Electric Corporation in Sunnyvale, California as a Project Engineer in the development of a shipboard missile launcher (Project "A") (Rockwell 1958). Her duties consisted of designing the controls, panels, cabling, instrumentation, and electrical hardware for the missile launcher. She also represented Westinghouse at various coordination meetings (Lund 1958). According to Rockwell, she was:

> …project engineer in charge of all electrical design work on a prototype launcher for the US Navy's Polaris missile project (the intermediate range ballistic missile to be launched from

a submarine). My particular part of the project is the design of a prototype launcher to be mounted on a ship of the US Navy (Rockwell 1956).

10.4 Later Life

How long Mabel MacFerran Rockwell (Fig. 10.2) worked for Westinghouse is not known. Rockwell studied philosophy starting in 1945 in her spare time as an avocation. She worked on a book on philosophy, science, and religion. Rockwell reported that she enjoyed small-boat sailing, swimming, and tennis. She built a sailboat (Rockwell 1958). Little is known about her life between when she received the Achievement Award in 1958 and when she died in 1979 (Ancestry.com n.d.; Engineering and Technology History Wiki—Mabel MacFerran Rockwell n.d.-b).

10.5 Professional Organizations

Rockwell became an Associate Member of AIEE (later the IEEE—Institute of Electrical and Electronic Engineers) in 1928 (Engineering and Technology History Wiki—Mabel MacFerran Rockwell n.d.-b). In 1934, Rockwell was one of only 10 women members (all Associate Members) of AIEE (Engineering and Technology History Wiki—History of IEEE Since 1984 n.d.-a).

Fig. 10.2 Mabel M. Rockwell (Photograph courtesy of Society of Women Engineers Records, Walter P. Reuther Library, Archives of Labor and Urban Affairs, Wayne State University)

Rockwell became a Member of AIEE in 1935. When Mabel and Edward married in 1935, they were both Members of the AIEE. They became the first couple to both be Members of AIEE. There were two other couples but all four were Associate Members (Electrical Engineering 1935-07).

10.6 Honors and Legacy

Mabel MacFerran Rockwell was the first and only woman to design, develop, and oversee the installation of the power-generating equipment for the Hoover Dam and the high-voltage power transmission lines and associated equipment from the Hoover Dam to points of use (Rockwell 1956). She earned awards for papers she wrote about that work from the American Institute of Electrical Engineers (AIEE) (Electrical Engineer 1935–07).

Rockwell and her husband Edward were the first couple to be full Members of AIEE (Electrical Engineer 1935–09). In 1942, AIEE had only three women full members. In 1948, Rockwell became one of the first three women Fellows in AIEE along with Edith Clarke, SWE Achievement Award 1954, and Vivien Kellems. The Fellow Award had been in existence since 1912 (Tietjen 2019).

The middle part of Rockwell's career was as the Engineer-in-Charge of Production Research for Lockheed Aircraft Company. Her work resulted in significant developments in processes for spot welding and deep draw forming which allowed Lockheed to increase the production rate for aircraft to support US efforts in World War II (Goff 1946).

Rockwell was recognized with the Society of Women Engineers Achievement Award in 1958 for work in the later part of her career developing an electrical control system for the Polaris missile launcher (Rockwell 1956). Edith Clarke, SWE Achievement Award Recipient 1954, was instrumental in making the award nomination happen. She commented about Rockwell, "She is the most brilliant woman I have ever known" (Clarke 1956).

References

Ancestry.com. n.d. California, US Death Index, 1940-1977. Accessed 15 May 2024.

Clarke, Edith. 1956. Letter to Joan Barriage, SWE Awards Committee Chair. February 7. Society of Women Engineers (SWE) Records (LR001539), Box 192, Folder 8, Walter P. Reuther Library, Archives of Labor and Urban Affairs, Wayne State University.

Electrical Engineering (American Institute of Electrical Engineers). 1935-07: Vol 54 Iss 7 p 787. https://archive.org/details/ieee-spectrum_1935-07_54_7/page/786/mode/2up. Accessed 10 Mar 2025.

Electrical Engineering (American Institute of Electrical Engineers). 1935-09: Vol 54 Iss 9 p 1019. https://archive.org/details/ieee-spectrum_1935-09_54_9/page/1018/mode/2up. Accessed 8 Mar 2025.

Engineering and Technology History Wiki n.d.-a "History of IEEE Since 1984," https://ethw.org/History_of_IEEE_Since_1984. Accessed 10 Mar 2025.

Engineering and Technology History Wiki. n.d.-b "Mabel MacFerran Rockwell." https://ethw.org/Mabel_MacFerran_Rockwell. Accessed 6 July 2023.

Goff, Alice C. 1946. Mabel MacFerran Rockwell. *Women Can Be Engineers*, 94-112. Ann Arbor, Michigan: Edwards Brothers, Inc.

Lund, H. 1958. Letter to Kay Broughton nominating Mabel Rockwell for Award. March 4. Society of Women Engineers (SWE) Records (LR001539), Box 192, Folder 8, Walter P. Reuther Library, Archives of Labor and Urban Affairs, Wayne State University.

US Patent Office. 1932. Patent Public Search Basic USPTO, Patent US-1889086A Protective Gap for Electrical Equipment. Roy B. Ashbrook, et al. n.d. https://ppubs.uspto.gov/dirsearch-public/print/downloadBasicPdf/1889086? requestToken=eyJzdWIiOiJmNmQ0Nzg0My0zZ-DQ3LTQ2Y2ItYTJlMi01MjgzZmUyNDFmNjAiLCJ2ZXIiOiJlYjViMTYxZS1lZTlmLTQ0OGEtYTc1ZC0wMWI2NDdhNzU1ODUiLCJleHAiOjB9. Accessed 20 Feb 2024.

Rockwell, Mabel. 1956. Civil Service Form 56/57. November 23. Society of Women Engineers (SWE) Records (LR001539), Box 192, Folder 8, Walter P. Reuther Library, Archives of Labor and Urban Affairs, Wayne State University.

Rockwell, Mabel. 1958. Handwritten letter to Kay Broughton. February 2. Society of Women Engineers (SWE) Records (LR001539), Box 192, Folder 8, Walter P. Reuther Library, Archives of Labor and Urban Affairs, Wayne State University.

Sigma Xi. n.d. https://www.sigmaxi.org. Accessed 5 March 2025.

The Pittsburgh Press. 1958. Ex-Pennsylvanian Named Woman Engineer of the Year. June 8. From the Society of Women Engineers (SWE) Records (LR001539), Box 192, Folder 8, Walter P. Reuther Library, Archives of Labor and Urban Affairs, Wayne State University.

Tietjen, Jill S. 2019. No Bent! All Women Were Given Was A Badge, The Bent of Tau Beta Pi, Winter, 8-13. https://www.tbp.org/pubs/Features/W19Tietjen.pdf. Accessed on 10 Mar 2025.

Tietjen, Jill, et al. 2023. Chapter 1—Those Electrifying Women! In *Women in Power*. 11-12. Women in Engineering and Science. https://doi.org/10.1007/978-3-031-29724-3_1

Chapter 11
Désirée le Beau, PhD (1909–1985) Achievement Award Recipient 1959

In recognition of her significant contributions to the field of rubber reclamation

Janet L. Williams

11.1 Introduction

Désirée Le Beau (Fig. 11.1) was a colloid chemist[1] with extensive experience and expertise in the reprocessing and reuse of synthetic and natural rubber. She was the first woman to hold the position of Director of Research for the Midwest Rubber Reclaiming Company in the rubber industry. She was responsible for developing materials from scrap rubber, mostly old tires, for reuse in new products. Le Beau earned five patents that covered chemical processes for rubber material reuse and, in one case, a pad for railroad ties made from reused rubber materials. In addition to receiving the 1959 Achievement Award from the Society of Women Engineers, the American Chemical Society honored her for service to the organization.

[1] Colloid chemistry refers to the study of the properties and behavior of colloidal particles, which are small particles suspended in a medium. It involves understanding the classification, shape, size, and dispersion of these particles (definition from Science Direct, https://www.sciencedirect.com/topics/chemistry/colloid-chemistry)

J. L. Williams (✉)
Albuquerque, NM, USA

C. D. Craig et al. (eds.), *Women Engineering Legends 1952-1976: Society of Women Engineers Achievement Award Recipients*, https://doi.org/10.1007/978-3-032-00224-2_11

Fig. 11.1 Désirée Le Beau, PhD (Photograph courtesy of Society of Women Engineers Records, Walter P. Reuther Library, Wayne State University)

11.2 Early Life and Education

Sidonie Désirée Le Beau was born in 1907 in Teschen in Austria-Hungary, now part of Poland. Her Polish mother gave birth while her father, a soldier in the Austrian Army, was serving in Dalmatia. Le Beau received her early education alternately in Austria and Sweden, and her family moved between these locations while she was growing up. She explored her aptitude in chemistry and music but decided that chemistry would be a better career choice. In the end, Le Beau elected to study pharmacy at the University of Vienna. However, a stroke of fate set her on a different path. Accidentally enrolling in the chemistry lab course intended for chemistry majors (rather than the less rigorous one for pharmacists), she fell in love with chemistry. Her mother was not keen on her daughter working in industry. However, she reluctantly agreed to pay for her daughter's PhD studies at the University of Graz, where Le Beau graduated with a degree in chemistry, as well as minors in physics and math, in 1931. As a result of her upbringing and education in Europe, she was fluent in French, Swedish, German, and Latin (Start 1959; SWE Records n.d.).

11.3 Early Career

Le Beau's first position was as a chemist with the Austro-American Rubber Works in Vienna, where she was first introduced to the field of rubber production. After 3 years, she accepted a job as a consultant for the Société de Progrès Techniques in

Paris. By 1935, Hitler's intentions for Europe were becoming clear, so she chose to move to the US in 1936. She declared her intent to become a US citizen in September 1938 and was granted citizenship in 1942 (National Archives 2025; Start 1959; SWE Records n.d.). After moving to the US, she began using the professional name Désirée S. Le Beau.

Le Beau worked as a research chemist for Dewey and Almy Chemical Co. in Cambridge, Massachusetts from 1936 to 1940, then as a research associate at the Massachusetts Institute of Technology (MIT) in the Department of Chemical Engineers and Division of Industrial Cooperation from 1940 to 1945. During this time, she worked on a research project set up by the Midwest Rubber Reclaiming Company in East St. Louis, Illinois. When the project ended, Le Beau accepted a job as their Director of Research (Start 1959).

11.4 Later Career

This career move leveraged Le Beau's extensive expertise in colloid chemistry. She was now fully immersed in the world of synthetic rubber recycling and reuse. At the time, Midwest Rubber Reclaiming was the largest rubber reclaiming company in the world. As the Director of research, Le Beau's job was to develop materials from scrap rubber, primarily old tires, to use in new products and applications. She was awarded five US patents from 1947 to 1965, some of which were also patented in other countries. They all involved improvements in processing for recycled synthetic rubber. In 1958, she developed a rubber tie pad for American railroads that was patented in 1963 (Fig. 11.2).

Le Beau became increasingly recognized as an expert in colloid chemistry as it pertained to the reuse of natural and synthetic rubber. She published many technical articles in the late 1940s through the 1950s. She authored a chapter on reclaiming of elastomers for a textbook on colloid chemistry and co-authored the chapter on the chemistry of clay minerals and films. And she was often an invited technical speaker in St. Louis and beyond (Dr. le Beau 1950; Group Told of Rubber Reclamation 1954; SWE Records n.d.).

In 1948, Le Beau was elected the chairman of the American Chemical Society (ACS), Colloid Chemistry division. Her role was organizing scientific meetings, publishing research, and participating in the leadership of ACS. She served as chair again in 1953. She was also active in the St. Louis Section of ACS, serving as Chairman in 1952. (Leaders in Chemistry 1952; Woman Chemist Honored by National Group 1948).

Le Beau had other technical interests outside of colloid chemistry. One of these topics, ceramics, led to meeting Henry W. Meyer, an engineer who had graduated from the Missouri School of Mines and Metallurgy and worked at General Steel Castings Corporation. This chance connection resulted in their marriage on August 6, 1955. She was 48 years old (Désirée Le Beau 1955; Start 1959).

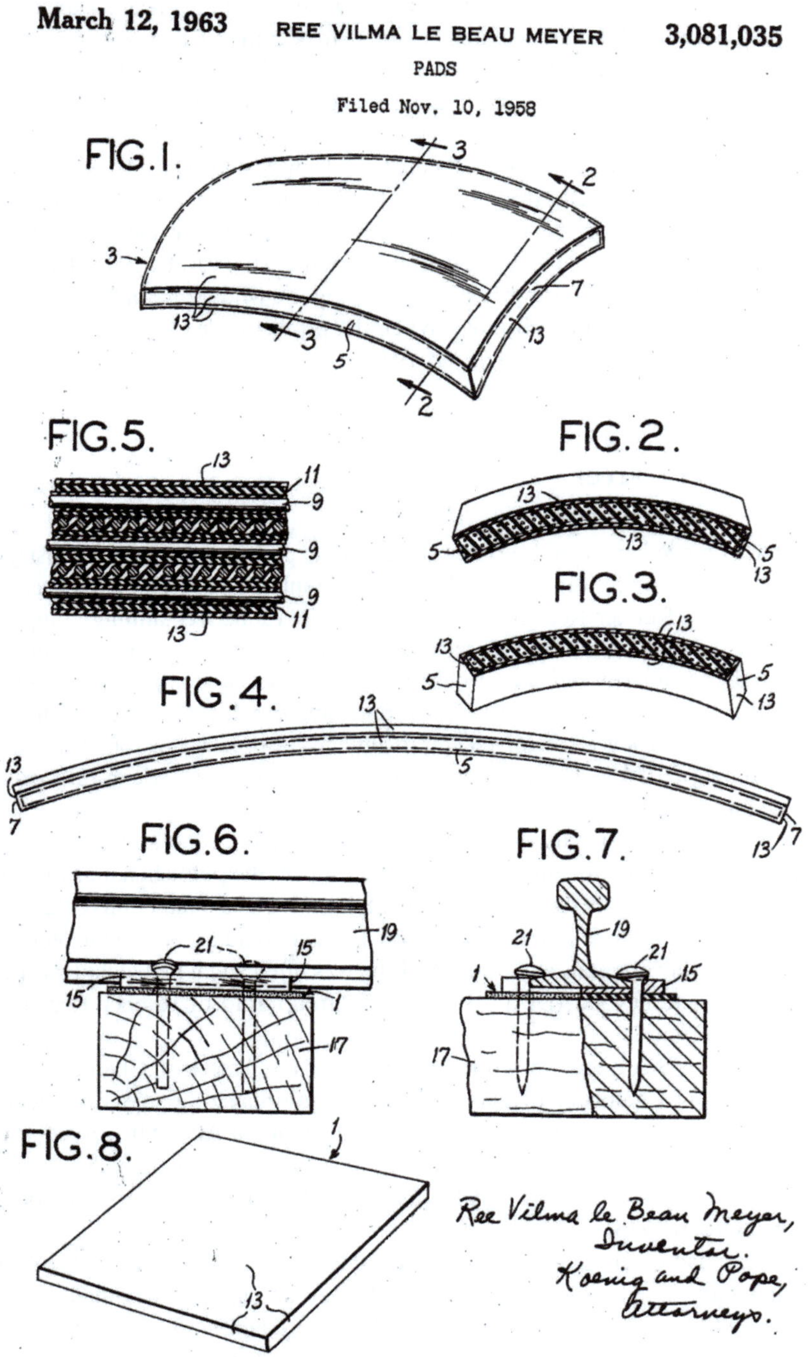

Fig. 11.2 Railroad Tie pad made from reclaimed synthetic rubber, US Patent No. 3081035

Le Beau continued as director of research at Midwest Rubber Reclaiming until her retirement. She and her husband moved to Cherokee Village, Arkansas in 1969. Henry Meyer passed away in 1991, and Le Beau followed him in death in 1993 (Dr. Désirée Le Beau Meyer 1993; Meyer 1991).

11.5 Honors and Legacy

Le Beau authored or co-authored dozens of papers related to colloid chemistry and its application to the rubber industry. She was selected as the 1950 Curie Lecturer at Penn State College (now Penn State University), speaking on "Elastic Polymers: Science and Economics of Rubber." As was true for so many of the early recipients of the Achievement Award, she rose to a position and stature in the technical community that was unusual for a woman in that era. In 1958, the *St. Louis Globe-Democrat* featured Le Beau as one of ten women selected and honored as "Women of Achievement" (SWE Records n.d.) (Fig. 11.3).

Fig. 11.3 Désirée Le Beau, left, receiving the 1959 Achievement Award (Photograph courtesy of Society of Women Engineers Records, Walter P. Reuther Library, Wayne State University)

Le Beau was honored with the SWE Achievement Award in 1959. In his recommendation letter for her Achievement Award nomination, the president of Midwest Rubber Reclaiming, Edward Bowles, pointed out that she was, to his knowledge, "the only woman to hold the position of Director of Research in the rubber industry" (SWE Records n.d.).

Le Beau also garnered honors through her technical society involvement. She was a Fellow of the American Institute of Chemists in recognition of the many ways she served that organization over the years. She belonged to the New York Academy of Science, The British Institute of the Rubber Industry, and the American Association for the Advancement of Science. (SWE Records n.d.).

After retirement, she remained active in her community, serving as a member of the board of directors of the Arkansas Wildlife Federation. In 1982, she was appointed by then Governor Bill Clinton to the Technical Advisory Committee on Hazardous Waste (Dr. Désirée Le Beau Meyer 1993).

Just as interesting and impressive as her technical achievements were her many outside interests, listed in her 1961 curriculum vitae as music, horseback riding, and interior decoration. In one newspaper article, she talked about making her own dresses, collecting records, laying parquet flooring, and hanging wallpaper in her home with her husband (Start 1959).

In an interview after receiving the SWE Achievement Award, Le Beau reminisced about the award's significance in light of her emigration story. When she applied to come to the United States in 1936, the man who handled her paperwork expressed doubt about the ability of a woman to be self-supporting in the engineering profession. "How I wish I could remember his name," she said. "I'd like to write to him now and say, 'See?'" (Start 1959).

References

Désirée Le Beau. 1955. Desiree Le Beau Becomes Bride of H.W. Meyer. *St. Louis Post-Dispatch*, August 7. St. Louis, Missouri.

Dr. Désirée Le Beau Meyer. 1993. Dr. Desiree Le Beau Meyer, 85, chemist, of Cherokee Village. *Arkansas Democrat-Gazette*, January 30.

Dr. le Beau. 1950. Dr. le Beau is Curie Lecturer. *Centre Daily Times*, October 27. State College, Pennsylvania.

"Group Told of Rubber Reclamation."1954. *Waterloo Region Record*, April 2. Kitchener, Ontario, Canada,

Henry W. Meyer. 1991. Henry W. Meyer. Obituaries. *Arkansas Gazette*, March 6.

"Leaders in Chemistry." 1952. *The Buffalo News* (Buffalo, New York), March 26.

National Archives. 2025. Waltham, Massachusetts; ARC Title: Petitions and Records of Naturalization, 8/1845 - 11/1911; NAI Number: 3000057; Record Group Title: Records of District Courts of the United States, 1685-2009; Record Group Number: RG 21, accessed January 2025.

Society of Women Engineers (SWE) Records (LR001539). n.d. *Désirée Le Beau*. Box 190, Folder 34, Walter P. Reuther Library, Archives of Labor and Urban Affairs, Wayne State University.

Start, Clarissa. 1959. She's Woman Engineer of the Year. *St. Louis Post-Dispatch,* July 7. St. Louis, Missouri.
"Woman Chemist Honored by National Group." 1948. *The Belleville News-Democrat,* November 6. Belleville, Illinois.

Chapter 12
Esther M. Conwell, PhD (1922–2014) Achievement Award Recipient 1960

In recognition of her significant contributions as a research physicist in the field of solid state research

Jill S. Tietjen

12.1 Introduction

Condensed matter theoretician Esther M. Conwell laid the theoretical foundation for a broad range of materials critical to the computer age—from transistors to integrated circuits to the wafers used for the manufacture of electronic devices. She worked in semiconductors, highly conductive quasi-one-dimensional organic crystals, conducting polymers, and DNA. Her work enabled the microelectronics industry, long-distance communication networks, advanced photocopying, solar cells, and light-emitting diodes (Garmire 2016). Conwell grew up in New York City. She got her AB in physics from Brooklyn College, her MS from the University of Rochester, and her PhD from the University of Chicago. She worked at Brooklyn College, Bell Laboratories, Sylvania (later GTE), Xerox, and the University of Rochester. The author of a seminal textbook, hundreds of papers, and the holder of four patents, Conwell died at age 92 while she was still working at the University of Rochester—and publishing papers—when she was hit by a car. The second woman elected to both the National Academy of Engineering and the National Academy of Sciences, she was the first woman to receive the Thomas Alva Edison Medal of the Institute of Electrical and Electronics Engineers (IEEE). Her many other honors include the National Medal of Science.

J. S. Tietjen (✉)
Technically Speaking, Inc., Greenwood Village, CO, USA

C. D. Craig et al. (eds.), *Women Engineering Legends 1952-1976: Society of Women Engineers Achievement Award Recipients*,
https://doi.org/10.1007/978-3-032-00224-2_12

12.2 Early Life and Education

Esther Marley Conwell grew up in New York City, the oldest of three daughters of immigrant parents. She and her sisters were exhorted by their father to be prepared to economically support a husband before they got married. Education was important in her family. "He impressed on his three daughters that in order to have a future, to be able to get married, we had to have jobs and be able to earn money, and he would prefer that we be professionals" (Ashrafi 2007). Conwell went to a good high school where she took biology and physics, but not chemistry. She majored in physics at Brooklyn College (BA, 1942, Phi Beta Kappa) intending to become a high school physics teacher, because that was all she had ever seen. She was generally the only woman in her physics classes (Ashrafi 2007; Braunstein and Stahl 1997; Howes and Herzenberg 2015; SWE Records n.d.).

During her college years, her physics teacher, Professor Kurrelmeyer, suggested that she apply to graduate school. Having half-heartedly taken a couple of education classes that held very little interest for her, Conwell took his suggestion seriously. She decided to accept the offer from the University of Rochester, which was closer to home than the University of Minnesota. At Rochester, she worked with Professor Weisskopf, although he was soon called to Los Alamos, leaving her to work on her research alone. The research topic for her master's thesis was impurity scattering with data obtained from physics personnel at Purdue. "A team there [Purdue] had obtained some data that they didn't know how to interpret on how resistance varies with temperature. Weisskopf came back with that as my thesis problem. He was around for a few months after that, and I was frantic to get something calculated to show him. So I worked hard and gave him my results the day before he left for Los Alamos." Semiconductors were just beginning to be studied—germanium at the time; silicon would come later. She derived the scattering of electrons by impurities in germanium from first principles (Ashrafi 2007; Garmire 2016; Polina 2022).

The summer of 1943, she worked on her master's thesis and at Western Electric. Although she started as an assistant engineer, after her employment started, she was informed that because there was no classification as assistant engineer for a woman, she would be retitled an engineer's assistant with an associated significant pay reduction (Howes and Herzenberg 2015). Back at Rochester, she tried cyclotron research but did not care for the work. She left to work at Columbia University doing separation of uranium isotopes, which she found very boring. (Howes and Herzenberg 2015).

She submitted her thesis to the University of Rochester. The University of Rochester was told by Purdue that her work had become classified. It couldn't be published or declassified until World War II was over. Without her knowing, the work became declassified when the war was over and the physics personnel at Purdue sent in two abstracts for the American Physical Society conference—one with her name on it—without her knowledge. Conwell saw her name on the abstract for the upcoming program and went to the conference (Ashrafi 2007; Braunstein and Stahl 1997; Krauss 2015; Polina 2022).

> My graduate thesis paper was listed as a 10-minute presentation, part of a group of presentations by several faculty members from Purdue University's physics department. This was news to me. As far as I knew the paper was still classified by the government. When I showed up, unannounced, and met the professor from Purdue who had planned to give the paper, I asked him why I had not been invited to present my own work. He seemed shocked and rather embarrassed, and agreed to let me give the paper (Polina 2022).

The Purdue physics personnel said they couldn't find her address or her telephone number and that they had tried to locate her. Today the theory that she proposed in her master's thesis is known as the Conwell-Weisskopf Theory, which is one of the fundamental properties for semiconductor physics and a prerequisite for modern electronics. It led to an understanding of how transistors operate. Transistors are the building blocks of integrated circuits, so her work enabled the computer age (Ashrafi 2007; Braunstein and Stahl 1997; Krauss 2015; Polina 2022).

When she arrived at the University of Chicago in 1944 to work on her PhD, she took Enrico Fermi's class on nuclear physics. Her PhD research on atomic physics was conducted under Subrahmanyan Chandrasekhar, who would win the 1983 Nobel Prize in Physics. She calculated the energy levels of the H-minus ion (hydrogen), also from first principles (Ashrafi 2007; Braunstein and Stahl 1997; Garmire 2016).

Conwell (Fig. 12.1) taught at Brooklyn College from 1946 to 1951, returning to Chicago to pass her orals and earning her PhD in 1948. "Having a PhD from the University of Chicago, any guy would figure on getting a nice professorship at a good university. But in physics in the early 1950s, there was no such thing for women and no such thing for me" (Ashrafi 2007; Braunstein and Stahl 1997; Garmire 2016).

Fig. 12.1 Esther M. Conwell Pach Brothers, New York (Courtesy AIP Emilio Segrè Visual Archives, Physics Today Collection)

12.3 Career

After teaching at Brooklyn College, Conwell took a leave of absence and spent 1 year at Bell Laboratories, where she worked with William Shockley, who would share a Nobel Prize for co-inventing the transistor. While there, she became familiar with hot electrons, a field in which she would work for many years after. She wrote a review paper that explained the properties of transistors at Shockley's request. She later said, "That article was really many people's introduction to semiconductors." That paper also led to her reputation as an expert on the complexity of electrons and holes in semiconductors. At the end of her year at Bell Labs, she did not want to return to Brooklyn College. Shockley helped her get a job at Sylvania which she began in 1952. She was thus able to continue her career as a condensed matter theoretician who researched fundamental properties of new electrically conductive materials (Ashrafi 2007; Garmire 2016).

Her initial projects at Sylvania (later GTE) pertained to research in semiconductors—theoretical analysis of germanium and silicon. Her focus was on carrier transport in p- and n-type materials, whether lightly or heavily doped, in high and low electric fields and at various temperatures, so-called "hot" or highly energetic electrons. "If you apply a high electron field, it changes the properties of electrons, what we call 'hot electrons.' I spent a lot of time theorizing about the properties of hot electrons, and comparing the theory with experiments" (Ashrafi 2007; ETHW 2018; Garmire 2016; Polina 2022; Rosengren 2024).

Conwell was pregnant with her son while at Sylvania and continued to work long after company policy said she should quit. She was sitting in the lunchroom one day when her boss came in. He asked her when her due date was. She said, "Yesterday!" He insisted she go home immediately. Her son, Lewis, was born the following weekend. She took maternity leave for three months and then worked part time for 3 months. Because there was no formal maternity leave policy, the company formally fired her and then rehired her. This permanently reduced her pension (Howes and Herzenberg 2015).

Conwell stayed at Sylvania for 20 years, although she did spend 1962–1963 as a Visiting Professor at the École Normal Superiéure in Paris, France. There she taught a graduate course in transport theory of electrons in solids and conducted independent research in the physics department. Her 1967 book, *High Field Transport in Semiconductors,* became a basic text in the field (Ashrafi 2007; ETHW 2018; Garmire 2016; Polina 2022; Rosengren 2024; SWE Records n.d.).

At Sylvania, she began her role as a mentor, encouraging young women to become scientists. One of those women was Millie Dresselhaus (SWE Achievement Award recipient 1977) who was an undergraduate intern at Sylvania (Ashrafi 2007; Garmire 2016).

Sylvania said of her work:

> We consider Dr. Conwell's work here to be of fundamental importance in the field of solid state research. The theoretical work she is doing in her chosen field represents an important advance in the understanding of the mechanism by which carrier conduction takes place in

> semiconducting materials. In particular, Dr. Conwell has concerned herself with a study of the mobility of carriers (both electrons and holes) in the material. These investigations have also led to the determination of values for the effective mass of electrons, since such measurement is necessary to a prediction of their mobility. Dr. Conwell's work in this field is an outstanding contribution to the understanding of these properties which are of fundamental importance in the study of such complicated devices as transistors, crystal diodes and photo-diodes.
>
> Dr. Conwell has been particularly active in presenting the results of her investigations in professional journals and before learned societies. Because of her outstanding reputation in solid state physics, Dr. Conwell has been frequently called upon to participate in meetings and colloquia and to lecture before graduate and student groups (SWE Records n.d.).

When GTE bought Sylvania, her research focus changed to support of telecommunications. Her research influenced the development of semiconductor electronics. She supported research underlying devices and then moved into integrated optics. When Sylvania phased out her lab in 1972, she needed to find another job (Ashrafi 2007; Garmire 2016; Krauss 2015).

Conwell reached out to her former intern Millie Dresselhaus (1977 SWE Achievement Award recipient) at the Massachusetts Institute of Technology who arranged for a temporary chair for Conwell at MIT. Conwell continued her research in integrated optics and looked for a job while she served as the Abby Rockefeller Mauzé Professor at MIT (Ashrafi 2007; Garmire 2016).

Conwell (Fig. 12.2) was hired by Xerox Corporation to work at their Rochester, New York lab in 1972. She discontinued work on integrated optics and started working on glassy one-dimensional organic materials. She worked with conducting polymers that radiated light when they were properly stimulated. A significant

Fig. 12.2 Esther Conwell Xerox Corporation Photographic Communications (Courtesy AIP Emilio Segrè Visual Archives, Physics Today Coll)

amount of physics is required in xerography—getting an image on the surface of a photoconductor and then transporting it to get it somewhere else - and she was involved with the transport mechanism as well. The materials and processes that she studied today are used in the manufacture of digital displays including those used in printers and cell phones. She worked at Xerox until 1998 and was named a Research Fellow in 1981 (Ashrafi 2007; Braunstein and Stahl 1997; Garmire 2016; NSTMF 2025).

In 1989, she helped bring the National Science Foundation Center for Photoinduced Charge Transfer to the University of Rochester. This was a collaborative effort of Xerox, Eastman Kodak, and the University of Rochester. In 1990, she began to serve as an adjunct professor at the University. In 1991 Conwell represented Xerox as associate director of the NSF Center she helped establish. The Center opened up new research avenues. Also in 1991, Conwell received Xerox's President's Award for outstanding achievement (Ashrafi 2007; Braunstein and Stahl 1997; Freile 2014; Garmire 2016; Rosengren 2024).

The press release issued when she received Xerox's President's Award explained what she had done in her career and the significance of her efforts (SWE Records n.d.):

> The world's understanding of the interactions between matter and energy would not be the same without Esther M. Conwell. She has made major contributions to theoretical condensed-matter physics since she began her career as a scientist more than 40 years ago.
>
> Dr. Conwell has been instrumental in furthering public and academic policies to increase the participation of women in science and engineering. Her stature as a scientist has enabled her to affect these policies through service on numerous committees and panels of the federal government's National Research Council and other influential scientific organizations.

After she retired from Xerox, Conwell maintained her affiliation with the University of Rochester, where she held a dual appointment in the chemistry and physics departments. She continued to serve as the Associate Director of the Center for Photoinduced Charge Transfer. She worked at the University and published papers until her death at age 92, when she was hit by a car. Her work at the University led to more knowledge of how electrical charges move through DNA; this knowledge could lead to an understanding of cancer-causing mutations (Ashrafi 2007; Garmire 2016; NSTMF 2025).

In 2003 she said, "My life is the story of women scientists making a place in the world." (Ashrafi 2007; Garmire 2016; NSTMF 2025) Conwell held four patents and published more than 250 scientific papers (ETHW 2018; Polina 2022).

12.4 Organizations and Advocacy for Women

Conwell performed extensive professional service and served on numerous governmental advisory boards. She was particularly active within the American Physical Society (APS), the American Institute of Physics (AIP), and the Institute of Electrical and Electronics Engineers (IEEE). She served as an officer, on the national board, or on the executive committee of the New York State Section of APS, APS Division of Condensed Matter Physics, APS Committee on Professional Concerns, APS Council, IEEE Awards Board, and the IEEE Foundation Board. In addition, she served on the Board of Editors of *Applied Physics Letters*, *Journal of Applied Physics*, and the *Proceedings of the IEEE* (SWE Records n.d.).

Her governmental service included the Defense Science Board, various committees of the National Research Council, the National Science Foundation, the National Academy of Engineering, and the National Materials Advisory Board on Critical Materials. Some of those activities were related to improving the representation of females in science and engineering careers (SWE Records n.d.).

12.5 Honors and Awards

Conwell received the SWE Achievement Award in 1960, "In recognition of her significant contributions as a research physicist in the field of solid state research" (Fig. 12.3). Conwell received many other honors and awards. She was the second woman to be elected to both the National Academy of Engineering (1980) and the National Academy of Sciences (1990). An IEEE Fellow (1980) "for contributions to semiconductor theory, particularly transport in both low and high electric fields," she was the first woman to receive the IEEE Thomas Alva Edison Medal (1997) "For fundamental contributions to transport theory in semiconductor and organic conductors, and their application to the semiconductor, electronic copying and printing industries." In 2002, she was named one of the 50 most important women in science by *Discover* magazine (Braunstein and Stahl 1997; Ediger 2019; ETHW 2018; Garmire 2016; Howes and Herzenberg 2015; Rosengren 2024; US NSF 2024).

Upon her receipt of the Edison Medal, Conwell said, (SWE Records n.d.)

> A dictum that has impressed me greatly comes from Thomas Edison, after whom the prize I am to receive is named, 'Genius is 1 percent inspiration and 99 percent perspiration.' My one time boss, Bill Shockley, impressed me with William Osler's words, 'In science, the credit goes to the man who convinces the world, not to the man to whom the idea first occurs.' But perhaps, more important than all those fine and truthful words were the acts of Marie Curie, who taught me that women could do what before only men were allowed.

The *Discover* magazine entry for her stated (Braunstein and Stahl 1997; Ediger 2019; ETHW 2018; Garmire 2016; Howes and Herzenberg 2015; Rosengren 2024; US NSF 2024),

> Half a century ago, Conwell's research on how electrons course through silicon and other semiconducting materials jumpstarted the Computer Age. Now, she studies the movement of electrical charges through DNA. The motion of charge in DNA can lead to mutations that can be cancer producing. And the DNA properties could be useful in assembling circuit elements in nano-electronic circuits.

Fig. 12.3 Betty Yost (l) and SWE Achievement Award recipient Dr. Esther Conwell (r) at the 1960 Society of Women Engineers National Convention in Seattle, Washington (Photograph Courtesy of the Society of Women Engineers Records, Walter P. Reuther Library, Archives of Labor and Urban Affairs, Wayne State University)

In 2009, she was awarded the National Medal of Science (Fig. 12.4) with a citation that read "For her broad contributions to understanding electron and hole transport in semiconducting materials, which helped to enable commercial applications of semiconductor and organic electronic devices, and for extending her analysis to studying the electronic properties of DNA." In the video produced for the awards presentation, Conwell said, "I like to solve puzzles, I like to think about things, things that are not familiar." The University of Rochester awarded her an honorary doctorate in 2011. She had earlier (1992) received an honorary degree from Brooklyn College (Braunstein and Stahl 1997; Ediger 2019; ETHW 2018; Garmire 2016; Howes and Herzenberg 2015; Krauss 2015; Rosengren 2024; US NSF 2024).

Fig. 12.4 Esther M. Conwell receiving the National Medal of Science from President Barack Obama in the East Room of the White House (Courtesy Ryan K Morris Photography and the National Science & Technology Medals Foundation)

Because she received so many honors, Conwell was asked to what she attributed her success. She said, (Rosengren 2024).

> A wise and encouraging father, good teachers and colleges, a helpful husband and son, and just plain tenacity and perseverance – what used to be called keeping the seat of one's pants glued to the seat of the chair.

12.6 Personal

Conwell married Abraham Rothberg in 1945, and they had a son, Lewis Rothberg, who was born in 1956. Lewis Rothberg became a tenured professor of chemistry and physics at the University of Rochester. Lewis Rothberg said that his father's "advice, encouragement, and support were an integral part of [his mother's] success" (Garmire 2016).

Conwell liked to cook and exercise. She started dancing in college—first modern dance and then ballet—and continued to take ballet classes up until her death. "I think the same things I love about ballet, I love about physics. They're both exact and require discipline. That sense of order is what really attracted me to physics in college." Her colleagues knew that she loved her ballet lessons on Friday morning and that she popped popcorn all day, every day in the office (NSTMF 2025; Polina 2022; Rosengren 2024; University of Rochester Newsletter 2014–2015).

12.7 Legacy

Conwell mentored young women for many decades. She was recognized for this dedication with the Dreyfus Foundation's Senior Scientist Mentor Program Award (2005), The Susan B. Anthony Award from the University of Rochester in 2006, and the American Chemical Society Award for Encouraging Women into Careers in the Chemical Sciences (2008). Conwell said, "Although it's not nirvana yet, women have come a long way in my working lifetime, and it gives me hope." Conwell is credited with inspiring numerous young women worldwide to pursue and grow in scientific and engineering careers (Garmire 2016; Rosengren 2024).

Todd Krauss, chairman of the University of Rochester chemistry department, said, "She was driven by her science. It really was at her core, her soul. She was all about discovery" (Freile 2014) Former chairman of that chemistry department Robert K. Boeckman, Jr., said, "Her theoretical models contributed greatly to our understanding of electron transport and charge redistribution in solids, which directly led to the practical devices we use every day" (Rosengren 2024).

Her contributions to scientific theory were summed up thusly: (Rosengren 2024)

- Charge motion in semiconductors, providing foundational research for developing the microelectronics industry.
- Light propagation physics, fiber optics, and lasers, providing foundational science for developing long-distance communication networks.
- Charge motion physics and quasi-one-dimensional organics—critical to developing advanced photocopying, solar cells and light-emitting diodes.

Millie Dresselhaus (1977 SWE Achievement Award recipient) said, "Esther was right there at a very exciting time in semiconductor research. Her work is the basis, the very fundamentals, of what we study today… Esther is a true role model for what it means to stay vital as a scientist and a woman. There is no denying that when Esther and I were first beginning, we faced many difficulties and obstacles as women scientists. But that climate is changing, and I'm proud of the role we've played" (Garmire 2016).

The Esther M. Conwell Graduate Fellowship Fund was established by her son Lewis Rothberg and his wife Shelby Nelson. It provides stipends for merit-based fellowships for advanced (third year or beyond) graduate students pursuing a PhD in Chemistry at the University of Rochester who show exceptional promise as researchers. The Fellowship was first awarded in 2019 and has been awarded annually since then (University of Rochester—Awards and Prizes 2025; University of Rochester Newsletter 2014–2015).

References

Ashrafi, Babak. 2007. "Interview of Esther Conwell." Niels Bohr Library & Archives. *American Institute of Physics*. January 22.

Braunstein, Rubin and Frieda Stahl. 1997. "Esther Marly Conwell." CWP at UCLA. Accessed January 19, 2025. http://cwp.library.ucla.edu/Phase2/Conwell,_Esther_Marly@841873837.html.

Ediger, Emily. 2019. "Happy Birthday, Esther M. Conwell." May 23. https://www.ccmsol.com/blogs/happy-birthday-esther-m-conwell,

Engineering and Technology History Wiki. 2018. "Esther Conwell: Biography." January 26. Accessed January 18, 2025. https://ethw.org/Esther_Conwell,

Freile, Victoria E. 2014. News: US Professor Esther Conwell remembered as a trailblazer. Rochester (NY) *Democrat & Chronicle,* November 17. https://www.democratandchronicle.com/story/news/2014/11/17/woman-struck-brighton-dies-injuries/19170219/.

Garmire, Elsa. 2016. "Esther M. Conwell 1922-2014." Memorial Tributes of the National Academy of Engineering (Volume 20). Accessed January 18, 2025. https://nap.nationalacademies.org/read/23394/chapter/13.

Howes, Ruth H. and Caroline L. Herzenberg. 2015. *After the War: Women in Physics in the United States.* San Rafael, California, Morgan & Claypool Publishers.

Krauss, Todd. 2015. Class Notes: Tribute—Esther Conwell '44 (MS): Lived and Breathed Science. University of Rochester. Accessed January 18, 2025. https://www.rochester.edu/pr/Review/V77N3/0704_conwell.html.

National Science & Technology Medals Foundation. 2025 "Esther M. Conwell." Accessed January 18, 2025. https://nationalmedals.org/laureate/esther-m-conwell/.

Polina, Ch. 2022. "Esther M. Conwell—the woman whose research helped pave the way for the technological revolution." *Medium.* November 4. https://medium.com/mycelium-network-media/esther-m-conwell-the-woman-whose-research-changed-the-face-of-modern-computing-7ed7c5cddad1.

Rosengren, Paul Lief. 2024. *Famous Women Engineers in History—Book 6: Esther Conwell: Theorist and Mentor.* IEEE-USA.

Society of Women Engineers (SWE) Records (LR001539). n.d. Box 187, Folder 38, Walter P. Reuther Library, Archives of Labor and Urban Affairs, Wayne State University.

US. National Science Foundation. 2024 Esther M. Conwell. Accessed January 18, 2025 https://new.nsf.gov/honorary-awards/national-medal-science/recipients/esther-m-conwell.

University of Rochester, Chemistry Department Newsletter. 2014–2015. Department Mourns the Loss of Esther Conwell. https://www.sas.rochester.edu/chm/assets/files/newsletter/newsletter15.pdf.

University of Rochester, Chemistry Department, Awards and Prizes. 2025 Accessed January 22, 2025. https://www.sas.rochester.edu/chm/news-events/awards.html.

Chapter 13
Laurel Van der Wal (1924–2009) Achievement Award Recipient 1961

In recognition of her significant contributions to the developing field of space biology

Linda A. G. Reed

13.1 Introduction

Laurel van der Wal's (Fig. 13.1) contributions to the newly burgeoning field of bioastronautics, specifically researching biological and medical aspects of human space flight, proved critical to the success of the human space program. Instrumental to that was a study she conducted on the physiological effects of space flight on mice in US rockets, called Project MIA. Additionally, she was involved in escape and recovery systems and human spacecraft design. Throughout her career, she pushed the boundaries and expectations of women in the workplace while broadening the understanding of mammalian tolerance to weightless conditions and the rigors of launch and reentry.

She was named the *Los Angeles Times's* "1961 Woman of the Year in Science" for her contributions to the field, and the Society of Women Engineers bestowed the Achievement Award the same year (Clifford 2009).

Van der Wal's commitment to the space program lasted throughout her career. She believed that space exploration is the greatest adventure of mankind's existence. She carried her dedication to safety into her work with the Los Angeles Board of Airport Commissioners from 1961 to 1967 and with the RAND Corporation during the 1970s (City of Los Angeles Officials database n.d.; Roennau 1969-1973). She designed a new transportation system for California, but RAND abandoned the plan when it changed its corporate focus (SWE Records n.d.).

L. A. G. Reed (✉)
San Miguel, CA, USA

C. D. Craig et al. (eds.), *Women Engineering Legends 1952-1976: Society of Women Engineers Achievement Award Recipients*,
https://doi.org/10.1007/978-3-032-00224-2_13

Fig. 13.1 Portrait of Laurel van der Wal, Society of Women Engineers Achievement Award recipient 1961 and Head of Bioastronautics at Space Technology Laboratories (Photograph courtesy of Society of Women Engineers Records, Walter P. Reuther Library, Wayne State University)

13.2 Early Life and Education

Van der Wal was born in San Francisco, the daughter of Lillian and Richard van der Wal, both from Spokane, Washington. Her mother was a former teacher and an alumna of the University of Washington; her father was a businessman. She graduated from prep school after studying to be an artist. She fell in love with airplanes and hoped to earn a pilot license. She went to Nevada to train in the Women's Air Force Service program during World War II. Ever resourceful, she paid for her education by working as a model, an art instructor, a deputy sheriff, a showgirl, a railroad switch tower operator, and a casino shill (Clifford 2009).

"I am impatient with people who do not make full use of all their capabilities," she explained in 1962 (Sweeney 1962).

The program was canceled before she earned her wings, and women were not flying professionally, so she turned her focus to aeronautical engineering. She worked as an aircraft mechanic at Hamilton Air Force Base. This experience led her to study mechanical engineering at the University of California Berkeley, where she earned a Bachelor of Science degree in 1949, with honors.

> Then, since I've always been a person who would rather do than watch, I got interested in the problems that kept people (which really meant "me") from going along on one of those exciting trips. And once I learned a little about bioastronautics, which really encompasses all aspects of manned space flight, I was hooked (SWE Records n.d.).

She pursued her graduate education at the University of California, Berkeley; UCLA - receiving the Howard Holmes Scholarship in 1947, and the Douglas Aircraft Scholarship in 1948. She worked as a reader and teaching assistant while finishing her degree.

In 1951, she earned a National Research Council fellowship that allowed her to travel to Stockholm, Sweden, and study aeronautics at the Royal Institute of Technology. She attended the Eighth International Congress of Theoretical and Applied Mechanics in Istanbul, Turkey, and the Third Congress of the International Astronautical Federation in Stuttgart, Germany. While there, she met international engineers and scientists and presented papers highlighting her work (SWE Records n.d.).

13.3 Career

During her university career (1947–1949) Van der Wal worked on an aerodynamic heating project that evaluated wind tunnel operation, model design and construction, and theoretical analysis. She then tackled a project that examined low-density subsonic and supersonic tunnels. (SWE Records n.d.).

In 1950, Van der Wal joined Douglas Aircraft Company, Inc. as a laboratory research analyst, working on servomechanisms in the Guided Missiles Division. She was involved in control system design and participated in firing program activities at White Sands Proving Ground, New Mexico for the NIKE program (SWE Records n.d.).

In 1953, Van der Wal moved to Rockaway, New Jersey, and began work at Reaction Motors Inc. in the Turbopump Group, Development Section. During her time with Reaction Motors, she worked on theoretical investigations of the starting time of self-sustaining turbopump rocket systems, evaluation of fuel-tank pressurization systems, determination of test area liquid oxygen losses, and boil-off loss reduction possibilities (SWE Records n.d.).

Later in 1953, Van der Wal joined the Rheem Manufacturing Company as a Design Engineer in the Government Products Division, Research and Development Laboratories in Downey, California. She focused on general performance analysis, subsonic and supersonic drones, and aerodynamics. She began working on tool design, proposals, and the design and testing of fuse components for fast-burning rockets before moving to the aircraft engineering department as an aerodynamicist. Here, she was involved in analyzing all phases of aerodynamics as applied to the design, performance, and operation of aircraft, missiles, and drones (SWE Records n.d.).

It was in 1956 that Van der Wal joined Ramo-Wooldridge, where she was involved in the preliminary design of advanced missile and space-probe systems. This launched her interest in the burgeoning field of bioastronautics. In 1958, Ramo-Wooldridge changed its name to Space Technologies Laboratories, Inc. (later TRW). Van der Wal (Fig. 13.2), as head of bioastronautics, proved to be instrumental in the drive to launch humans into space and successfully bring them home (SWE Records n.d.).

Fig. 13.2 Laurel van der Wal works on equipment at Space Technology Labs, circa 1959–1960. (Photograph courtesy of Society of Women Engineers Records, Walter P. Reuther Library, Wayne State University)

Her research included a revolutionary experiment called Project MIA. Project MIA, or Mice-In-Able, placed white mice in the nose cones of Thor-Able rockets to measure their heartbeats. This data was telemetered to Earth and provided information on the physiological effects of space flight. The mice, named Benji and Laska, traveled in a "Mouse House" that cradled the animals and held food, water, and dehumidifying chemicals. Another cylinder contained air purifiers and a blower system that circulated air through the Mouse House. The total weight of all the MIA equipment was ten pounds. Although the nose cones were not located by recovery ships, telemetry proved that the mice had normal heart rates while traveling at speeds exceeding 15,000 miles per hour, reached a record altitude of 14,000 statute miles, and withstood decelerations well over 60 g's and nose cone skin temperatures of several hundred degrees Fahrenheit. Both mice survived reentry. Her work paved the way for human space flight in 1961. (SWE Records n.d.).

She also focused on the design of spacecraft capsule escape and recovery systems for astronauts.

When asked if she would want to go into space herself, she answered, "I'd go in a minute, if they'd let me" (City of Los Angeles Officials database n.d.). The first American woman astronaut wouldn't happen until Sally Ride in 1983. That didn't stop Van der Wal from helping to ensure that astronauts would be safe.

By the early 1970s, she had written several reports for the Rand Corporation about planning a more effective transportation system for California (Roennau 1969-1973). When Rand shifted focus and the project was dropped, Van der Wal needed a job. She was chosen for a position to develop a transportation system for Nigeria. In 1974, now divorced, she packed up her two children and headed for Africa. When she arrived, she learned there had been a coup; the government that had hired her was no longer in power. She and her children escaped Nigeria under the cover of night (SWE Records n.d.).

When she returned to the States, she worked for the Southern California Association of Governments. As a noted activist on behalf of slow growth in Santa Monica, California, she served as an advisor to the Santa Monica Coalition for a Livable City and worked with the Community Redevelopment Agency. In later years, she would join the Ocean Park Community Organization.

13.4 Consulting and Civic Involvement

Van der Wal was the Director of the Southern California chapter of the American Rocket Society, where she partnered with Explorer Scouts, an organization that introduces children to different career paths. Van der Wal had a keen interest in teaching children about space and aeronautics.

In 1961, Van der Wal was appointed to the Los Angeles Board of Airport Commissioners and worked as a commissioner until 1967. She was the first woman appointed to the board in its history. In 1968, she served as the planner for the Los Angeles International Airport (City of Los Angeles Officials database n.d.).

13.5 Awards and Recognition

Van der Wal received many awards and honors. While working toward her degree, she received a National Research Council Fellowship in Aeronautics to study at the Royal Institute of Technology in Stockholm. This allowed her to travel to Europe, attend conferences, and meet leaders in the European space industry (SWE Records n.d.).

In 1961, the Los Angeles Times honored her as "Woman Scientist of the Year," and the Aerospace Medical Association Wives' Wing named her "Outstanding Women Scientist."

Also, in 1961, the Society of Women Engineers awarded her their highest honor, the Achievement Award, for her contributions to bioengineering and bioastronautics. (SWE Records n.d.).

13.6 Legacy

Van der Wal's pioneering work with mice in rocket nose cones paved the way for human space flight. As the space program gained popularity in the 1960s, she was a popular speaker at women's clubs, teachers' meetings, and other events. As she explained in 1964, "The sky is no longer the limit" (Keele 1964).

In addition to her groundbreaking work in the field of bio-astronautical engineering, Laurel van der Wal was the first woman commissioner of the Los Angeles Board of Airport Commissioners in 1961 and served until 1967.

As a tall, beautiful woman, she garnered attention for her focus on her career. Many articles about her stressed her "willowy" form and "wide blue eyes." Reporters couldn't resist mentioning that she'd been a model. Many articles at the time stressed her choice of career over family, even though in 1961, she married fellow engineer William Henry Roennau in Arlington, Virginia. They went on to have two children but later divorced.

She strongly believed in a role for women in the workplace. In her 1962 speech at the Air Force Association Sixteenth Convention and Aerospace Panorama in Las Vegas, Nevada, she said,

> First of all, both in time and importance, I am a woman, and I intend to keep on being more and more so for the rest of my life. Last year I also became a wife and, if all goes well, in the not too distant future I shall become a mother…I would especially like to point out that these three factors, although of prime importance to the fulfillment of any woman as a person, should not and need not interfere in any way with a full and gratifying schedule of activities in the professional, civic and cultural fields (SWE Records n.d.).

References

City of Los Angeles Officials Database. n.d. https://cityclerk.lacity.org/chronola/index.cfm?fuseaction=app.FacultyDetail&OfficeHolderID=3340. Accessed 09 March 2025.

Clifford, Peggy. 2009. Laurel Roennau Dies Thursday. Archived 2015-04-02 at the Wayback Machine. *Santa Monica Dispatch*, August 14.

Keele, Beth. 1964. Well, She Started As a Model. *The Philadelphia Inquirer Magazine*. June 21.

Roennau, Laurel V. 1969-1973. Research & Commentary. RAND Corporation. http://www.rand.org/pubs/authors/r/roennau.laurel_V.html. Accessed 08 March 2025.

Sweeney, Joan. 1962. Work by Woman Scientist Helped Prepare Way for Manned Flights. *Franklin Evening Star.* June 21.

Society of Women Engineers (SWE) Records (LR001539). n.d. *Van der Wal, Laurel.* Box 193 Folder 20, Walter P. Reuther Library. Archives of Labor and Urban Affairs. Wayne State University. Accessed 09 March 2025.

Part III
1962–1966

Chapter 14
Setting the Stage: 1962–1966
The Space Race and Civil Rights Dominate the Headlines

Holly J. Teig and Janet L. Williams

14.1 Introduction

The space race between the USA and the Soviet Union drove engineering innovation and technology during the mid-1960s. Technology in computing, materials science, petrochemicals, and communications developed at an ever-faster rate. The USA began a slow shift from industrial work toward an economy of white-collar and service industries. Environmental activism emerged after the publication of Rachel Carson's *Silent Spring* and conservation-focused publicity by the Sierra Club.

The civil rights movement, led by Dr. Martin Luther King, Jr. and others, achieved legal victories with the passing of the Civil Rights Act of 1964, the Economic Opportunity Act of 1964, and the Voting Rights Act of 1965. These events changed the way society viewed future opportunities for minorities and women. The small number of women engineers graduating and reaching the workforce leveraged the gains of the civil rights movement to advance their own ambitions. For girls growing up in the 1960s, just as "the sky became the limit" in space, it also represented new possibilities in careers—whether it was aspirations to become a lawyer, a scientist, a judge or an engineer.

H. J. Teig (✉)
Peoria, IL, USA

J. L. Williams
Albuquerque, NM, USA

C. D. Craig et al. (eds.), *Women Engineering Legends 1952-1976: Society of Women Engineers Achievement Award Recipients*,
https://doi.org/10.1007/978-3-032-00224-2_14

In 1962, John Glenn became the first American to orbit Earth. He flew aboard a Mercury spacecraft dubbed Friendship 7, propelled into space by an Atlas rocket.

Laurence Delisle Pellier received the Achievement Award in recognition of her significant contributions to the field of metallurgy. Then she developed a technique for applying electron microscopy to metallurgical problems and a patent for gold-plated surgical needles.

AT&T's Telstar I satellite was launched in 1962, enabling the first transatlantic television transmission linking France and the USA.

1962

Rachel Carson's *Silent Spring*, published in 1962, revealed the devastating effects of DDT and other pesticides on wildlife and human health. She inspired a whole future generations' efforts to protect the planet's flora, fauna, and people.

Telephones: In 1963, the first electronic push-button system with touch-tone dialing was offered by Bell Telephone to customers in Pittsburgh, Pennsylvania.

Beatrice Hicks received the Achievement Award in recognition of her significant contributions to the theoretical study and analysis of sensing devices under extreme environmental conditions, and her achievements in international technical understanding, professional guidance, and engineering education.

Grace Murray Hopper, PhD received the Achievement Award in recognition of her significant contributions to the burgeoning computer industry as an engineering manager and originator of automatic programming systems. She designed the first compiler, opening the way for modern data processing.

1963

In 1963, Betty Friedan's book, *The Feminine Mystique*, was published. It described widespread dissatisfaction amongst women in American society.

1964

The civil rights movement achieved legal victories with the passage of the Civil Rights Act of 1964, and the Voting Rights Act of 1965.

The Harris County Domed Stadium, later dubbed the Astrodome, was the first enclosed, air-conditioned, multi-purpose stadium in the world. Opened in Houston, Texas, in 1965, it was called the "Eighth Wonder of the World".

Martha J. B. Thomas, PhD received the Achievement Award in recognition of her significant contributions to the science of chemistry, as an engineer, educator, and administrator, while fulfilling her duties as a wife and mother. She developed natural white phosphor for improved fluorescent lights.

In 1965, Stephanie Kwolek invented Kevlar®, a polyamide fiber five times stronger than steel. The bullet-proof vest, made from layers of Kevlar® fabric, protects law enforcement officials and soldiers every day.

1965

- Women represented 2.6% of the members of the 88th Congress, with two senators and 12 representatives.
- The participation of women in the workplace increased to 39%.

In 1966, NASA created temper foam, also known as memory foam, to improve passenger protection in airplane crashes. It is now used in mattresses, pillows, safety equipment, aircraft seats, and more to cushion the human body.

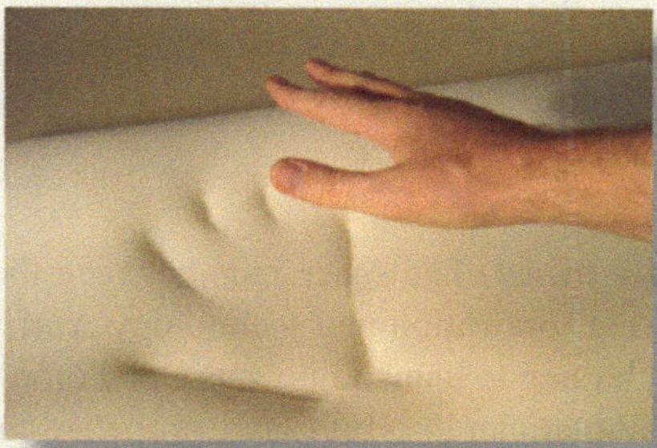

In 1966, aerospace companies started using computer-aided design and drafting (CADD) tools to depict component dimensions and fit with other parts of assemblies.

Dorothy Martin Simon, PhD received the Achievement Award in recognition of her significant contributions to space engineering, especially in the fields of combustion and ablative coatings. She isolated an isotope of calcium and developed a process to manufacture the predecessor of Orlon®.

1966

In 1966, the National Organization of Women (NOW) formed, intent on gaining a full expansion of women's rights as documented in the proposed Equal Rights Amendment (ERA). Among NOW's 28 founders was feminist Betty Friedan.

14.2 State of Technology

14.2.1 In Space

Space travel The space race between the USA and the Soviet Union drove engineering innovation and technology. Yuri Gagarin of the Soviet Union became the first human to launch into space and orbit in 1961. In 1962, John Glenn became the first American to orbit Earth, propelled into space by an Atlas rocket in a Mercury spacecraft (Uri 2022). The flight plan during the first orbit maintained an optimal spacecraft attitude for radar tracking and communication checks. The flight lasted 4 h, 55 min, and 23 s before the *Friendship 7* spacecraft splashed down in the ocean (Ostovar 2024). Dorothy Martin Simon (Achievement Award 1966) was credited with advancement in aerospace fuel and combustion principles, leading to improved spaceflight engine designs (ETHW n.d.). These were critical steps in the US manned flight program, proving the ability of a human to function in space and NASA's ability to bring them safely back to earth.

Satellites American Telephone and Telegraph Company's *Telstar 1* satellite, launched in 1962, was the first television transmission linking the USA and France. The satellite received microwave signals from ground stations and retransmitted them across vast distances back to Earth, laying the groundwork for all subsequent communications satellites. Telstar's electronics became compromised through exposure to Van Allen Belt radiation,[1] resulting in deactivation within a year (Smithsonian 2025).

14.2.2 Computer Hardware and Software

Mainframe computers The first supercomputer was the Control Data Corporation *CDC 6600* released in 1964, designed by Seymour Cray, Jim Thornton, and a team of engineers in Wisconsin (Computer History Museum 2025a). The core processing technology was fast—featuring 400,000 transistors and more than 100 miles of wiring. It also boasted ten peripheral processors that simultaneously performed housekeeping and input/output tasks, with user interaction through punch cards, tape, or direct terminal input (Computer History Museum 2025a). The price was seven million USD, with the first unit delivered to the Lawrence Livermore National Laboratories in September 1964 (Computer History Museum 2025b).

Programming languages In 1964, a software language called *Beginner's All-purpose Symbolic Instruction Code* (BASIC) was invented by mathematicians John Kemeny and Thomas Kurtz at Dartmouth College. BASIC made computing accessible to non-technical students (and faculty), who wanted to solve numeric prob-

[1] Van Allen Belt radiation is a zone of intense radiation held by Earth's magnetic field that can damage satellites and spacecraft passing through it.

lems but did not want to learn a complex programming language like FORTRAN (Kurtz 2025). BASIC emerged as the standard programming language loaded on early personal home computers in the 1970s.

Computer Aided Design (CAD) In 1957, Dr. Patrick Hanratty created the first numerical control system, which would later become Computer Aided Design. Evolving in step with the computer industry, in 1966 CAD took a leap forward when IBM and Lockheed began using Computer Graphics-Augmented Design and Manufacturing (CADAM) software to create their aerospace designs. McDonnell Douglas and Ford quickly followed suit with their own software, enhancing the precision, versatility, and portability of engineering drawings. Designing in software revolutionized the engineering, architecture, and manufacturing landscape, obsoleting drafting boards by the 1980s and largely the staff who had handcrafted technical drawings (Beck 2017).

Liquid-crystal display Electrical engineer George Heilmeier showed off the first liquid-crystal display (LCD) at RCA Laboratories in 1964. For several years, he had run lab experiments on this class of compounds, called liquid crystals, which flowed like a liquid but retained the optical properties of a crystalline solid. Heilmeier and his colleagues filed numerous patents as they advanced the technology, envisioning it used in advertising displays, automobile rear-view mirrors, and even an LCD wristwatch. Executives at RCA were not convinced of Heilmeier's vision. When the company suffered a severe business contraction in 1976, the technology was sold to Timex. (Gross 2012). Concurrently, James Fergason was pursuing the practical usage of LCDs while working at the Liquid Crystal Institute at Kent State University. Fergason patented the first low-power LCD in 1971 (NIHF 2025). His invention made the LCD commercially viable and led to its widespread usage for televisions and computer monitors.

Compact cassette tape Starting in 1958, RCA and its competitors embarked on the development of magazine-style tapes in all shapes and sizes. The technology allowed people without technical training to simply insert the tape case into a tape player and press a button to listen. Philips inventor Lou Ottens introduced in 1963 a small cassette version, featuring a reversible tape housing that allowed 30–45 min of stereo music per side. In 1965, after protracted negotiations with Sony, Philips made the technology available free to manufacturers worldwide, thus guaranteeing their design would become the common standard (Vintagecassettes.com 2012). Tape recording and playback devices quickly shrank to take advantage of the compact size.

14.2.3 Communications

Telephones In 1963, the first electronic push-button system with touch-tone dialing was offered by Bell Telephone to customers in Pittsburgh, Pennsylvania. The system used audible tones for each digit on the push-button keypad, called dual-tone multi-frequency (DTMF) technology. When a button was pressed, the dial generated a combination signal of the frequencies, which a switching center decoded to route the call. Engineers tested 15 different keypad layouts before settling on the modern configuration with the numeric 1 at the top left and 0 in the bottom row (MacNeil 2019).

14.2.4 Home and Safety

Kevlar fabric In 1965, Dupont chemist Stephanie Kwolek discovered a new synthetic polymer that could be spun into fiber many times stronger than steel. The Kevlar fibers are so tightly woven that it is difficult to separate them, making it ideal for applications to prevent cuts and punctures involving sharp objects, such as industrial gloves. Over time, Kevlar fiber became the standard in many safety-related products, including bulletproof vests used by police and military personnel. When a projectile hits Kevlar, the fibers catch the projectile while absorbing and dissipating its energy (Dupont 2025). Kwolek was inducted into the National Inventors Hall of Fame in 1974.

Temper foam Known as memory foam, NASA created this material in 1966 to improve crash protection for airline passengers. As a shock absorber, a three-inch foam absorbs the impact of a ten-foot fall by an adult (NASA Spinoff 1981). It is now used in mattresses, pillows, safety equipment, aircraft seats, and more to cushion the human body.

14.2.5 Architectural Engineering

Event stadiums The Houston Astrodome, the world's first roofed, air-conditioned stadium, opened to the public in 1965. Buckminster Fuller's geodesic dome design completely captured the essence of the space age and Houston's role at the center of NASA's human spaceflight program. The stadium boasted no columns to impede fans' view of events and had the first private suites. The groundskeeping staff wore spacesuit-inspired gear to tend the three acres of Bermuda grass on the playing field, although the indoor grass struggled to grow, prompting the invention of synthetic *astroturf* the next year. In 1973, the Astrodome hosted a high-profile moment in gender politics, called the Battle of the Sexes, when Billie Jean King defeated Bobby Riggs in tennis (Houston Chronicle 2015).

14.2.6 Energy

Nuclear Power In 1963, The Jersey Central Power and Light Company announced the purchase of a 515-megawatt nuclear power plant from General Electric to be built in New Jersey. It was the first nuclear power plant selected on purely economic grounds without government aid and in direct competition with a conventional fossil-fueled facility (US Department of Energy 2025a). Ada I. Pressman (Achievement Award 1976) developed emergency safety systems for fossil-fuel and nuclear power plants (ETHW n.d.). By 1965, the System for Nuclear Auxiliary Power (SNAP) program had launched the first nuclear reactor in space, SNAP-10A, with the aim of providing power to research satellites (US Department of Energy 2025b).

14.3 State of Women

In 1961, President John F. Kennedy named Eleanor Roosevelt to lead the *President's Commission on the Status of Women*, hoping to get insights into female discrimination in education, workforce, and Social Security (Britannica 2016). He also tapped Esther Eggertsen Peterson to serve as executive vice chairman. Kennedy had appointed Peterson as head of the Women's Bureau in the Department of Labor at the beginning of his term, and promoted her to assistant secretary of labor in 1963 (Little 2024). The outcome of the commission was the *Peterson Report*, published in 1963, documenting widespread workplace discrimination against women and including a call to end to all legal restrictions on married women's right to own property, to enter into business, and to make contracts; equal opportunity in employment; and greater availability of child-care services (Britannica 2016). At the time, women had extremely limited roles and influence in American government:

> Although women comprised about half of the nation's voters, there were no female Supreme Court justices, federal appeals court justices, governors, cabinet officers, or ambassadors. Only 2 out of 100 US senators and 15 out of 435 representatives were women. Of the 307 federal district judges, 2 were women. Of the 7700 members of state legislatures, 234 were women. Only two American women had ever been elected governor, only two had ever served in a president's cabinet, and only six had ever served as ambassador (digitalhistory 2021).

Peterson was the highest-ranking woman in Kennedy's administration and the driving force behind the passage of the Equal Pay Act of 1963, legislating *equal pay for equal work* (Little 2024).

In 1963, Betty Friedan published *The Feminine Mystique*, which described widespread dissatisfaction amongst women in American society. She coined the phrase *feminine mystique* to describe the societal assumption that women only find fulfillment through housework, marriage, sexual passivity, and child rearing, and had no desire for higher education, careers, or a political voice. Friedan argued that women who worked during the wartime years were pressured to resume traditional feminine roles at home, regardless of their preferences. She advocated for harmonious equality between men and women (Churchill 2024). The book catalyzed the women's movement in the late 1960s and caused many women to rethink their educational goals and role in society.

Also in 1963, the American physicist Maria Goepper-Mayer shared the Nobel Prize for discoveries concerning atomic nuclei shell structure (NobelPrize.org

2025a). Raised in Germany, she had emigrated to the USA in the 1930s with her American husband, who was employed at Johns Hopkins University. Job scarcity during the Great Depression and nepotism rules prevented the university from offering her a position since her husband held a job there. Undeterred, she continued working, "just for the fun of doing physics" (Nobel Lectures 1972). She completed significant work in nuclear physics at Argonne National Laboratory and other employers (NobelPrize.org 2025b).

Women were engaged in all levels of the civil rights movement, performing organizational tasks, leading local civil rights organizations, conducting voter registration campaigns, and serving as lawyers on school segregation lawsuits (Library of Congress 2025). Their efforts were often behind the scenes of the men who were the well-known faces of the movement, depicted in the headlines. Congress passed the Civil Rights Act of 1964, the Economic Opportunity Act of 1964, and the Voting Rights Act of 1965 in response to pressure from the civil rights movement and from President Lyndon Johnson (Britannica 2025a).

Women realized they could benefit from the new legislation aimed at racial minorities. It was not a guarantee of full equality with men, but provided a foundation to enable them to speak up for additional women's rights. In 1966, the National Organization of Women (NOW) formed, intent on gaining a full expansion of women's rights as documented in the proposed Equal Rights Amendment (ERA). Among the 28 founders of NOW was Betty Friedan (NOW 2025). The NOW statement of purpose read:

> The purpose of NOW is to take action to bring women into full participation in the mainstream of American society now, exercising all privileges and responsibilities thereof in truly equal partnership with men (NOW 2025).

14.4 State of SWE[2]

In 1962, SWE settled into the headquarters at the United Engineering Center, finally recognized amongst the peer group of other engineering societies. Membership statistics compiled in 1965 reported SWE having 828 members, with virtually all the growth since 1960 coming from 259 student members (SWE Records 2000). SWE was foremost still an organization focused on professional working women engineers. It was not a presence—yet—on most college campuses. Barbara Faust, SWE Fellow and Life member, recalled, "We didn't have enough women engineering students to establish a viable SWE section at Penn State at the time" (personal correspondence, 2025).

Technical schools mounted a concerted effort to attract women students. In 1962, the Massachusetts Institute of Technology built a new women's dorm. Colorado School of Mines followed suit in 1964. Paula Loring Simon, SWE Fellow and Life member, commented,

[2] This section includes personal recollections about the time period and SWE from a survey of SWE Fellows and Late Career/Retiree Affinity group members conducted in February 2025.

> Women at Rensselaer Polytechnic Institute did not have a dorm until my junior year. We lived at Russell Sage College, an all-female school with a dress code. Therefore, we had a difficult time integrating with the men in our class after class hours (personal correspondence 2025).

Rutgers announced a re-entry program for women in math and chemistry, and Caltech announced it would start admitting women in the fall of 1970 (SWE Records 2000). By 1964, the engineering schools finally saw an upward trend in enrollment. This also increased the demand for engineering educators.

Career guidance was more important than ever in order to fill enrollment at these colleges. In 1964, future SWE Fellow Amy Spear created a slideshow featuring women engineers at work for junior and senior high schools as part of SWE's career guidance activities. For the first time, SWE focused on expanding the image of engineers using real-life role models in their work, informing the public about careers in engineering. The *Shining Star Award* for section excellence in career guidance activities was given for the first time that same year to the Los Angeles SWE Section. By 1965, NASA contacted SWE about opportunities for women as astronauts. At the time, candidates were required to have a PhD, yielding very few qualified candidates from the ranks of women engineers. Later, the requirements for all candidates (male and female) were lowered to the baccalaureate (Kindya 1990).

Women engineering students during this time were typically the lone female in their engineering classes. They used slide rules for calculations and received their first exposure to computer systems and software. Most students did not have access to a SWE section on their campus, but received encouragement to complete their degrees from professors and engineering deans. Their unique presence stood out in large lecture halls and laboratories. They were considered the oddity amongst the rest of the engineering students.

Women engineers in the workforce during this era wore professional suits and heels, but were often mistaken as clerical staff, asked to take notes, and given less prestigious assignments. Employers assumed they would quit work once they became pregnant. Many reported receiving less pay than their male counterparts doing the same job. Connecting with other women engineers through SWE sections provided a counterbalance to the isolation many faced as the sole female in their workplaces. Within the workplace, Barbara Faust recollected,

> Mainframe computers were coming into use for aerodynamic and other complex simulations. The computer calculation speed required programs to be run overnight, after the engineer or computer programmer transported the code and input parameters to the computer room via punch cards. A genuine fear was dropping the box of punch cards (personal correspondence, 2025).

In 1964, SWE applied for associate membership in the Engineers' Council for Professional Development (ECPD), joining 32 other professional societies. The Engineering Manpower Commission tracked facts and figures on employment, demand, supply, and other statistics vital to the planning effort of SWE. In 1966, the Engineering Joint Council coordinated efforts to create continuing education courses to counteract technical obsolescence. This was particularly relevant for women engineers, who often graduated, worked for a while, and then returned home to raise a family. Those who opposed seeing women in the engineering profession made much of this assumption (Kindya 1990).

In 1964, active SWE member Ruth Shafer organized the first International Conference of Women Engineers and Scientists (ICWES) in New York. The conference was held at the United Engineering Center, attended by 529 participants representing every state in the USA and 35 countries. The theme was *Focus on the Future* with lectures on topics including future needs for food, shelter, clothing, and energy, and the consequent need for engineers and technologists. The conference also included technical tours and a visit to the New York World's Fair. The conferences were planned to be held every 3 years in the future (Kindya 1990).

In 1966, the national convention held in Phoenix featured exhibit booths for the first time. These included a large-scale model of the Lower Colorado River Basin by the US Department of the Interior and a Motorola display tracing the growth of integrated circuits (Kindya 1990).

14.5 Laws and Culture

The space race John Glenn and NASA checked a milestone for the USA by launching a human into space and orbiting the Earth in 1962. In 1965, the National Academy of Engineering was formed within the National Academy of Sciences as an advisor to the government in matters of science and technology (Kindya 1990).

Societal change Profound social change, marked by the civil rights movement, resulted in instability in all regions of the USA. Organized nonviolent protests led by activists raised awareness of pervasive and systemic discrimination for non-white Americans. Dr. Martin Luther King, Jr., led a march on the Washington Mall in 1963, advocating for African American rights with the iconic *"I Have a Dream"* speech. In the 1964 Freedom Summer voter registration drive, sponsored by the Student Non-Violent Coordinating Committee (SNCC), volunteers faced violent resistance by the police and some community members. The Black Panthers organization formed, signaling an impatience with the pace of change for those involved.

Environmental movement Rachel Carson published the book *Silent Spring* in 1962, documenting the ecological harm to birds and wildlife caused by the indiscriminate use of the pesticide Dichlorodiphenyltrichloroethane, commonly known as DDT (PBS 2014). Carson's accusations of disinformation from the chemical industry and public officials sparked the environmental awareness movement. In 1966, the Sierra Club publicly opposed plans to build two dams inside the Grand Canyon and the Internal Revenue Service retaliated by suspending the tax deductibility of contributions to the club. The public mobilized and convinced Stewart Udall, Secretary of the Interior, to change his mind. In a major victory for conservationists, Congress moved to prohibit dams in the Grand Canyon and expand the national park (PBS 2025).

Summary. Table 14.1 provides examples of these changes.

Table 14.1 Culture and legal highlights for 1962–1966

Date	Event
1962	**Space Race:** John Glenn became the first American to orbit Earth, propelled into space by an Atlas rocket in a Mercury spacecraft (Uri 2022)
1962	**Cuban Missile Crisis:** The Cuban Missile Crisis was a tense confrontation between the USA and the Soviet Union after the Soviet Union placed nuclear missiles in Cuba. For weeks, there was extreme tension as the two superpowers hovered close to the brink of nuclear war. A negotiated return of the weapons eventually diffused the situation (Britannica 2025b)
1963	**The Equal Pay Act:** Prohibits employers from paying different wages to employees of different sexes for substantially equal work (Little 2024). Despite the law, women and minority races continued to earn less than men in nearly all fields (Fugiero 2024)
1963, 1970, 1977, 1990	**Clean Air Act:** Regulated all sources of air emissions. Congress also ratified the Limited Nuclear Test Ban Treaty, ending testing of nuclear weapons in the atmosphere, underwater, and in space (PBS 2014)
1963	**I Have a Dream:** On Aug. 28, 1963, the Rev. Martin Luther King Jr. gave his *I Have a Dream* speech to more than 200,000 people on the Washington Mall to demand equal rights for African-Americans (NPR 2010). King declared in words the equality and respect that African-Americans in the civil rights movement were fighting for: "I have a dream that my four little children will one day live in a nation where they will not be judged by the color of their skin but by the content of their character (NPR 2010)"
1964	**Civil Rights Act:** Signed by President Lyndon B. Johnson, the Civil Rights Act outlawed discrimination in the USA based on race, color, religion, sex, or national origin. Aimed at addressing racial discrimination of the time, it established a legal basis for women to bring lawsuits regarding unfair treatment in the workplace (Britannica 2025c)
1964	**Economic Opportunity Act:** Federal legislation establishing social programs aimed at improving education, health, employment, and welfare for impoverished Americans. It was a landmark of the Great Society and War on Poverty programs signed into law by President Lyndon B. Johnson (Britannica 2023)
1964	**The Wilderness Act:** Congress set aside 9.1 million acres to be preserved in perpetuity, *where man is a visitor but does not remain*. Congress added national forest, national park, and Bureau of Land Management lands to the system, which grew to 100 million acres (PBS 2014)
1965	**Voting Rights Act:** Legislation signed by President Lyndon B. Johnson to remove legal barriers at the state and local levels that prevented minorities from exercising their right to vote under the Fifteenth Amendment (Britannica 2025c)
1965–1972	**Contraception:** In 1965, *Griswold v. Connecticut,* the Supreme Court ruled that state bans on birth control violated marital privacy, overturning remaining access limits to married couples. In 1972, *Eisenstadt v Baird* extended birth control access to unmarried people. (Justia 1965; Justia 1972)
1965	**Medicare and Medicaid:** As part of the *Great Society* program**,** President Lyndon B. Johnson signed into law federal health insurance for the aged (Medicare) and for poor families (Medicaid), bringing the first medical safety net to vulnerable populations (Isserman and Kazin 2004)

References

Beck, Adam. 2017. 60 Years of CAD Infographic: The History of CAD since 1957. CADENAS. September 25. https://www.3dfindit.com/en/engiclopedia/60-years-of-cad-infographic-the-history-of-cad-since-1957. Accessed March 17, 2025.

Churchill, L.B. 2024. The Feminine Mystique. *Encyclopedia Britannica*. October 5. https://www.britannica.com/topic/The-Feminine-Mystique. Accessed February 14, 2025.

Computer History Museum. 2025a. CDC 6600's Five Year Reign. https://www.computerhistory.org/revolution/supercomputers/10/33. Accessed February 14, 2025.

Computer History Museum. 2025b. CDC 6600 CPU cabinet. https://www.computerhistory.org/collections/catalog/X1385.97F. Accessed February 14, 2025.

Digitalhistory. 2021. Feminism Reborn. https://www.digitalhistory.uh.edu/disp_textbook.cfm?smtID=2&psid=3341, Digital History ID 3341. Accessed February 14, 2025.

Dupont. 2025. What is Kevlar? https://www.dupont.com/what-is-kevlar.html. Accessed February 10, 2025.

Editors of Encyclopaedia Britannica. 2016. President's Commission on the Status of Women. *Encyclopedia Britannica.* October 21. https://www.britannica.com/topic/Presidents-Commission-on-the-Status-of-Women. Accessed February 14, 2025.

Editors of Encyclopaedia Britannica. 2023. Economic Opportunity Act. *Encyclopedia Britannica*. March 20. https://www.britannica.com/topic/Economic-Opportunity-Act. Accessed February 17, 2025.

Editors of Encyclopaedia Britannica. 2025a. Civil Rights Act. *Encyclopedia Britannica.* February 15. https://www.britannica.com/event/Civil-Rights-Act-United-States-1964. Accessed February 17, 2025.

Editors of Encyclopaedia Britannica. 2025b. Cuban missile crisis. *Encyclopedia Britannica*. February 15. https://www.britannica.com/event/Cuban-missile-crisis. Accessed February 16, 2025.

Editors of Encyclopaedia Britannica. 2025c. Voting Rights Act. *Encyclopedia Britannica*. January 17. https://www.britannica.com/event/Voting-Rights-Act. Accessed February 17, 2025.

Engineering and Technology History Wiki (ETHW). n.d. SWE Awards. https://ethw.org/SWE_Awards. Accessed March 22, 2025.

Fugiero, M. 2024. Equal Pay Act of 1963. *Encyclopedia Britannica*. October 18. https://www.britannica.com/topic/Equal-Pay-Act. Accessed February 16, 2025.

Gross, Benjamin. 2012. How RCA lost the LCD. November 1. https://spectrum.ieee.org/how-rca-lost-the-lcd. Accessed February 14, 2025.

Houston Chronicle. 2015. How the Astrodome was ahead of its Time. April 3. https://www.chron.com/news/houston-texas/slideshow/How-the-Astrodome-was-ahead-of-its-time-changed-106803.php. Accessed February 16, 2025.

Isserman, Maurice and Kazin, Michael. 2004. *America Divided: The Civil War of th*e 1960s. Second Edition. Oxford University Press. ISBN 0-19-516046-0

Justia. 1965. Griswold v. Connecticut, 381 U.S. 479. https://supreme.justia.com/cases/federal/us/381/479/. Accessed March 7, 2025.

Justia. 1972. Eisenstadt v. Baird, 405 U.S. 438. https://supreme.justia.com/cases/federal/us/405/438/. Accessed March 7, 2025.

Kindya, Marta. 1990. Four Decades of the Society of Women Engineers. Society of Women Engineers Records (LR001539). Box 180, Folder 9, Walter P. Reuther Library, Wayne State University. Accessed January 23, 2025.

Kurtz, Thomas E. 2025 Thomas Kurtz. https://cis-alumni.org/TKurtz.html. Accessed February 16, 2025.

Library of Congress. 2025. Women in the Civil Rights Movement. https://www.loc.gov/collections/civil-rights-history-project/articles-and-essays/women-in-the-civil-rights-movement/. Accessed February 14, 2025.

Little, Becky. 2024. Esther Peterson: The Woman Behind the Equal Pay Act. March 11. https://www.history.com/news/equal-pay-act-esther-peterson. Accessed February 14, 2025.

MacNeil, Jessica. 2019. Touch Tone Phones are Invented. November 18. https://www.edn.com/tone-dialing-telephones-are-introduced-november-18-1963/. Accessed February 10, 2025.

NASA Spinoff. 1981. Temper Foam. https://spinoff.nasa.gov/node/9021. Accessed February 10, 2025.

NIHF. 2025. James L. Fergason. *National Inventors Hall of Fame*. https://www.invent.org/inductees/james-l-fergason. Accessed February 18, 2025.

NobelPrize.org. 2025a. Maria Goeppert Mayer—Facts. Nobel Prize Outreach. https://www.nobelprize.org/prizes/physics/1963/mayer/facts/.Accessed February 14, 2025.

NobelPrize.org. 2025b. Maria Goeppert Mayer—Biographical. Nobel Prize Outreach. https://www.nobelprize.org/prizes/physics/1963/mayer/biographical/. Accessed February 14, 2025.

NPR. 2010. Read Martin Luther King Jr.'s 'I Have a Dream' speech in its entirety. January 18. https://www.npr.org/2010/01/18/122701268/i-have-a-dream-speech-in-its-entirety. Accessed February 10, 2025.

Nobel Lectures. 1972. Physics 1963-1970. Elsevier Publishing Company, Amsterdam. https://www.nobelprize.org/organization/publications/. Accessed February 14, 2025.

NOW. 2025. When and how was NOW founded. https://now.org/faq/when-and-how-was-now-founded/. Accessed February 14, 2025.

Ostovar, Michele. 2024. Mercury-Atlas 6: Friendship 7. NASA. February 12. https://www.nasa.gov/mission/mercury-atlas-6-friendship-7/. Accessed February 10, 2025.

PBS. 2014. Timeline of Environmental Movement and History. April 15. https://www.pbs.org/wnet/americanmasters/a-fierce-green-fire-timeline-of-environmental-movement/2988/. Accessed February 18, 2025.

Smithsonian. 2025. Telstar. https://airandspace.si.edu/collection-objects/communications-satellite-telstar/nasm_A20070113000. Accessed February 10, 2025.

Society of Women Engineers (SWE) Records (LR001539). 2000. SWE 50 Years 1950-2000. Box 180, Folder 9, Walter P. Reuther Library, Archives of Labor and Urban Affairs, Wayne State University. Accessed January 10, 2025.

Uri, Jon. 2022. 60 Years Ago: John Glenn, the First American to Orbit the Earth aboard Friendship 7. NASA. Feb 18. https://www.nasa.gov/history/60-years-ago-john-glenn-the-first-american-to-orbit-the-earth-aboard-friendship-7/. Accessed February 10, 2025.

US Department of Energy. 2025a. Timeline of Events: 1951 to 1970. https://www.energy.gov/lm/timeline-events-1951-1970. Accessed February 10, 2025.

US Department of Energy. 2025b. System for Nuclear Auxiliary Power (SNAP) Overview. https://www.energy.gov/etec/system-nuclear-auxiliary-power-snap-overview. Accessed February 10, 2025.

Vintagecassettes.com. 2012. History of Compact Cassettes. http://vintagecassettes.com/_history/history.htm. Accessed February 14, 2025.

Chapter 15
Laurence Delisle Pellier (1904–2004) Achievement Award Recipient 1962

In recognition of her significant contributions to the field of metallurgy

Holly J. Teig

15.1 Introduction

Laurence Delisle Pellier's contributions to metallurgical engineering include developing techniques for applying electron microscopy to metallurgical problems, researching alloy manufacturing methods for fine instruments and wires, and establishing a breadth of metallurgical testing procedures throughout her career. Pellier was honored by the Society of Women Engineers as the 1962 Achievement Award recipient.

Pellier (Fig. 15.1) was a specialist in physical metallurgy, particularly light and electron microscopy. She was a prolific author of technical papers and a US patent holder for the gold plating of surgical needles. Her participation as a member of the American Society of Testing Materials (ASTM) subcommittee on electron microstructure of metals generated micrographs of alloys and a multitude of publications about their metallurgical structures.

H. J. Teig (✉)
Peoria, IL, USA

C. D. Craig et al. (eds.), *Women Engineering Legends 1952-1976: Society of Women Engineers Achievement Award Recipients*,
https://doi.org/10.1007/978-3-032-00224-2_15

Fig. 15.1 Laurence Delisle Pellier, Portrait (circa 1955–1962). (Photograph courtesy of Society of Women Engineers Records, Walter P. Reuther Library, Archives of Urban and Labor Affairs, Wayne State University)

15.2 Early Life and Education

Born in Paris, France, in 1904, Laurence Pellier and her husband, Christian (Rene) came to the US in the 1920s (SWE Records n.d.). Described as very quiet and having a thick accent, she used her French language skills in 1933 as a bilingual secretary in the office of the French Commercial Attaché in New York, perhaps to fund her educational goals.

Pellier earned a Bachelor of Science in Chemical Engineering (cum laude) from the College of the City of New York (CCNY) in 1939 when she was about 35. What is interesting about this accomplishment is that CCNY did not admit women at the time, with the reason given that no sanitary facilities existed for women. Pellier later explained that she attended night school to earn her degree, which allowed a bypass of the bathroom barrier of the era (US Woman Engineer 1990). She obtained a Master of Science degree in Metallurgy from the Stevens Institute of Technology of New Jersey in 1942 (SWE Records n.d.).

While at Stevens Institute, she had her first experiences in metallurgical investigations as a research associate and research fellow for the International Nickel Company. For 6 years, she was engaged in studying and designing alloys by powder metallurgy (SWE Records n.d.). She also taught at Stevens in the Department of Powder Metallurgy. During her early career, Pellier continued working toward a PhD in physical metallurgy at Columbia University.

15.3 Early and Mid-Career

In 1946, Pellier became a senior metallurgical engineer at Sylvania Electric Products in Bayside, NY. She was responsible for developing techniques for applying electron microscopy to problems in physical metallurgy, with emphasis on deformation, recovery, and recrystallization in tungsten and copper (SWE Records n.d.).

Six years later, Pellier accepted a role as a metallurgist with the American Cyanamid Company in Connecticut. Here, she focused on research studies of corrosion. She was engaged in specifications of metals used in the construction of chemical plants, particularly stainless steels and titanium alloys (SWE Records n.d.). She concentrated on electron metallography, employing light and electron microscopes in her research. There was also work with electro and electroless plating techniques (SWE Records n.d.).

Pellier's work at American Cyanamid Company led to the invention of gold plating of surgical needles, awarded US patent number 2,865,376 (Fig. 15.2) in December 1958 (US Patent Office 1958). It was common for surgical needles at the time to be made of carbon steel and pre-threaded with sutures stored in various conditioners to keep the sutures pliable for surgery. However, the conditioning fluids had undesirable corrosive effects on the needles. Gold had been regarded as impossible to plate onto the tiny wire surfaces of the surgical needle until Pellier found that an electroless plate of gold from an alcoholic solution readily formed on the surface of the steel needle (US Patent Office 1958).

Over the next 5 years, Pellier moved between three different companies. Her job title varied, but consistent was her ongoing work involving alloy design and development of metallurgy techniques using electron microscopes. In the summer of 1956, Pellier joined Sigmund Cohn Corporation of Mount Vernon, NY as a research metallurgist. She again applied her expertise to the design and processing of precious metal alloys of high corrosion and wear resistance, with a wide range of electrical properties. The new alloys were used in fine instrumentation products, including potentiometers, resistance thermometers, and thermocouples (SWE Records n.d.).

A year later, Pellier switched to Philips Electronics, Inc. as an application engineer for electron optics. Here, she worked with prospective customers to perform services in electron microscopy and diffraction. It clearly was not a fit for her, as seven months later she was back at International Nickel Company in New Jersey as a senior scientist. She performed electron metallography analyses of high temperature alloys, high strength steels, stainless steels, and nickel powders. Pellier used different techniques to understand the microstructures, such as extraction replicas, direct examination of thinned sections, and electron diffraction. She had time to conduct her own research in addition to performing work that others in the company required (SWE Records n.d.).

15.4 ASTM Subcommittee and Publications

Throughout the decade of the 1950s, Pellier was very involved with the ASTM subcommittee on electron microstructure of metals, taking part in documentation work and authoring publications from 1950 through 1957. Examples of her work include: *Electron Microstructure of Steel* (1950), *Techniques for Ferrous Electron*

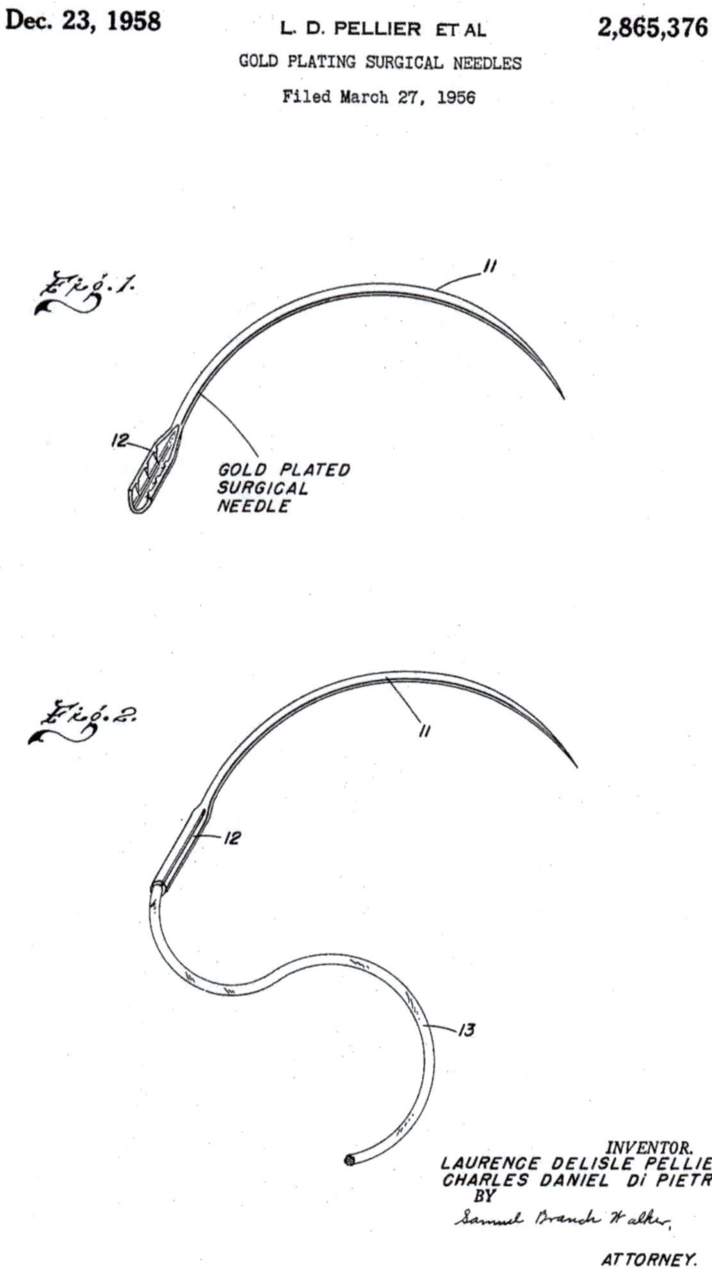

Fig. 15.2 Gold Plating of Surgical Needle patent drawing, Laurence Delisle Pellier, inventor, U. S. Patent 2,865,376

Metallography (1953), *Electron Microstructure of Tempered Bainite and Tempered Martensite* (1954), and many others.

Pellier described the significance of the invention of the electron microscope to her work as a metallurgist:

> In the physics of metals, for instance, it had been postulated and shown indirectly by X-ray diffraction analysis that metals were imperfect crystals and that their imperfections were responsible for the several orders of magnitude difference between their actual strength and that, calculated theoretically. It was now possible to observe these defects.
>
> It had been shown theoretically that line defects, called dislocations, greatly facilitated slip in metals under stress; these dislocations could be observed directly, they could be made to move and multiply under the effect of mechanical or thermal stress applied directly to the samples in the [electron] microscope, their reaction with inclusions and grain boundaries could also be followed.
>
> In other words, a wealth of direct evidence had become available for the formulation of mechanisms of deformation and fracture of metals. Such progress steps were perhaps minor when one thinks of the release of nuclear energy; they were enormous to a metallurgist whose field was not yet a science; they were fascinating to me when I had the opportunity to be an observer (ICWES 1964).

Keeping current technically was very important to Pellier. Beyond ASTM, she was a member of the American Institute of Mining and Metallurgical Engineers (AIME), the Metal Science Club, the Electron Microscope Society of America, Société Française de Métallurgie, the Society of Women Engineers (SWE), and several other technical organizations. She attended the International Metallurgical Society meetings held in 1951 (Detroit), 1955 (Europe) and 1957 (Chicago).

Pellier began publishing technical papers in 1944 with “Why Not a 50-50 Cobalt-Copper Solid Solution Alloy?” in the *Transactions of the Electrochemical Society* (SWE Records n.d.). More papers followed in that periodical as well as other technical research topics featured in the AIME Technical Publication, *Metal Transactions Journal* and *Proceedings of ASTM*. She co-authored an article with Gordon A. Davis entitled “A Method of Examination of Sections of Fine Metal Powder Particles with the Electron Microscope” that was included in the McGraw-Hill reference book *Physics of Powder Metallurgy* in 1952 (SWE Records n.d.). ASTM exhibits featured her electron micrographs of copper, tungsten, nickel powder, and various alloys. The publications allowed Pellier to reach a technical credibility level that opened doors within the metallurgical community.

15.5 Later Career

In 1960, her husband, Christian, a mechanical engineering consultant, was sent by his company to oversee the installation of machinery in a plant in Belgium. Pellier left her employment and moved with him to Europe for a year. She used her time there productively, attending technical meetings at Cambridge, England; Delft, Holland; and Paris, France. She also followed research work in metallurgy conducted by the Division of European Research Associates in Brussels, Belgium, on refractory metals, such as tantalum (SWE Records n.d.).

Returning to the US from Europe in 1961, she took on a metallurgist role at the Burndy Corporation in Norwalk, Connecticut. Burndy manufactured electrical and electronic connectors for utilities and industrial use and provided testing and certification of tools and connectors to standards.

Pellier attended the first International Conference of Women Engineers and Scientists (ICWES) in June 1964, held in New York City. She and other past SWE Achievement Award recipients engaged with students at a luncheon held in their honor. Pellier provided a written contribution to the proceedings related to her work with electron microscopes and metallurgy.

> Throughout history, the practical demonstration of a few major ideas made progress advance suddenly by leaps and bounds…Persons my age have had the privilege to follow the burst of progress that such ideas have generated during the present century (ICWES 1964).

15.6 Post Retirement

After retiring from corporate work, Pellier privately purchased an electron microscope and installed it in a small, specially built house in her yard in Westport, Connecticut. In the lab surrounded by a lovely garden, her husband Chris, a machinist by trade, assisted Pellier by meticulously preparing the metal specimens and developing the photomicrographs (SWE CT Newsletter 2006). Their small business was called the Pellier - Delisle Metallurgical Laboratory, as noted in the 1982 Massport's *Directory of Women in Business*. Services that were offered included: optical and electron metallography, physical metallurgy, powder metallurgy, consulting, and mechanical engineering (her husband's specialty) (Massport Directory 1982). Pellier continued to work and report results to several clients until her vision declined as she neared age 90 (SWE CT Newsletter 2006).

15.7 Honors and Legacy

In 1962, Pellier (Fig. 15.3) received the Achievement Award of the Society of Women Engineers *in recognition of her significant contributions to the field of metallurgy*, exemplified by her continuous leadership in research and development of

Fig. 15.3 Laurence Delisle Pellier, Portrait (circa 1955–1962). (Photograph courtesy of Society of Women Engineers Records, Walter P. Reuther Library, Archives of Labor and Urban Affairs, Wayne State University)

products, physical metallurgy testing and manufacturing methods. The award chair cited her "talent, courage, and perseverance" in achieving her many accomplishments (SWE Records n.d.). Ten years into the awarding process, SWE was also progressing forward by including an accomplished SWE member on the award judging panel for the first time.

STEEL - The Metalworking Weekly magazine featured Pellier as a guest in the *Men in the News* column following the announcement of the SWE Achievement Award. Pellier stated in the article,

> The role of women in engineering is the same as that of men. The timeworn prejudices and hearsays leveled at women have long been proved false as more of them have entered the profession. Women are subjected to the same exacting training as men; they take the same examinations; receive the same degrees and - on the job - fulfill the same assignments (STEEL 1962).

Pellier was one of the 92 Society of Women Engineers charter members in 1950 (SWE Records 2000). By 1962, when she was recognized with the Achievement Award, the Society had grown to more than 700 members. Pellier reflected, "In industry, women have achieved such a reputation for competence that the number of opportunities for them in engineering has expanded tremendously in the last decade" (STEEL 1962). She noted that the number of engineering schools admitting women students or women to their teaching ranks was growing (STEEL 1962).

With her Achievement award, Pellier was presented with an honorary life membership in SWE. Conscious of the limited resources of the fledgling SWE organization, she continued to pay dues annually designated as a donation to SWE, with half directed to headquarters operations and half to the Connecticut SWE section (SWE Records n.d.). It is clear that her career and SWE meant a great deal to Laurence in her life. Laurence and her family also generously donated in her name to the SWE

Fig. 15.4 Laurence Pellier featured in the 2023 National Inventors Hall of Fame Museum exhibit in Alexandria, Virginia (Courtesy of the National Inventors Hall of Fame)

Connecticut Jean R. Beers Endowed Scholarship Fund, which supports women pursuing engineering education who lack financial resources (SWE CT Newsletter 2006).

> Although Laurence and Rene did not have children, they adored the children and grandchildren of their friends (I am one of the latter) and were very interested in our education, wrote Robin Dibner, a family friend. I think it would have thrilled Laurence to know that my daughter, who she knew from babyhood, will be pursuing a career in math or the physical sciences (Dibner, personal correspondence 2006).

In 2023, Pellier was one of six featured pioneers honored for Women's History month (Fig. 15.4) at the National Inventors Hall of Fame Museum (NIHF) housed in the US Patent and Trademark Office. "Because of our partnership with SWE on this exhibit, we wanted to specifically focus on women in engineering. We wanted to represent members of SWE as well as NIHF inductees, so that gave us a great group of women to choose from," stated Sophia Johnson, museum manager (SWE Magazine 2023). The museum has inducted 12 SWE-affiliated women, the first being Yvonne Brill (1986 SWE Achievement Award recipient), SWE Fellow and a rocket and jet propulsion engineer, inducted in 2010, and most recently biophysicist and researcher Alice Stoll (1969 SWE Achievement Award Recipient), inducted in 2024 (SWE Magazine 2023).

References

International Conference of Women Engineers and Scientists (ICWES). 1964. *Proceedings of the First International Conference of Women Engineers and Scientists.* University of Illinois Urbana-Champaign. Accessed March 26, 2025.

Massport Directory of Women in Business. 1982. Massport Directory 1982. Accessed January 6, 2025.

STEEL. 1962. Bias Against Women in Engineering Waning. *STEEL - The Metalworking Weekly.* Accessed 20 December 2024.

Society of Women Engineers (SWE) Records (LR001539). n.d. *Pellier, Laurence Delisle*. Box 191 Folder 34, Walter P. Reuther Library. Archives of Labor and Urban Affairs. Wayne State University.

Society of Women Engineers (SWE) Records (LR001539). 2000. Five Decades of the Society of Women Engineers. Walter P. Reuther Library, Archives of Labor and Urban Affairs, Wayne State University.

SWE Connecticut (CT) Newsletter. 2006. In Memory of Laurence Delisle Pellier (1904-2004). July. Accessed 20 December 2024.

SWE Magazine. 2023. Women Inventors Grab Spotlight in National Exhibit. Society of Women Engineers. Spring. *SWE Magazine*. Accessed 06 January 2025.

US Patent Office. 1958. Patent US-2865376, Gold Plating Surgical Needles issued December 23, 1958. Patent Public Search Basic USPTO. Accessed November 06, 2024.

US Woman Engineer. 1990. Celebrating the Pioneer Spirit. September/October. Accessed November 06, 2024.

Chapter 16
Beatrice Hicks (1919–1979) Achievement Award Recipient 1963

In recognition of her significant contributions to the theoretical study and analysis of sensing devices under extreme environmental conditions, and her substantial achievements in international technical understanding, professional guidance, and engineering education

Jill S. Tietjen

16.1 Introduction

Beatrice A. Hicks (Fig. 16.1) decided to be an engineer when she was 13 years old after she learned that the Empire State Building and the George Washington Bridge were the work of engineers. The first President of the Society of Women Engineers (SWE) and one of its co-founders, Hicks invented a gas density sensor that could sense the actual amount of gas, instead of the pressure produced by the gas, within a container. Her sensor has been used in space, in aircraft, in communications, to monitor nuclear weapons, and in equipment for the electric utility industry. Her many awards include the Society of Women Engineers Achievement Award (1963), election to the National Academy of Engineering (1978), and posthumous induction into the National Women's Hall of Fame and the National Inventors Hall of Fame.

16.2 Early Life and Education

Beatrice A. Hicks was born in Orange, New Jersey and demonstrated an aptitude for math and science from when she was very young. At age 13, inspired by the Empire State Building and the George Washington Bridge, and learning that engineers were

J. S. Tietjen (✉)
Technically Speaking, Inc., Greenwood Village, CO, USA

C. D. Craig et al. (eds.), *Women Engineering Legends 1952-1976: Society of Women Engineers Achievement Award Recipients*,
https://doi.org/10.1007/978-3-032-00224-2_16

Fig. 16.1 Beatrice Hicks (Photograph courtesy of Society of Women Engineers Records, Walter P. Reuther Library, Archives of Labor and Urban Affairs, Wayne State University)

responsible for those infrastructure miracles, she told her engineer father that she, too, would become an engineer (Candee 1957; Fields 1963). Although her high school classmates, school counselors, and some of her teachers tried to discourage her, pointing out that engineering was not a proper field for women, her parents did not stand in her way (Barth 1960). Hicks said, "However, while father didn't actually discourage me, he certainly didn't do any cheering for my being an engineer" (Candee 1957).

After her high school graduation in 1935, she entered the Newark (New Jersey) College of Engineering. At first, she worked retail at a store in New York City and later at night on switchboards to earn the money she needed to pay for her college expenses. She also worked for the library abstracting technical articles. Her extracurricular activities in college included editing the yearbook, and she was a student member of the Institute of Radio Engineers, the American Institute of Electrical Engineers, and the American Institute of Chemical Engineers. In 1939, she received a BS in chemical engineering, one of two women in a class of 900, and took a position as a research assistant at the College (Barth 1960; Candee 1957; New York Mirror 1961).

16.3 Career

From 1939 to 1942, Hicks served as a research assistant at Newark College. In that position, she researched the history of the inventions of Dr. Edward Weston. She paid particular attention to the dynamo, the electric lamp, and instruments for electrical measurements. Her research was foundational for a book published in 1949 by McGraw-Hill, written by David O. Woodbury and titled *A Measure For Greatness*. Weston had bequeathed his instruments to the college, and one of her responsibilities associated with her work in this area was setting up a museum containing the models of Weston's inventions (Barth 1960; Candee 1957; SWE Records 1961).

In 1942, Hicks got a job with the Western Electric Company—the first woman hired as an engineer. She also took graduate classes in electrical engineering during this time at the Newark College of Engineering. She worked first in the test set design department where she set up tests for telephone equipment that was used in toll circuits. She then worked in the quartz crystal department, where she was involved with the research, manufacture, and testing of quartz crystal oscillators. Of her three and a half years at Western Electric, Hicks said, "They were good years because I was putting into practical application all the things I had been so long learning. I was being well paid and was getting ahead, but it was fundamentally a man's world there. So I left and started on my own, designing on special jobs" (Candee 1957; Hicks n.d.-a).

Hicks joined Newark Controls Company, the company her father had founded, as chief engineer in 1945. When her father died in 1946, she became vice president as well. Newark Controls Company specialized in environmental sensing devices. She married engineer Rodney D. Chipp in 1948. She studied at night while employed and, in 1949, earned an MS in physics from Stevens Institute. Subsequently, she undertook further graduate work at Columbia University. Hicks pioneered the theoretical study, analysis development, and manufacture of sensing devices, patented a molecular density scanner, and developed an industry model for quality control procedures (Candee 1957; SWE Records 1961).

In 1955, Hicks bought control of the company and became president. One of the major products of the company at that time was low-water cutoffs and other devices to protect people from their own forgetfulness, often sold through mail-order companies. Hicks said, "Americans are in the habit of going off for the weekend—or even to Florida—and never bothering to turn off the furnace. They expect everything to take care of itself—so we make several devices which look after things while they are away." During her years with the Company, it manufactured specialized electro-mechanical devices such as liquid level controls, pressure controls, and altitude switches for aircraft and space vehicles, where extreme reliability under severe environmental conditions was required (Candee 1957).

In 1959, she was awarded patent 3,046,369 for a molecular density scanner or gas density switch (see Fig. 16.2). This type of switch is a key component in systems using artificial atmospheres. The gas density switch (today called a gas density sensor, controls, or monitor) was designed to monitor gas leakage, particularly for artificial atmospheres around electronic equipment. The artificial atmospheres include dry air, nitrogen, sulfur hexafluoride (SF_6), and fluorochemicals. These gases are suitable as insulators and heat dissipators for a broad range of applications, including sealed electronic equipment (non-airborne), sealed airborne

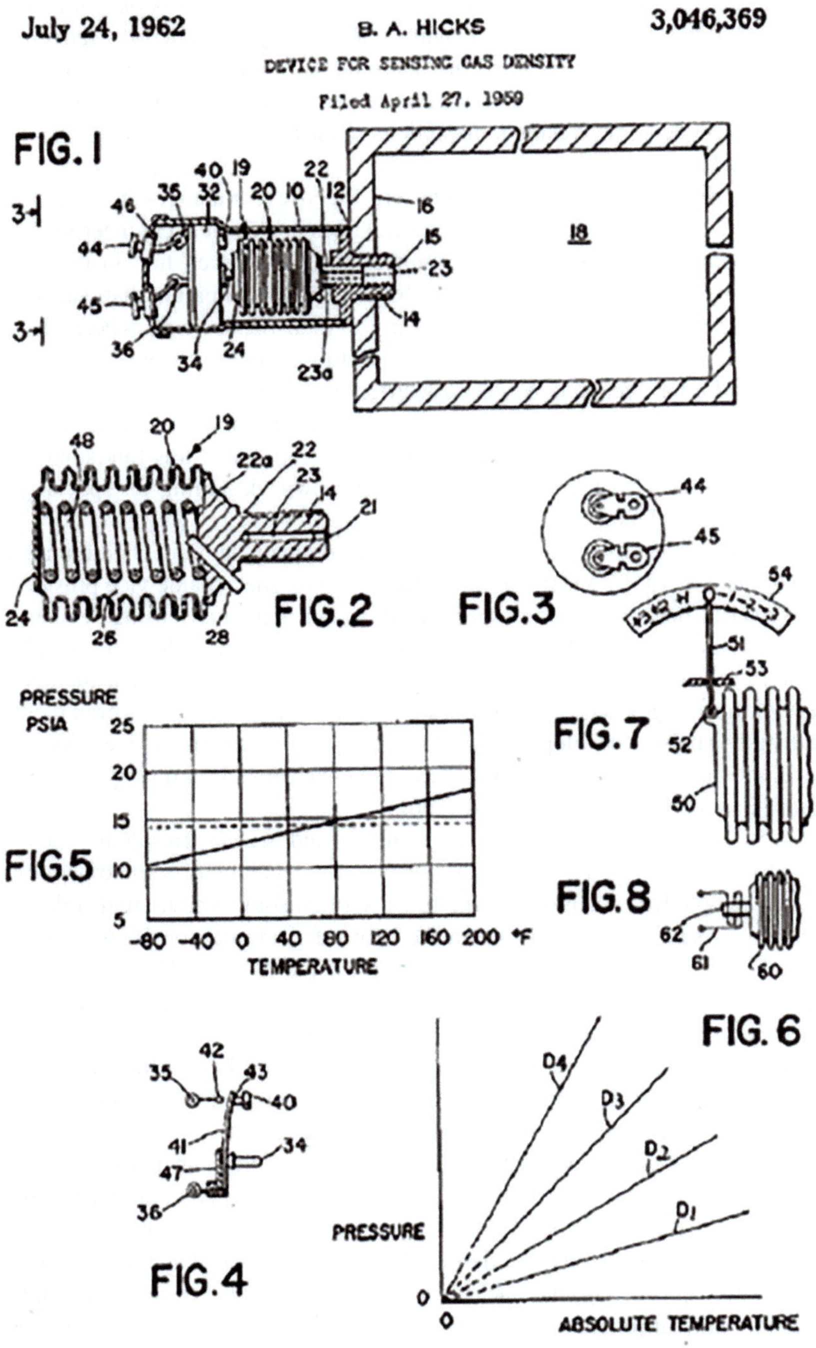

Fig. 16.2 Beatrice Hicks' Patent 3,046,369—Gas Density Sensor

electronic equipment, power transformers, switchgear, X-ray units, pressurized power cables, waveguides, and coaxial cables.

At the time of her invention, no mechanism existed to monitor the gas leakage and then signal to the equipment operator that action was warranted. Her gas density switch monitors the number of molecules per unit volume and was capable of indicating leakage at a pressure value that varied with temperature. Density, not pressure or temperature, is the important variable in a sealed-in atmosphere, because it determines arc resistance and heat transfer. Her switch was a unique bellows-type switch. It could be calibrated to protect throughout the entire temperature range. In addition to its function of signaling critical density, the switch could also be used to activate control circuits of systems, for example, opening a valve or starting a pump.

The sub-miniature gas density switch was in use in a number of applications, particularly in aircraft and missiles. The Saturn V rockets that launched the Apollo moon missions used her gas density switch in their ignition systems. The Boeing 707 used a gas density switch to monitor the nitrogen atmosphere around the high voltages in a sealed communications antenna coupler. On both the Boeing 707 and the Hustler B-58, the antenna couplers handle very high voltages and would arc-over in the upper atmospheres. A gas density switch mounted in each antenna coupler indicated remotely at the pilot's panel board if a problem developed. The sensor is also used to monitor nuclear weapons in storage (National Inventors Hall of Fame 2023; Tietjen n.d.).

Fire extinguishers use gas density switches to indicate which units are ready for recharging. Power transformers using gas and vapor cooling are monitored by gas density switches. At the time the switch was developed, it allowed early detection of leakage with an associated remote indicator that prevented critical loss of the gases and permanent damage to the transformers. Gas density switches are of value when power transformers are in shipment or storage, at the time they are put in service, as well as when they are in operation.

Bellows-type gas density switches are still manufactured, in use, and marketed today. One of the most significant applications for these types of gas density switches today is for high voltage SF_6 breakers. The monitors ensure that the breaker will adequately perform its function of properly interrupting faults. The monitor alarms and then trips the breaker when there is no longer adequate SF_6 to extinguish a fault. Monitors are manufactured in the US and abroad. Manufacturers today include Trench Limited, Solon Manufacturing Company (with the largest market share in the USA), Trafag, AKM, and WIKA (Tietjen n.d.).

In addition to her gas density sensor, Hicks developed other sensors that monitored pressure, fuel levels, and flow rates for both liquids and gases. One sensor that she developed set off an alarm when a rocket, missile, or plane exceeded the speed that was safe for structural components, as they could lose their integrity above the threshold speed (National Inventors Hall of Fame 2023).

Her marketing efforts included technical papers, presentations, and articles written for or presented to the American Society of Heating and Air Conditioning Engineers, *Electronic Equipment Engineering*, *Military System Design*, *Aerospace Electronics*, *Product Engineering*, *Space Aeronautics*, *American Engineer*, and *Power Engineering* (SWE Records 1961).

In 1963, the New York *Daily News* reported that Newark Controls made five different sensors (Fields 1963):

- The Mach sensor sounds the alarm when the speed of the rocket, missile or plane exceeds the safe structural stresses on the frame.
- Another sensor watches the level of liquids in containers like a balloon or gas tank or the fuel system of a jet, rocket or missile.
- A third warns when switches assigned to the job aren't properly controlling the pressure of liquids or gases, like fuel for a rocket or oxygen for a pilot.
- A fourth watches the rate of flow of such liquids and gases and alerts everyone concerned if the desired rate isn't achieved.
- A fifth senses when the density of a gas or liquid approaches the danger point, too low for the operation of the jet or missile, too low for the men in them.

16.4 Society of Women Engineers

A co-founder and the first President (1950–1952) of the Society of Women Engineers (SWE), Hicks (Fig. 16.3) knew there was a great future for women in engineering. She gave numerous talks about the opportunities for women in engineering to high schools, college classes, and organizations. Following one such talk, after the principal congratulated her, he said, "It didn't really matter just what you said—the important thing was for the children to see a woman engineer wasn't some sort of freak" (Candee 1957).

Fig. 16.3 Beatrice Hicks at the United Engineering Center dedication dinner, 1961 (Photograph courtesy of Society of Women Engineers Records, Walter P. Reuther Library, Archives of Labor and Urban Affairs, Wayne State University)

When SWE was founded, it had 60 members. Within a few years, that membership had grown to several hundred, with interest demonstrated from coast to coast. Hicks also chaired the professional guidance and education committee and served as Chair of the Board of Trustees. She said, "When a woman engineer starts on a job, people are usually a little uncertain and keep a close watch on you. But after you have done two or three things which turn out OK, they will accept you without question from then on. Only when there is an administrative position to be filled does a woman engineer find there is prejudice against her" (Barth 1960; Candee 1957: SWE 1963). Hicks chaired the Steering Committee for the First International Conference of Women Engineers and Scientists held in New York City in 1964 (SWE Records).

Her love for her SWE colleagues and for the organization was evident in her acceptance speech for SWE's 1963 Achievement Award (Hicks "Acceptance Speech" 1963):

> The award which your generosity and kindness has bestowed on me bears the name of a society which is a distinguished pioneer in American engineering. A society that has demonstrated through the accomplishments of its individual members and through their combined efforts that education and work in engineering is both the responsibility and the privilege of <u>all</u> whose talents are scientific in nature – a society that has given freely of its knowledge, time, and financial resources to build broader based and more capable engineering resources for the United States – a society whose deep and sympathetic understanding has inspired younger women to have the courage to obtain an education and to go on to accomplish outstanding engineering jobs – a society composed of those people for whom I hold the greatest respect, admiration, and love.
>
> To have been selected to receive the Society of Women Engineers Award is not only a high honor, but a challenge to work unceasingly for its high values, dedicated effort, and foresighted goals.
>
> In the beginning it looked like a lonely road strewn with neglect and disinterest, but it hasn't been. From the very beginning we constantly found new friends with the same dedication and foresightedness, both women and men. This group has swollen so now our members are almost 800, but our friends who work with us are in the thousands.
>
> The Society of Women Engineers started only thirteen years ago with a few people and the important goal to bring engineering education and opportunity and women with this talent together. Today it is active in all of the United States and many countries. Its aims and ideals are becoming a reality. This Denver conference is a fine testimony to its accomplishments, and next year in 1964 we look forward to another outstanding achievement, our first International Conference at the United Engineering Center in New York City.
>
> There are at least a thousand women making outstanding contributions to engineering, many of them to whom I look for advice, help, and teaching, and who have far greater claim to this award than I. I can only accept this award as a representative of the very many outstanding women engineers whose accomplishments, courage, and loyalty to the realistic needs of our people and civilization have been promoted and inspired by the work of this society, of which this award is an example of its vision and its faith; for as Edna St. Vincent Millay said:

The world stands out on either side
No wider than the heart is wide;
Above the word is stretched the sky,
No higher than the soul is high.
The heart can push the sea and land
Farther away on either hand;
The soul can split the sky in two,
And let the face of God shine through.

16.5 Honors, Awards, and Activities

Hicks received many honors and awards over her lifetime in addition to SWE's Achievement Award. In 1952, she was named "Woman of the Year in business" by *Mademoiselle* magazine. Also in 1952, *Charm* magazine ran a three-page article about her. In 1961, she was the first woman engineer appointed by the US Secretary of Defense to the Defense Advisory Committee on Women in the Services. Also in 1961, she was named the Alumna of the Year by Newark College (Hicks n.d.-b). She was the first woman to receive an honorary doctorate (DEng) from Rensselaer Polytechnic Institute (1965) (Fig. 16.4). She also received honorary degrees from Hobart and William Smith Colleges (ScD, 1958), Stevens Institute of Technology (1978), and Worcester Polytechnic Institute (1978). In 1978, she was elected to the National Academy of Engineering, the sixth woman to be elected. She was a Fellow of the International Academy of Law and Science (Candee 1957; Ogilvie and Harvey 2000; Proffitt 1999; Stanley 1995).

Her awards have continued posthumously. In 2001, she was inducted into the National Women's Hall of Fame. In 2013, Hicks received the Advancement of Invention Award from the New Jersey Inventors Hall of Fame. In 2017, she was inducted into the National Inventors Hall of Fame (National Inventors Hall of Fame 2023; National Women's Hall of Fame 2024; SWE Magazine 2014).

Hicks was a member of the American Society of Mechanical Engineers, The National Society of Professional Engineers, The New Jersey State Society of Professional Engineers, Montclair Society of Engineers, the Institute of Radio Engineers (later part of the Institute of Electrical and Electronics Engineers), the American Society of Heating and Air Conditioning Engineers (ASHAE), the Women's Engineering Society of Great Britain, and Eta Kappa Nu. She was licensed as a professional engineer in New Jersey, New York, Pennsylvania, and the District of Columbia (Barth 1960; Busignies 2024; Hicks n.d.-b; SWE 1961; SWE Records n.d.).

The National Society of Professional Engineers selected Hicks and her husband to represent the engineers of North America in a fact-finding and goodwill tour of South America. Her selection for Project Ambassador (1961) was in recognition of "her achievements as an engineer-executive, a member of many engineering societies, and an experienced world traveler (Barth 1960; SWE, "Information Release"

Fig. 16.4 Beatrice Hicks after receiving her honorary doctorate from Rensselaer Polytechnic Institute, 1965 (Photograph courtesy of Society of Women Engineers Records, Walter P. Reuther Library, Archives of Labor and Urban Affairs, Wayne State University)

1963). Hicks served as a delegate from the US to the International Management Conference multiple times including in Sao Paulo, Brazil in 1954; in Paris in 1957, and in Sydney, Australia in 1960 (Barth 1960).

Hicks also served as a Trustee of Newark College and as a member of the Newark College of Engineering Alumni Association. She served as President of the Newark College of Engineering Alumni Association 1958–1959. She was a member of the Orange Lawn Tennis Club. Hicks was listed in *Who's Who*, *Who's Who in Engineering*, and *Who's Who in American Women*. (Hicks n.d.-b.) She indicated that her outside interests were reading, music, and travel (Barth 1960; Hicks n.d.-b: SWE 1961; SWE Records).

16.6 Legacy

"There's hardly a space shot from Cape Canaveral [Kennedy Space Center] that does not carry an environmental sensor designed and made by Beatrice Hicks" (Fields 1963). Hicks' sensor was "an innovation that made possible the development of advanced technologies of the time and was a critical breakthrough to enabling space travel" (National Inventors Hall of Fame 2023).

In late 2024, the Society of Women Engineers had more than 40,000 members, over 100 professional member sections, and over 300 collegiate member sections. In addition, SWE offers activities for international members and high school students (SWE 2024).

References

Barth, E.C. 1960. Dr. Beatrice Hicks. From Society of Women Engineers Records (LR001539). Box 189, Folder 18. Walter P. Reuther Library, Archives of Labor and Urban Affairs, Wayne State University.

Busignies, Henri G. "Beatrice Hicks 1919-1979," National Academy of Engineering—Memorial Tributes. Accessed December 1, 2024. https://www.nae.edu/215725/BEATRICE-HICKS-19191979.

Candee, Margorie Dent, ed. 1957. *Current Biography*. New York, New York: The H. W. Wilson Company.

Fields, Sidney. Only Human – Space Trouble Shooter. *Daily News* [New York, New York], December 10, 1963. From Society of Women Engineers Records (LR001539). Box 189, Folder 18. Walter P. Reuther Library, Archives of Labor and Urban Affairs, Wayne State University.

Hicks, Beatrice. n.d.-aBiographical Sketch, From Society of Women Engineers Records (LR001539). Box 189, Folder 18, Walter P. Reuther Library, Archives of Labor and Urban Affairs, Wayne State University.

Hicks, Beatrice A. n.d.-b. Curriculum Vitae. From Society of Women Engineers Records (LR001539). Box 189, Folder 18. Walter P. Reuther Library, Archives of Labor and Urban Affairs, Wayne State University.

Hicks, Beatrice A. 1963. Society of Women Engineers Award Acceptance Speech. From Society of Women Engineers Records (LR001539). Box 189, Folder 18. Walter P. Reuther Library, c Wayne State University.

National Inventors Hall of Fame. Beatrice Hicks. Accessed June 24, 2023, https://invent.org/inductees/beatrice-hicks.

National Women's Hall of Fame. "Beatrice A. Hicks," Accessed December 1, 2024, https://www.womenofthehall.org/inductee/beatrice-a-hicks/.

New York Mirror (author unknown). 1961. Beauty and the Rockets. From Society of Women Engineers Records (LR001539). Box 189, Folder 18. Walter P. Reuther Library, Archives of Labor and Urban Affairs, Wayne State University.

Ogilvie, Marilyn and Joy Harvey, eds. 2000. *The Biographical Dictionary of Women in Science: Pioneering Lives from Ancient Times to the Mid-20th Century*. New York, New York: Routledge.

Proffitt, Pamela, Editor. 1999. *Notable Women Scientists*. Gale Group, Farmington Hills, Michigan

Society of Women Engineers. 1961 Convention. Dr. Beatrice A. Hicks, President, Newark Controls. From Society of Women Engineers Records (LR001539). Box 189, Folder 17. Walter P. Reuther Library, Archives of Labor and Urban Affairs. Wayne State University.

Society of Women Engineers. 2024. wAbout SWE. Accessed November 30, 2024. https://swe.org/about-swe/.

Society of Women Engineers. 1963. June 19 Information Release. From Society of Women Engineers Records (LR001539). Box 189, Folder 17. Walter P. Reuther Library. Archives of Labor and Urban Affairs. Wayne State University.

Society of Women Engineers Records (LR001539) n.d.. Box 189, Folder 17. Walter P. Reuther Library. Archives of Labor and Urban Affairs. Wayne State University.

Stanley, Autumn. 1995. *Mothers and Daughters of Invention: Notes for a Revised History of Technology*. New Brunswick, NJ: Rutgers University Press.

SWE Magazine. Beatrice Hicks recognized by New Jersey Inventors Hall of Fame. Winter 2014. Accessed May 24, 2015, http://www.nxtbook.com/nxtbooks/swe/winter14/index.php#/22.

Tietjen, Jill S. n.d. Unpublished nomination.

Chapter 17
Admiral Grace Murray Hopper, PhD (1906–1992) Achievement Award Recipient 1964

In recognition of her significant contributions to the burgeoning computer industry as an engineering manager and originator of automatic programming systems

Jill S. Tietjen

17.1 Introduction

Admiral Grace Murray Hopper's development of the computer compiler—having the computer translate the letters of the alphabet into machine code—laid the foundation for the computer age as we know it today. It was a giant step toward the widespread usage of computers that now affects almost every aspect of our lives. She received numerous awards and honors in recognition of her efforts during her lifetime, as well as posthumous awards.

"Amazing Grace" successfully combined careers in academia, business, and the US Navy while making history in the computer field. Hopper is referred to as the mother of computerized data automation in the naval service. Her pioneering computer compilers enabled the computer user to communicate with the computer using human languages, not just mathematical symbols. Furthermore, she loved taking credit for finding the first computer bug—it was a moth stuck in the relay of a computer.

When asked about the many accomplishments in her life, Hopper (Fig. 17.1) said "If you ask me what accomplishment I'm most proud of, the answer would be all of the young people that I've trained over the years, that's more important than writing the first compiler."

J. S. Tietjen (✉)
Technically Speaking, Inc., Greenwood Village, CO, USA

C. D. Craig et al. (eds.), *Women Engineering Legends 1952-1976: Society of Women Engineers Achievement Award Recipients*,
https://doi.org/10.1007/978-3-032-00224-2_17

Fig. 17.1 Admiral Grace Murray Hopper (Courtesy of United States Navy)

17.2 Family, Education, and Early Career

The oldest of three children, Grace Murray Hopper was born in 1906 in New York City to a father who was an insurance broker and a mother who was the daughter of a civil engineer. Her father's legs were amputated by the time Hopper was in high school—the standard treatment for hardening of the arteries at the time. He told his children that if he could walk with two wooden legs and a cane, then they could do anything. He would say, "You can overcome things if you want to." His handicap was an extra incentive for the children to excel—and to bring home A's from school, which they did. His handicap was also a reason for their mother to take care of the family's financial matters, so that Hopper learned a fascination with numbers and with getting answers! (Billings 1989; Whitelaw 1995).

She also loved gadgets from the beginning, a trait that would serve her well in the computer industry:

> When I was about seven years old, each room in our Wolfeboro home had an alarm clock. Those round clocks had a bell on top and two feet that shake apart when the alarm goes off. When we were going on a summer hike or a trip, mother would go around and set the clocks for her children and all of the cousins who were visiting. She went around one night and found all the clocks taken apart. What had happened was that I had taken the first one apart and couldn't get it together . . . and then I'd opened the next one and the next one, and . . .But I'd always loved a good gadget. I'm afraid that in my heart, when I saw the first Mark I, my first thought was "Gee, that's the prettiest gadget I ever saw." Which probably would not have pleased Howard Aiken, had he heard it (Tropp 1984).

Her father's handicap and his response were not the only philosophies that formed Hopper's life and career. Her father also believed that his daughters should have the same educational opportunities as his son. He wanted all his children to attend college, and he wanted his daughters to work for at least 1 year after they finished college to ensure they knew how to fend for themselves economically (Billings 1989).

After attending private schools in New York City and New Jersey, she passed the entrance exams and enrolled at Vassar in 1924. Hopper studied mathematics and physics while she was at Vassar. She was elected to Phi Beta Kappa, graduated with a bachelor of arts (1928) and won a Vassar College Fellowship which provided funding for further education. She used the money to enroll at Yale University and earned her master's degree in mathematics (1930) (Billings 1989; Whitelaw 1995).

She then married and taught mathematics at Vassar College. In addition, she completed her PhD in mathematics from Yale University in 1934. Hopper (Fig. 17.2) stayed in the Mathematics Department at Vassar from 1931 to 1943. She and her husband divorced in 1945—no children had been born of the marriage (Billings 1989; Whitelaw 1995).

Fig. 17.2 Grace Murray Hopper (Courtesy of Library of Congress)

17.3 World War II and Harvard

With World War II raging, Grace Murray Hopper wanted to join the military as had many members of her family before her. When she tried to enlist, the Navy rejected her—too old (36 years of age), too light (she only weighed 105 pounds and the minimum for her height of 5′6″ was 121 pounds), and she was a math teacher—whom the Navy believed needed to stay at their institutions to teach future soldiers and sailors. But she persisted! She got a six-month leave of absence and associated permissions from Vassar College and waivers for her age and weight. In December 1943, Grace Murray Hopper was sworn into the United States Navy as part of the Women Accepted for Voluntary Emergency Service—the WAVES. To her relief, as she would later say, "There was a war on! It was not unusual for a woman at that time to join the Navy; there were 30,000 to 40,000 women there at the time." Besides, she said, "It was the only thing to do." Off she went to Midshipman's School for Women at Smith College, where almost everyone else was the age of the students she had recently taught in her classes at Vassar (Billings 1989; WAVES 2019; Whitelaw 1995; Yale 2019).

Upon graduating at the top of her class from Midshipman's School, she was commissioned a lieutenant junior grade and assigned to work at the Bureau of Ordnance Computation Project at Harvard University. And, thus Hopper became, as she called herself, the "third programmer" on the Mark I, the world's first large-scale, automatically-sequenced digital computer. This new computing machine was a joint project between IBM and the Harvard Research Facility. It was officially dedicated in August of 1944 and leased by the US Navy for the remainder of World War II. In addition to rocket trajectories and range tables, Hopper and her colleagues programmed Mark I to calculate the area covered by a mine-sweeping detector towed behind a ship and the top secret mathematical simulation of the shock waves that were expected from the explosion of the first atomic bomb. The first microprocessors and the first subroutines originated at Harvard with the work that Hopper was doing (Billings 1989; Whitelaw 1995; Yale 2019).

The first computer bug (Fig. 17.3) also resulted from their work at Harvard:

> In the summer of 1945 we were building Mark II; we had to build it in an awful rush—it was wartime—out of components we could get our hands on. We were working in a World War I temporary building. It was a hot summer and there was no air-conditioning, so all the windows were open. Mark II stopped, and we were trying to get her going. We finally found the relay that had failed. Inside the relay—and these were large relays—was a moth that had been beaten to death by the relay. We got a pair of tweezers. Very carefully we took the moth out of the relay, put it in the logbook, and put scotch tape over it.
>
> Now, Commander Howard Aiken had a habit of coming into the room and saying "Are you making any numbers?" We had to have an excuse when we weren't making any numbers. From then on if we weren't making any numbers, we told him that we were debugging the computer. To the best of my knowledge that's where it started. I'm delighted to report that the first bug still exists (Tropp 1984).

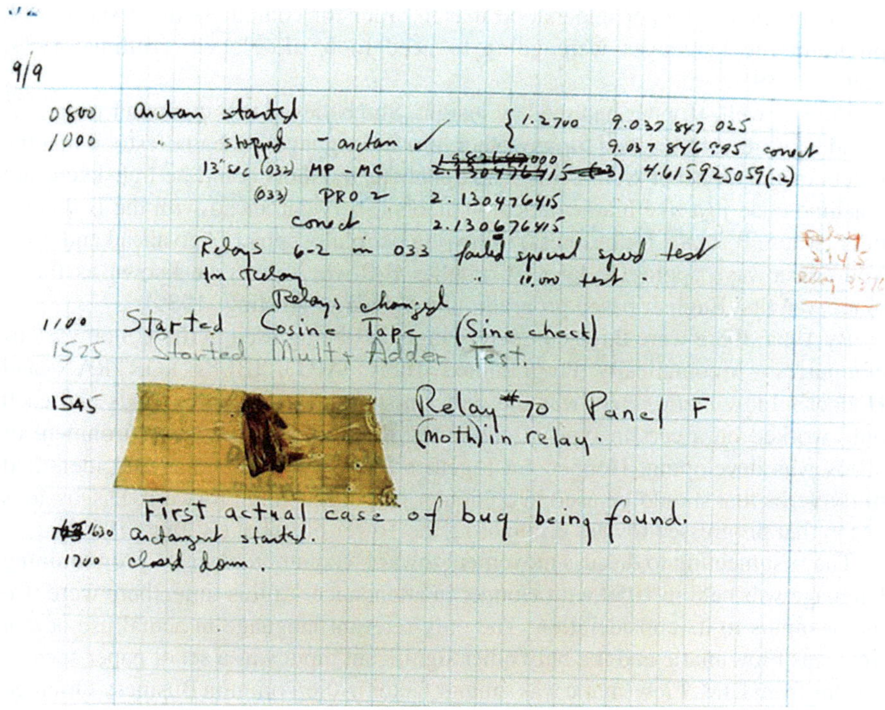

Fig. 17.3 First Computer Bug (Courtesy of Smithsonian Institution)

17.4 After Harvard: The Computer Compiler

Hopper left Harvard in 1949, as there were no permanent positions for women, to join the Eckert-Mauchly Computer Corporation in Philadelphia as a senior mathematician. She worked with many of the early computers. She believed that a computer could be used to write its own programs and that in order for computers to be used widely, they would need to be accessible to a much wider audience than just computer specialists (Billings 1989; Whitelaw 1995; Yale 2019).

The first completed compiler was Hopper's A-0 system developed at Remington Rand, a successor company to the Eckert-Mauchly Corporation, in 1952. Her original paper "The Education of a Computer" in the Proceedings of the Association of Computing Machinery Conference, Pittsburgh, May 1952 is regarded by programming language compiler developers and writers as the primordial presentation on this topic. It was also the first of more than fifty papers that she published on programming languages and computer software (Billings 1989).

A-0 was followed by A-1 and A-2, which were both improved versions. A-2 was completed in 1955 and is said to be the first compiler to be used extensively. Computers were starting to be sold to businesses both in the USA and abroad, and they needed a language for businesses that was easy to use—that did not require mathematicians. Hopper understood that in order for computers to be successful, programming languages were going to need to be developed. (Billings 1989; Tietjen 1989).

Here as well, Hopper had the key insight. She believed that computer programs could be written in English (or any other human language). She understood that the letters of the alphabet were symbols. Alphabetic symbols could be translated into machine code just like mathematical symbols. She went ahead with the B-0 compiler (where B stands for business) following her philosophy of "go ahead and do it. You can always apologize later." Compiler B-0 would later be known as Flow-matic, the first English-based programming language. (Billings 1989).

By 1957, there were three computer languages being used in the US on various computers—Automatically Programmed Tools (APT), IBM's FORTRAN and Hopper's Flow-matic. Hers was the only one using English commands and each only worked on a certain type of computer. Recognizing that an environment of chaos was developing, Hopper put together 500 typical programs and identified thirty verbs that should be used in computer languages going forward. Discussions began that progressed toward eliminating the Tower of Babel (Billings 1989).

The first meeting to discuss movement toward a common business programming language was held in 1959 with Hopper in attendance. At this time, there were two major inputs to its consideration. The only relevant language in actual use at that time was Flow-matic and the only other significant input was a set of paper specifications from IBM. Flow-matic was a major factor in the Common Business-Oriented Language (COBOL) committee's deliberations. As one of the members of the committee stated, "Flow-matic was the only business-oriented programming language in use at the time COBOL development started… Without Flow-matic we probably never would have had a COBOL" (Billings 1989; Hopper 1994; Tietjen 1989).

The committee that was established, with Hopper as one of two technical advisers, set itself very aggressive goals. In 1960, the Government Printing Office issued the COBOL 60 Report. Although IBM didn't officially accept COBOL until 1962, many other companies announced in 1960 that they would be developing COBOL compilers for their computers and users. COBOL then became the standard language used in a wide variety of computer applications. By the 1970s, COBOL was the "most extensively used computer language" in the world (Billings 1989; Hopper 1994; Tietjen 1989; Yale 2019).

17.5 Naval Service

Hopper had been affiliated with the US Naval Reserve since 1946, promoted to lieutenant commander in 1952 and promoted to commander in 1966. However, also in 1966, she received a letter from the Navy requesting that she retire. She complied with the request and on December 31, 1966, which she recalls as "the saddest day of my life," Hopper retired. That retirement lasted for 7 months (Billings 1989; Whitelaw 1995).

Hopper was recalled by the Navy to standardize COBOL for the Navy and to persuade everyone across the Navy to use this higher-level language. She returned to "temporary" active duty on August 1, 1967, taking a leave of absence from Remington Rand. Later, "temporary" was amended to "indefinite." Indefinite lasted for almost 20 years—"The longest six months I've ever spent" (Billings 1989).

In 1973, Hopper was promoted to captain by President Richard M. Nixon. In 1983, she became a Commodore (Fig. 17.4) by a special act of Congress (because she was too old to otherwise be promoted). In 1985, when she was elevated to rear admiral, she became the first female rear admiral in the history of the US Navy (Billings 1989).

In 1986, when she was 79, Admiral Hopper retired again from the US Navy. She asked that the retirement ceremony be held on the deck of the *USS Constitution* in Boston Harbor, and the Navy honored her request. This ship is the oldest commissioned ship in the Navy, and Hopper said, "We belong together. After all, I'm the oldest sailor in the Navy." When she died in 1992, she was buried with full military honors at Arlington National Cemetery (Whitelaw 1995).

Fig. 17.4 Grace Murray Hopper being promoted to Commodore (Courtesy of United States Department of Defense)

17.6 Awards and Recognitions

Hopper received numerous awards and honors in recognition of her efforts during her lifetime, including more than 40 honorary doctorates, as well as some honors awarded posthumously. She said that the Naval Ordnance Development Award in 1946 was particularly meaningful to her as it indicated "You're on the right track. Keep up the good work" (Billings 1989; Yale 2019).

The Society of Women Engineers recognized her accomplishments with its 1964 Achievement Award. In 1969, she received the first computer science "Man of the Year" award from the Data Processing Management Association. In 1973, she was elected to membership in the National Academy of Engineering. Also in that year she received the Legion of Merit from the USA, which was created by Congress in 1942 and is given for the performance of outstanding service. Hopper was also named a Distinguished Fellow of the British Computer Society in 1973, the first woman and first person from the USA to receive that distinction. (Billings 1989).

In 1984, her induction citation into the Engineering and Science Hall of Fame read:

> "In tribute to your superior technical competence and mathematical genius; in tribute to your creative leadership, vision, and commitment as a computer pioneer; in tribute to your setting a foremost example as an author and inventor; in tribute to your dedication to human and humane elements of teaching, learning, and scholarship; and in tribute to your insights and innovations in meeting the challenges of rapidly changing times, we induct you – Commodore Grace Murray Hopper – into the Engineering and Science Hall of Fame" (Billings 1989).

In 1988, she received The Emanuel E. Piore Award from the Institute of Electrical and Electronics Engineers, which is awarded for outstanding achievement in the field of information processing in relation to computer science, deemed to have contributed significantly to the advancement of science and to the betterment of society.

In 1991, she received the National Medal of Technology presented by President H.W. Bush. Hopper was the first individual woman to receive this Medal. The citation for the National Medal of Technology reads, "For her pioneering accomplishments in the development of computer programming languages that simplified computer technology and opened the door to a significantly larger universe of users."

Hopper was inducted posthumously into the National Women's Hall of Fame (1994). In 1996, the *USS Hopper* (DDG 70), an Arleigh Burke-class guided missile destroyer, was launched. It was commissioned in 1997. This destroyer was only the second to be named for a woman who had served in the Navy.

In 2016, President Barack Obama posthumously awarded her the Presidential Medal of Freedom, the nation's highest civilian honor in recognition of her "lifelong leadership in the field of computer science" (Fuller 2019; Yale 2019).

Additional posthumous awards include The Grace Hopper Achievement Award, presented in April 2021 by the Institute of Electrical and Electronics Engineers (IEEE) Philadelphia Section, the IEEE Computer Society and the IEEE Women in

Engineering Affinity Group. The plaque for this award was the first placed on the Philadelphia Women's Walk of Fame, on a sidewalk in Philadelphia. In May 2024, an IEEE Milestone plaque was affixed to the Moore Building of Engineering at the University of Pennsylvania. It was placed in the room next to the Electronic Numerical Integrator and Computer (ENIAC) (McDevitt and McGettigan 2024).

17.7 Legacy

Hopper had a long and distinguished career in the computer field with major contributions to the development and use of software, to the encouragement and leadership of workers in the field, and to promoting the software field. Hopper was one of the most influential persons in the field of computer science in the promotion of the development of programming languages and the expansion of our knowledge and expertise in language translation (Tietjen 1989).

Hopper early recognized the need for better programming languages and techniques, understood also the close interaction between hardware design and software design and used a firm grounding in computer operation and principles of computation to take the innovative technical steps necessary for new language development. Without her pioneering efforts, there would have been little software to run on computers, their basic efficiency of operation would have been severely limited, and their ease of use would have been at a standstill. She contributed significantly to the more effective use of computers through the development of major new software tools: Tools that were essential to the realization of the potential of the emerging technology for computer hardware (Tietjen 1989).

Hopper's role in the emergence of the computer field was an extremely broad one, and one that extended over a period of more than 40 years. From the beginning, Hopper challenged the industry to accept new technological advances and to avoid the "but we've always done it that way" mentality. She confirmed computers could be used for more than just mathematics, and computer users owe her a debt of gratitude for proving computers could be programmed in English, not just through the use of symbols. In addition to the development of essential new technology, she had the vision and persistence to exert overall leadership. She was an effective and ubiquitous spokesperson for the Navy, for the computer field, and for its career opportunities for young people. Jean Sammet says in *Programming Languages* (Prentice Hall, 1969), "In my opinion, Dr. Grace Hopper did as much as any other single person to sell many of these concepts from an administrative and management, as well as a technical point of view" (Tietjen 1989).

Hopper distinguished herself as a Naval officer in the service of her country, as a computer pioneer in the evolution of the high technology industry, and as an educator who inspired generations of young people. She was not only a pioneer in the development of data processing and computer languages she was also a significant pioneer and role model for women in the Navy, and the whole nation. For many decades, she inspired generation after the generation of engineers and naval officers and enlisted to take ideas and turn them into real accomplishments. Grace Hopper not only achieved great things herself but also inspired hundreds of others to great accomplishments of their own (Tietjen 1989).

Former Secretary of the Navy John Lehman sums up her contributions this way: “Military success is in no small measure due to the efficiencies and innovations provided by the pioneering introduction of signal and data processing, led by Grace Hopper. These accomplishments are of heroic proportions (Tietjen 1989).

Hopper’s legacy is also honored through the Grace Hopper Celebration, an annual conference hosted annually by anitab.org. In 2010, Cray launched its new supercomputing system—the Grace Hopper XE6 system. Nvidia offers its Grace Hopper Superchip for giant-scale artificial intelligence and high-performance computing. The Superchip combines Nvidia’s Grace and Hopper architectures (NERSC 2010; Nvidia 2024).

Author’s Note This chapter is abridged from a chapter written by Jill S. Tietjen titled “Amazing Grace—Computer Pioneer Admiral Grace Murray Hopper,” published in Smith, Alice E., Editor, *Women in Computational Intelligence: Key Advances and Perspectives on Emerging Topics*, Springer Nature: Cham, Switzerland, 2022.

References

Billings, Charlene W. 1989. *Grace Hopper: Navy Admiral and Computer Pioneer.* Hillsdale, New Jersey: Enslow Publishers, Inc.

Fuller Rear Admiral John. US Navy Toughness: ‘Amazing Grace and Namesake USS Hopper. Navy Live. https://navylive.dodlive.mil/2017/03/17/u-s-navy-toughness-amazing-grace-and-namesake-uss-hopper/. Accessed September 1, 2019.

Hopper, Grace Murray. 1994. Grace Hopper Celebration of Women in Computing 1994 Conference Proceedings. http://www.cs.yale.edu/homes/tap/Files/hopper-story.html, Accessed August 28, 2019.

McDevitt, Kathleen and Doreen McGettigan. 2024. Grace Hopper Earns IEEE Milestone at Last. *IEEE Women in Engineering Magazine*, December. pp. 38-40.

NERSC. 2010. Grace Hopper Powers Science on NERSC’s New Cray XE6. https://www.nersc.gov/news-publications/nersc-news/nersc-center-news/2010/grace-hopper-powers-science-on-nersc-s-new-cray-xe6/. June 4. Accessed January 7, 2025.

Nvidia: Cloud and Data Center. 2024 Nvidia Grace Hopper GH200 Superchip. https://www.nvidia.com/en-us/data-center/grace-hopper-superchip/. Accessed January 7, 2025.

Tietjen, Jill. 1989 [unpublished]. Nomination and Supporting Letters for the National Medal of Technology, Submitted December 4, 1989.

Tropp, Henry S. 1984. Grace Hopper: The Youthful Teacher of Us All. *Abacus* 2(1) (Fall): 7-18.

The WAVES 75th Birthday. Naval History and Heritage Command. https://www.history.navy.mil/browse-by-topic/wars-conflicts-and-operations/world-war-ii/1942/manning-the-us-navy/waves_75th.html. Accessed August 29, 2019.

Whitelaw, Nancy. 1995. *Grace Hopper: Programming Pioneer.* New York, New York: W. H. Freeman and Company.

Yale University. Biography of Grace Murray Hopper. Office of the President. https://president.yale.edu/biography-grace-murray-hopper. Accessed August 28, 2019.

Chapter 18
Martha J. B. Thomas, PhD (1926–2006) Achievement Award Recipient 1965

In recognition of her significant contributions to the science of chemistry, as an engineer, educator, and administrator, while fulfilling her duties as a wife and mother

Marie Aloia

18.1 Introduction

Dr. Martha Jane Bergin Thomas (Fig. 18.1) spent her four-decade career at GTE Sylvania in Salem, Massachusetts, as a scientist and chemical engineer, earning her masters and doctorate degrees while working full time. While leading her research group at GTE Sylvania, Thomas made significant contributions to electric lighting, chemistry of phosphors, and the development of lightbulbs, setting up and running two pilot plants for the manufacture of new types of phosphors. Her work in phosphor chemistry to color balance fluorescent lighting to a more natural white light set an industry standard and made these lights much more commercially viable. It also opened the door for other "spin off" technologies including plant lights, black lights, light boxes to treat seasonal affective disorder and even to the phosphor technology needed for the first commercially viable cathode-ray tube color television sets. Thomas is the sole inventor or co-inventor on 22 US patents and 13 foreign patents for improvements to phosphors and electric lighting, and published several dozen papers in technical journals. She was also happily married to her husband Dr. George Thomas for forty years, and the mother of four daughters.

M. Aloia (✉)
Morris Plains, NJ, USA

C. D. Craig et al. (eds.), *Women Engineering Legends 1952-1976: Society of Women Engineers Achievement Award Recipients*, https://doi.org/10.1007/978-3-032-00224-2_18

Fig. 18.1 Martha JB Thomas, portrait. (Photograph courtesy of Society of Women Engineers Records, Walter P. Reuther Library, Wayne State University)

18.2 Early Life and Education

Martha Jane Bergin Thomas approached her life and career with a simplicity and modest matter-of-fact attitude that made her remarkableness her normal. Born Martha Jane Bergin to John and Augusta Bergin, she entered Radcliffe College with the class of 1946, but an accelerated program allowed her to graduate in 1945. Thomas graduated cum laude, at only 19, with an AB degree in chemistry. Her parents, both teachers, advised her to become a teacher like they were. Instead, she applied to and was accepted at several medical schools because she was thinking that she would pursue a career in medicine. For whatever reason, Thomas decided to go to work instead and she accepted a job at GTE Sylvania Inc. She joked that she accepted the position because they were looking for someone with a scientific background for their patent office, but "No one seemed to know what to do with me so I ended up in the laboratory as a chemist" (Fowler 1965; SWE Records n.d.).

At GTE Sylvania Thomas found her calling as a chemist and engineer. With World War II winding down and men coming back to take industry jobs, many women were leaving, but she stayed. In some of her job interviews she was actually told that she should be ready to give up her job for a returning veteran, but GTE Sylvania did not require that of her (King 1972; SWE Records n.d.).

18.3 Career

At GTE Sylvania, her starting position was junior technician because that "was the level they would take a woman scientist on in 1945". Within a year she was a junior engineer (Andrews 1991). Taking advantage of GTE Sylvania's tuition reimbursement program, Thomas attended Boston University at night while working full time and earned her masters in 1950 and her PhD in 1952, both in chemistry. While she was earning her degrees, she was promoted to senior engineer and she held a National Research Fellowship at Boston University. Thomas didn't leave Boston University, however, when she finished her PhD. She taught chemistry in the evening division until 1970, as "a wonderful way to keep scientifically up to date". Later, she taught chemistry as an adjunct professor at the University of Rhode Island from 1974–1978 (King 1972; SWE Records n.d.).

In her early years at GTE Sylvania Thomas (Fig. 18.2) was actively working on research that led to her first patent, US2566615, filed just after her 21st birthday. It described a chemical process for etching tungsten filament wire to a finer diameter than was mechanically possible without compromising the integrity of the wire. This patent facilitated the manufacture of incandescent lamp filaments.

After 1952, Thomas (Fig. 18.3) began a very productive period of research into phosphors and electric lamp construction. She started with fluorescent lights, which work when an electrical current passes through a mercury vapor, excites the atoms, and produces ultraviolet light at the core of the lamp. A phosphate coating on the

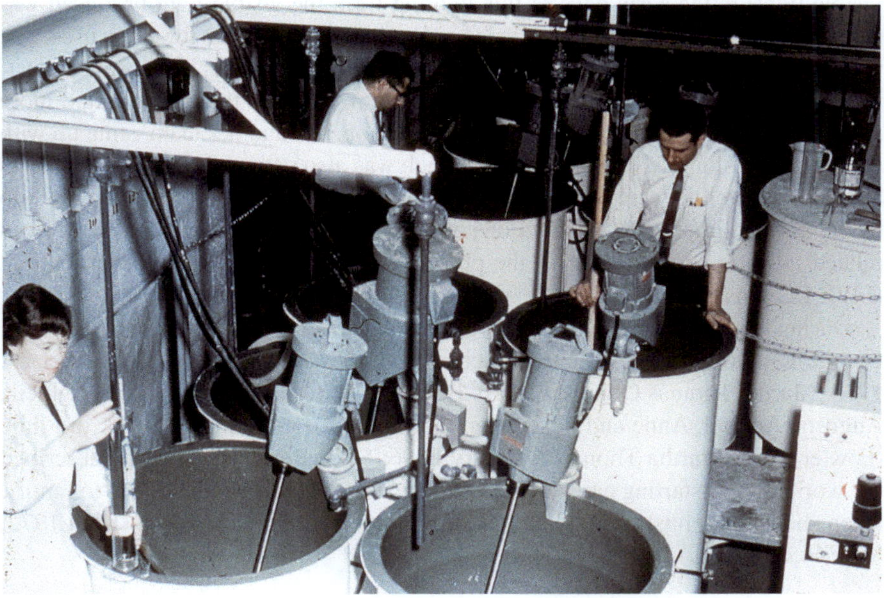

Fig. 18.2 Martha Thomas at work at the Sylvania Lighting Products Division. (Photograph courtesy of Society of Women Engineers Records, Walter P. Reuther Library, Wayne State University)

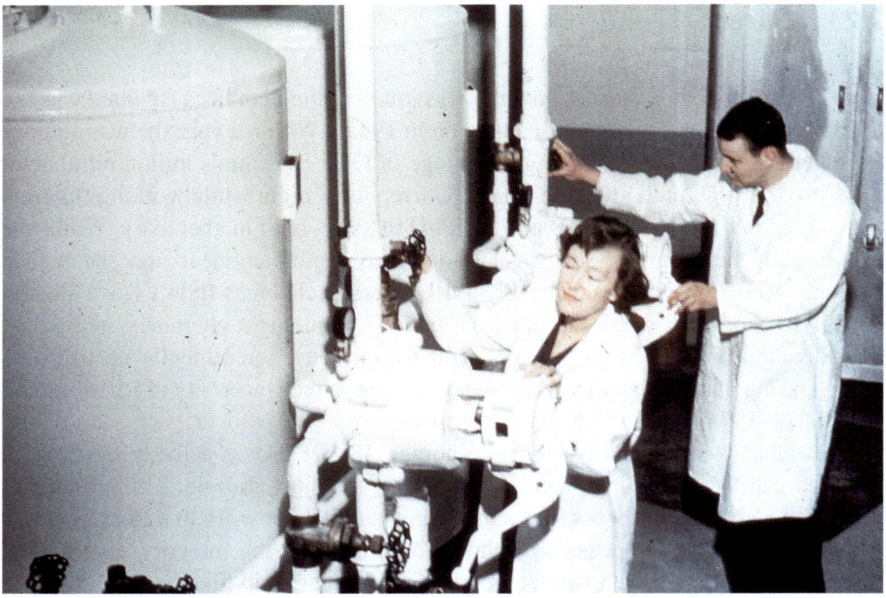

Fig. 18.3 Martha Thomas working on phosphor development at the Sylvania Lighting Products Division plant. (Photograph courtesy of Society of Women Engineers Records, Walter P. Reuther Library, Wayne State University)

inner surface of the tube surrounding the core absorbs the ultraviolet light and emits visible light. In the early days of this lighting form the visible light produced was not very efficient and poorly color balanced making it unsuitable for many uses. She provided novel solutions that led to improvements to the blends of phosphate coating compounds, and the processes to create them, to make the light emitted more natural by correcting the color balance. In addition, she and her team came up with new methods of applying the coatings, which was also patented. Additional patents along the same line were issued for phosphor blends, methods of applying them to the lamps, and improved constructions of those lamps. One patent that she received related to optimizing the size of the phosphor particles for best light distribution. (SWE Records n.d.)

During her time in graduate school at Boston University she met her husband Dr. George R. Thomas, a government scientist, and they were married on October 29, 1955 (Martha Thomas Obituary 2006). Between 1956 and 1963 her four daughters, Augusta, Abigail, Anne and Susan were born, and she continued to work full time (Fowler 1965; Martha Thomas Obituary 2006; SWE Records n.d.). She described her work days as starting early and ending late, but the weekend belonged to family, when she and her husband looked forward to time to unwind together. (King 1972; SWE Records n.d.).

The first commercially viable color televisions in the 1950s were cathode-ray tubes using screens with phosphor coatings to produce the colors. Sylvania unveiled their first one in 1964, at a time when all of Martha's patents on phosphor manufacture were still active (Sylvania Develops New Color-Tv Tube 1956). Her patent US3668140, a Process for Yttrium Orthovanadate Phosphors, specifically deals with the handling of some phosphors being used in these first television screens (SWE Records n.d.).

In 1958 Thomas was promoted from Senior Engineer to Group Leader and Engineering Manager, which she held till 1972. During this time she led her research group and supervised the startup of two pilot plants for the manufacture of phosphors (SWE Records n.d.). Some of her further patents, issued during the 1960s and 1970s show entire industrial schemes for creating phosphor blends that were likely applied in the pilot plants she started up.

In 1972 Thomas (Fig. 18.4) was promoted from Engineering Manager to Manager of the Technical Assistance Laboratory, taking her away from actual hands-on research. As a manager she felt the need to learn about management so she returned to night school earning an MBA at Northeastern University in 1981 (Andrews 1991; SWE Records n.d.). She completed her career with one more promotion from Manager to Director of the Technical Assistance Laboratory, a position she held until her retirement in 1990.

Fig. 18.4 Martha JB Thomas, portrait. (Photograph courtesy of Society of Women Engineers Records, Walter P. Reuther Library, Wayne State University)

18.4 Awards and Recognitions

Thomas received a number of awards in addition to the 1965 SWE Achievement Award "In recognition of her significant contributions to the science of chemistry, as an engineer, educator, and administrator, while fulfilling her duties as a wife and mother." In 1966, she received the Golden Plate Award as "Woman Engineer of the Year" in the category of Science and Exploration. Thomas was named a Fellow of the American Institute of Chemists in 1969. (SWE Records n.d.).

Elected to Sigma Xi, the science honor society, Thomas also received two Distinguished Alumna Awards—from her high school (1978) and from Boston University (1983) (SWE Records n.d.). In 1991 she received the New England Inventor Award from the Boston Museum of Science, for "an individual whose application of science and technology, creativity, and independent thought has positively impacted society" (King 1972).

She was listed in three compilations of "Who's Who": "Who's Who of American Women", "Who's Who in Engineering" and "The World Who's Who of Women". Thomas is also listed in the *American Men and Women of Science*, making the list before 1971 when it was still called *American Men of Science*. (SWE Records n.d.).

18.5 Organizations and Community

During her long career Martha was active in the scientific and engineering professional community. She joined her local section of SWE, the Boston Section, in 1951 in the society's founding days. She held a number of offices and served on committees, including Boston Section Chairman, National Recording Secretary, National Correspondence Secretary, and Board of Trustees. In 1968–1969 she served SWE on the Scholarship Awards Committee (SWE Records n.d.).

She was active in the local section of the American Chemical Society, holding offices of Treasurer, Auditor, and Section Director. She published some of her papers as a member of the Electrochemical Society (SWE Records n.d.). She also served as a Director of the American Institute of Chemists.

18.6 Legacy

Thomas made it clear throughout her life that work-life balance mattered. When asked about her ability to manage her family and job she responded with a great simplicity that many other women are doing this. She further explained that "While the rewards are tremendously satisfying…there are several prerequisites for success". For her these were the complete support of her husband and children, quality care for her children, and the support of her employer. These were things that men could always presume to have, but not women, making the point that if more women did have these "prerequisites" more could succeed as she did. She used her moment in the spotlight to speak out for women. (SWE Records n.d.).

The work that Thomas did for fluorescent light color balancing had an impact not just on the industry that makes the light bulbs, but in phosphor chemistry in other applications as well. Her research also added to the understanding of how humans and nature are affected by light.

Her research in the chemistry of phosphors not only found the phosphor compounds that emitted light in the "missing colors" in early fluorescent light to make it more naturally white, but proposed ways to economically manufacture them, blend them, and apply them to the lamps (SWE Records n.d.). Color balancing made fluorescent lamps much more commercially acceptable, and set an industry standard for them. Tuning the color balance also diversified the applications of fluorescent lighting to products like "black lights" and plant lights like GTE's Gro-Lux light, which is still available today. (SWE Records n.d.). Several newer technologies came to life built on the knowledge base of phosphors that she established. Today, the active ingredient in most "glow in the dark" products is a phosphor.

References

Andrews, Rebecca. 1991. Boston Museum Honors GTE Chemist With 1991 New England Inventor Award. The Scientist, Vol:5, #6, pg. 22. March 18.

Fowler, Elizabeth M. 1965. Radcliffe Girl's Success Formula; A Chemist Since '45, She Has 7 Patents and 4 Children. *New York Times*, Section Business Financial, p 43. August 4.

King, Mary Sarah. 1972. Women in Business, 2 engineers, 2 approaches. *The Boston Globe*, Section Financial, p 64. July 20.

Martha Thomas Obituary. 2006. Dr. Martha Jane Bergin Thomas, *The Boston Globe*. https://www.legacy.com/us/obituaries/bostonglobe/name/martha-thomas-obituary?id=25840450. October 18.

Society of Women Engineers (SWE) Records (LR001539). n.d. *Thomas, Martha J.B.* Box 193 Folder 12, Walter P. Reuther Library. Archives of Labor and Urban Affairs. Wayne State University.

Sylvania Develops New Color-Tv Tube. 1964. *The New York Times*. June 5.

Chapter 19
Dorothy Martin Simon, PhD (1919–2016) Achievement Award Recipient 1966

In recognition of her significant contributions to engineering administration and space engineering, especially combustion and ablative coatings

Linda A. G. Reed

19.1 Introduction

Dr. Dorothy Martin Simon's hands-on contributions to space engineering, especially with aerospace combustion and the development of polymers, were critical to the success of the Apollo space program. She made important contributions while at the National Advisory Committee for Aeronautics (NACA) regarding heat shield construction and improvement of rocket engine designs. Her work with the kinetics of polymerization on vinyl compounds, especially the vinyl cyanide polymer, was instrumental to the mass production of Orlon and Dacron at DuPont (NASA Oral Histories Database 2012).

Dr. Simon (Fig. 19.1) rose to the level of Vice President at Avco Corporation. The company noted her "international reputation for her research contributions, particularly in the field of combustion and in addition, in radiochemistry, vinyl polymerization, and aerothermochemistry" (SWE Records n.d.).

19.2 Early Life and Education

Dorothy Martin was born in 1919, in Harwood, Missouri, to Laudell and Robert W. Martin (United States Federal Census 1920). Her parents valued education and achieving big things. Dr. Simon said of herself, "I was born into science. I knew the

L. A. G. Reed (✉)
San Miguel, CA, USA

C. D. Craig et al. (eds.), *Women Engineering Legends 1952-1976: Society of Women Engineers Achievement Award Recipients*,
https://doi.org/10.1007/978-3-032-00224-2_19

Fig. 19.1 Dr. Simon rolls up the sleeves of her lab coat circa 1960. (Photograph courtesy of Society of Women Engineers Records, Walter P. Reuther Library, Wayne State University)

word chemistry before I learned to talk" (Scientific American 1964). Dr. Simon graduated from Greenwood Grade and High School as class Valedictorian. There, she also won the highest sports honor, the Natalie Briggs Medal.

Strongly influenced by her father, a chemist and college professor, she earned her AB degree from Southwest Missouri State University, where her father was head of the science department. With a major in chemistry and minors in mathematics and French, she was the first graduate from that institution to achieve a perfect score (4.0 GPA) (SWE Records n.d.).

In 1940, she attended the University of Chicago for 1 year before transferring to the University of Illinois, where she worked as a Teaching Assistant in the Chemistry department. Her dedication to the study of chemistry earned her the Strong Fellowship in 1944. In 1945, she completed her PhD in physical chemistry with thesis work regarding radon. Her thesis research on active deposits from radon and thoron gas was the first work on fallout. In her testing, she found evidence of the formation of aggregates of isotopes of lead and bismuth, the products of radioactive disintegration of the gases (NASA Oral Histories Database 2012).

19.3 Career

Following graduation in 1945, Dr. Simon joined DuPont de Nemours & Company, Rayon Division, Pioneer Research Lab, in Buffalo, New York, as an associate Chemist. Her work focused on the development of new catalysts to change the properties of synthetic polymers. This work led to the development of two polymers, which would later be named Dacron and Orlon (NASA Oral Histories Database 2012).

While in Buffalo, she met and married fellow scientist Dr. Sidney L. Simon, a fellow graduate student from the University of Chicago, and took his surname. As was common at the time, she subordinated her career to his and followed him wherever his work took him. This did not thwart her ability to excel wherever she landed, accruing a work history that led to higher responsibility and challenges (SWE Records n.d.).

Sidney Simon was on loan from the National Advisory Committee for Aeronautics (NACA now NASA) Lewis Laboratory to the Atomic Energy Commission (AEC). In 1946, Dr. Dorothy and Dr. Sidney moved to Chicago, where both worked for the AEC Argonne Laboratory. Dr. Dorothy Simon was an associate chemist. There, she made discoveries in the field of radiochemistry while studying the triple fission of uranium. Dr. Dorothy Simon also became the first to discover a new isotope of calcium (SWE Records n.d.).

In 1949, both returned to NACA Lewis Laboratory where in her role as Assistant Chemistry Branch Chief, Dr. Simon began her research in the field of combustion at the Lewis Research Center Fuels and Combustion Division. She advanced the understanding of flames by relating fundamental flame properties to each other through heat and mass transfer principles with chemical reactions (SWE Records n.d.).

She conducted research regarding flame velocities and types of fuel for aerospace applications, including studies to determine the minimum diameter of tubing necessary to maintain an engine flame. This led to Dr. Simon being credited in numerous technical publications and research papers and led to improvements in engine designs for aeronautics and spaceflight purposes (SWE Records n.d.).

As a result of her work at NACA, Dr. Simon was given the Rockefeller Federal Service Award, which entailed a $10,000 (~$110,000 in 2022) grant from the Rockefeller Foundation. She used this award to travel to Europe to further her education at the University of Cambridge in Cambridge, England. She studied with Ronald G. W. Norrish, who later won the Nobel Prize in Chemistry (Proffitt 1999). Dr. Simon toured around Europe, where she met with scientists from Germany, England, France, and Italy to examine their research regarding physical chemistry and radiochemistry. Upon returning from her tour, she was promoted to assistant chief of NACA's Chemistry Branch (SWE Records n.d.).

Dr. Simon left Lewis in 1955 and briefly worked for the Magnolia Petroleum Company in Dallas, Texas. Her role there was Group Leader of the Fundamental Combustion Branch. Her work focused on oil sands and ways to profitably remove the oil, separating it from the sand underground (NASA Oral Histories Database 2012).

In 1956, Dr. Simon joined Avco Corporation, an aviation conglomerate in New England. There, she applied her previous experience with polymer construction and physical chemistry to develop the technology for early aerospace heat shields, which were critical to NASA's crewed spacecraft endeavors. This technology would go on to be used most notably during NASA's Apollo program and later was applied to ballistic missiles at the height of the Cold War.

Her success led to her appointment as Director of Corporate Research (Fig. 19.2) in 1962, where she directed advanced research in six Avco divisions: Aerospace Structures, Lycoming; Electronics; Ordinance; Avco-Tulsa; and Research and Advanced Development (NASA Oral Histories Database 2012).

Fig. 19.2 Dr. Simon indicates an advanced propulsion system during a presentation, circa 1966. (Photograph courtesy of Society of Women Engineers Records, Walter P. Reuther Library, Wayne State University)

In 1964, she was promoted to Vice President of these six groups, becoming the company's first female corporate officer, and in 1965, she added Director of Economic Systems Corporation to her portfolio. This subsidiary operated the largest girls' Job Corps Training Center at Poland Springs, Maine (SWE Records n.d.). The program ran until 1969. Hundreds of women from disadvantaged areas of the country received vocational training (Poland Springs Timeline n.d.).

In 1983, Dr. Dorothy Simon testified before the Senate Subcommittee on Science, Technology, and Space, stating that fire-related costs of building construction in the United States could be reduced by 40% through the use of modern fire technology (SWE Records n.d.).

Her management skills were applauded by the American Management Association, and they invited her to speak at several seminars on research management and technical planning. From time to time, various other organizations requested her help in resolving problems arising from government-industry research and development interfaces (SWE Records n.d.).

19.4 Consulting and Civic Involvement

Upon retiring from Avco Corporation, Dr. Simon operated her consulting firm, Simon Associated, from her home in Chapel Hill, North Carolina. She served on numerous committees and advisory boards for government agencies, universities, and corporations. The most notable positions include the Department of Commerce Committee for the National Bureau of Standards and the National Medal of Science Committee during the Carter administration. Dr. Simon served as the first female trustee at Worcester Polytechnic Institute, where she set up the Dorothy M. Simon Endowed Fund for Fire Safety Studies (WPI 2016).

She also served on the National Research Council's National Materials Advisory Board, the Department of Defense's Policy Advisory Committee, and the Department of Commerce's Statutory Committee for the National Bureau of Standards (Proffitt 1999).

19.5 Awards and Recognition

Dr. Dorothy Martin Simon received numerous awards and recognition throughout her career. She was named an "Outstanding Alumnus" by her alma mater, the SMSU Alumni Association, in 1957. Businessweek listed her as one of the top 100 corporate women in 1976, and she was listed in *American Men of Science, Who's Who in the East*, and *Who's Who of American Women* (SWE Records n.d.).

Dr. Simon (Fig. 19.3) was awarded the Society of Women Engineers Achievement Award in 1966 "In recognition of her significant contributions to space engineering, especially in the fields of combustion and ablative coatings" (SWE Records n.d.).

In 1971, Dr. Simon received an Honorary Doctor of Science degree from Worcester Polytechnic Institute (WPI). The citation read, "Perhaps the most important woman executive in American industry today…" specifically calling out that she oversaw the wide-ranging research interests of Avco Corporation as Vice President and Director of Corporate Research (WPI 2016).

She served on the boards of directors for five companies and as trustee for two universities. She was made a fellow of the American Institute of Aeronautics and Astronautics and the American Institute of Chemists (WPI 2016).

19.6 Legacy

Dr. Simon considered the most important contribution she made was to the integrity of the heat shields used on spacecraft. As she explained in a NASA Oral History, "... it was a very, very vital point. You would not have had a space program if they couldn't come home. Even if they had been to Mars, they couldn't get back. So it was very successful. I got certain congratulations from people, but I wasn't included in the report. At first I was worried about it, but then I thought I've done something important" (NASA Oral Histories Database 2012).

In addition to her scientific contributions, Dr. Simon made her mark as a respected speaker on the role of management in technological enterprises and as an advocate of technical training and professional advancement for women (SWE Records n.d.). She wrote on women's struggle for executive authority:

Fig. 19.3 Formal portrait of Dr. Dorothy Martin Simon circa 1966. (Photograph courtesy of Society of Women Engineers Records, Walter P. Reuther Library, Wayne State University)

> Men tend to lead by asserting authority, but women manage through the participative mode, by helping people develop their capabilities and solve problems. Now, men are beginning to do the same. In science and engineering you have to manage participatively because your staff knows more than you do. (Scientific American 1964).

As a social pioneer, she was one of the leaders of women in science and engineering. Dr. Simon saw the work of scientists and engineers as critical to humankind's evolution on planet Earth.

> …in our time, we are attempting to perpetrate a discontinuity in man's evolution, to cause a change for man which is as great as that which nature accomplished when she caused a land animal to go into the sea and become a whale… (Rudd 1963).

Throughout the 1950s, she served as an advocate for women in science and education, appearing in several radio shows and magazines, including *The Saturday Evening Post*, *Missouri State University Magazine*, and the Junior Engineering Technical Society—*Jets Journal*.

Among her many firsts, Dr. Dorothy Martin Simon played a critical role in the development of synthetic polymers, leading to Dacron and Orlon, and was the first to discover a new isotope of calcium. She became Avco's first female corporate officer and was the first woman trustee at Worcester Polytechnic Institute (WPI 2016).

References

NASA Oral Histories Database. 2012. National Aerospace and Space Administration. http://www.historycollection.jsc.nasa.gov. Accessed March 10, 2025.

Proffitt, Pamela, editor. 1999. *Notable Women Scientists*. Farmington Hills, Michigan: Gale Group.

Poland Springs Timeline. n.d. https://polandspringps.org. Accessed January 30, 2025.

Rudd, Dorothy, editor. 1963. *Jets Journal* Volume XI, November.

Scientific American. 1964. The womanly art of directing research on turbine engines.

Society of Women Engineers (SWE) Records (LR001539). n.d. *Simon, Dorothy Martin*. Box 192 Folder 39, Walter P. Reuther Library. Archives of Labor and Urban Affairs. Wayne State University. Accessed 09 March 2025.

United States Federal Census. 1920. https://*www.ancestry.co.uk*. Accessed March 10, 2025.

Worcester Polytechnic Institute (WPI). 2016. In Memoriam: Dorothy Martin Simon, Trustee Emerita. 26 July 2016. Accessed February 2, 2025.

Part IV
1967–1971

Chapter 20
Setting the Stage: 1967–1971
The Turmoil of 1968 and Landing a Human on the Moon

Holly J. Teig and Janet L. Williams

20.1 Introduction

During the late 1960s, the USA was in upheaval. The Vietnam War was ongoing, sparking protests on college campuses across the nation. The struggle of the civil rights movement boiled over in April 1968 when an assassin killed Dr. Martin Luther King, Jr. in Memphis, Tennessee. Senator Robert F. Kennedy was assassinated two months later. Women engineers were joining the workforce in increasing numbers but were still struggling to gain recognition and respect for their work. They benefited indirectly from the gains of the civil rights movement, and through legislation on affirmative action and equal pay.

Engineers invented the foundational technologies of the modern computer age during this period. Two of the most impactful were the development of the microprocessor chip and dynamic random access memory (DRAM) technology in semiconductors, enabling miniaturization of the core components of today's computers, mobile phones, and consumer electronics. Networking of computers in different geographical locations and electronic communications emerged, ushering in a newly connected landscape. For the next decades, engineers enhanced and tuned these breakthroughs to enable ever more powerful computing solutions.

The US space program experienced the high emotion of putting a human being on the surface of the moon with Apollo 11 along with the low point of an averted space flight disaster with Apollo 13. Both events increased engineers' and scientists' understanding of the physics of space travel, vital information for the booming space industry.

H. J. Teig (✉)
Peoria, IL, USA

J. L. Williams
Albuquerque, NM, USA

C. D. Craig et al. (eds.), *Women Engineering Legends 1952-1976: Society of Women Engineers Achievement Award Recipients*,
https://doi.org/10.1007/978-3-032-00224-2_20

In 1968, Robert Dennard changed the world with the patent of the one-transistor memory cell. It was the basis of the DRAM (dynamic random-access memory) and other types of modern memory devices used in servers, PCs, and mobile devices.

Marguerite M. Rogers, PhD received the Achievement Award for her outstanding contributions to the field of air-delivered tactical weapons, resulting in changing US Navy emphasis on nuclear weapons to improved conventional weapons.

Isabella L. Karle, PhD received the Achievement Award in recognition of her significant contributions to the development of unique procedures for crystal structure analysis, using electron and X-ray diffraction to study the structure of molecules.

1967

1968

President Johnson signed Executive Order 11375 extending Executive Order 11246 to include gender as a protected category. This now required affirmative action on behalf of women as well as minorities so that hiring was in line with gender and racial proportions in the relevant labor pools.

In 1969, after an enormous effort, the Apollo11 mission took three astronauts to moon orbit. Neil Armstrong and Buzz Aldrin rode the lunar module to the surface and set foot on the moon.

Alice Stoll received the Achievement Award in recognition of her significant contributions in the development of fire-resistant fibers and fabrics, based on her pioneering studies of heat transfer by flame contact.

With a single stretch, scientist Bob Gore discovered a revolutionary new material, expanded polytetrafluoroethylene (ePTF). This 1969 discovery became the basis for thousands of life-improving innovations from medical devices and protective fabrics to air-cleaning filters and cables used to explore the universe.

1969

In the late 1960s/ early 1970s:

- Long-time barriers for women in education began to fall. Princeton and Yale began admitting women in1969, with Brown University following in1971, and Dartmouth in 1972.

Erna Schneider Hoover worked on the Bell Labs development team that built the first large-scale computerized electronic telephone switching system in the mid-1960s. Subsequent to the deployment of the 1ESS switch, she devised and patented key software improvements to the system that avoided freeze-up when the switch was inundated with calls, resulting in dramatic improvements in phone service. The patent was approved in 1971.

After developing a network of Internet Message Processors (IMP), Ray Tomlinson sent the first email across computers in 1971. He is also credited with the use of @ to direct emails to the intended recipient.

Irmgard Flügge-Lotz received the Achievement Award for her significant contributions to the field of fluid mechanics, in particular, wing theory and boundary layer theory.

1970

- Women represented 2% of the members of the 91st Congress, with one senator and 10 representatives — less than the prior Congress.
- The participation of women in the workplace was 43%, up from prior decades.
- 9.3% of doctors were women while 3% of engineers and attorneys were women.

IBM released the first read-only, 8-inch flexible diskette drive system (floppy disk) for sale in 1971. It could hold 80 kilobytes of data. This spurred rapid development of read-write capabilities and smaller disk formats.

Alva T. Matthews Solomon, PhD received the Achievement Award in recognition of her significant contributions to the field of engineering mechanics and applied mathematics in the areas of shock analysis, elasticity, and structural design.

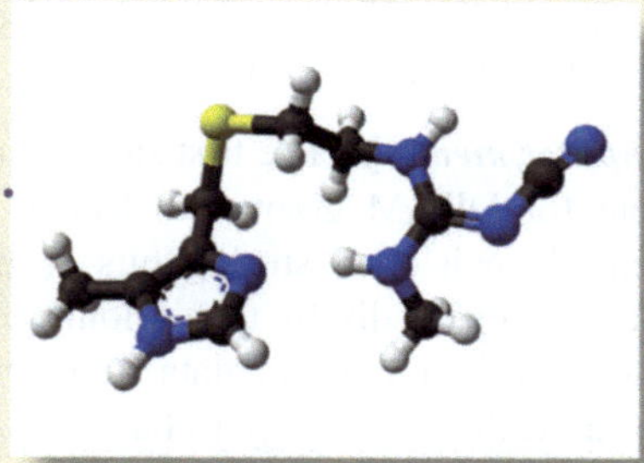

The drug Cimetidine was developed in 1971. Later released as Tagamet®, it was the first targeted medication for stomach ulcers, replacing bland diets, antacids, and surgery as primary treatments.

1971

In the early 1970s, educated women had more career opportunities but were still considered secondary wage earners.

20.2 State of Technology

20.2.1 Computer Hardware and Software

Microprocessors The first commercially available microprocessor (4-bit) was the Intel 4004 chip released in 1971. Containing the equivalent of 2300 transistors in a single integrated circuit, it featured data processing logic and control on a single chip using cutting-edge silicon-gate technology (Peddie 2021). Jon Peddie, of Jon Peddie Research, eloquently summarized the magnitude of this innovation in 2011:

> No one knew at the time the Intel 4004 was developed what it could or would evolve into. Nothing less than brains of everything we use, from our cars to our TVs, phones, dishwashers, and the airplanes we travel in. The microprocessor and subsequent technological revolution was the biggest change to ever happen and far surpassing the industrial revolution in terms of wealth creation, quality of life, communications, and understanding. Jon Peddie, October 2011 (Intel 2011).

Computer memory The first step in miniaturizing computer memory came from a patent for DRAM granted to Robert Dennard and IBM in 1968 (IBM 2025c). Dennard envisioned storing bits as positive or negative charges on capacitors, refreshed repeatedly by tiny amounts of energy. The charging process enabled the capacitor to maintain the data, and it made the memory system *dynamic*. In 1970, Intel used the technology to build a 1-kilobit DRAM chip using a three-transistor cell design. The simplicity, low cost, and low power consumption of DRAM made earlier magnetic technologies immediately obsolete (IBM 2025a). DRAM exponentially increased the capacity of digital information storage, sparking a revolution in computers, consumer electronics, and mobile phones.

Birth of the internet The Advanced Research Projects Agency's (ARPA) *ARPANET* created in 1969 was the first wide area packet switching network. It grew into the Internet we all use today. Bolt, Beranek & Newman (BBN) delivered an interface message processor to create the first network node in the Network Measurements Center at University of California, Los Angeles. Shortly thereafter, a second node was set up at the Stanford Research Institute and connected over a phone line provided by the AT&T telephone company. By the end of 1972, there were 24 sites on the ARPANET, including the Department of Defense, the National Science Foundation, NASA, and the Federal Reserve Board (Hauben 2025).

Birth of email In 1971, Ray Tomlinson at BBN sent the first network email between host computers, using an @ sign separating the name from address for the first time. Widely considered the first electronic mail between two different computers, the use of email led to a huge shift in business and personal communications (Tomlinson 2001).

Personal computer In 1971, John Blankenbaker commercially released the *Kenbak-1*, acknowledged as the first personal computer. With a target audience of teachers and educators, it came fully assembled at the high cost of $750, which limited sales. Built before the advent of microprocessors, the logic engine incorporated integrated circuits mounted on a single printed circuit board (Blankenbaker 2010). Switches keyed the input and lights displayed the output. The rapid pace of microprocessor development resulted in the *Kenbak-1* being a short-term invention, replaced by the cheaper and higher-functioning *Altair 8800* in 1975.

Programming languages Software was a burgeoning by-product of the computer industry due to the variety of computing hardware systems emerging. Software classes appeared in college curriculums by the mid-1960s. Niklaus Wirth released the first Pascal compiler in 1967 (Hemmendinger 2024). In 1969, Ken Thompson, Dennis Ritchie and a group of other developers at Bell Labs created UNIX, an operating system that made large-scale networking of diverse computing systems possible. UNIX programmers leveraged a small number of included programs to conduct network monitoring and housekeeping tasks (Ford 2014). UNIX runs today's web servers and is contained within billions of phones. In 1983, the Association for Computing Machinery honored Thompson and Ritchie for their innovation with the Turing Award, the most significant award in computing (Ford 2014). In 1970, the software FLOW, developed by Jef Raskin, mechanized the coding language entry by auto-correcting syntax errors while the user typed (Raskin 1974).

Word processing and publishing Len Deighton wrote the first novel with a word processor on an IBM Magnetic Tape/Selectric Typewriter in 1970. This machine included a magnetic tape drive and playback controls (Kirschenbaum 2013). A year later, Michael S. Hart produced and posted the first e-book, a copy of the United States Declaration of Independence, on the University of Illinois' mainframe computer, which was also the origin of Project Gutenberg (Hart 1992). Hart was given a virtually unlimited computer time budget and no specific project to work on, so he decided his own scope of use. Hart announced, "the greatest value created by computers would not be computing, but would be the storage, retrieval, and searching of what was stored in our libraries" (Hart 1992). His prediction was way ahead of its time, as the first internet search engines would not come along until 1990.

20.2.2 Computer Accessories

Data Storage In 1967, David L. Noble and engineers at IBM developed portable magnetic data storage using a thin, flexible disk that was eight inches in diameter. Named *floppy disks*, they became widely available in 1971. Quickly adopted by others in the fledgling computer industry, floppy disks were used to transfer data, store data, and run programs on a multitude of hardware (IBM 2025b).

Computer mouse Douglas Engelbart and William English were trying to augment human interaction with information displays when they stumbled upon the idea for the computer mouse. After testing out light pens, joysticks, and other mechanisms, they determined that the most efficient device was their prototype mouse, in which the human hand did not have to raise up in order to engage with the information display. The original design patented in 1967 used two discs to indicate an X-Y position and had one button. William English devised the ball mouse in 1972, allowing the freedom of motion in all directions users today still appreciate (Abby 2024).

Dot Matrix Printers By 1970, both Digital Equipment Corporation and Centronic introduced the first dot matrix printers (Webster 2000). A microprocessor in the printer operated a pin head to form characters out of small ink dots, creating words and numbers on paper (Britannica 2025b). Known for their speed, but not necessarily their print quality, dot matrix printers became the economical choice for early personal computer users lasting well into the 1980s (Lewis 1985).

20.2.3 In Space

Satellite navigation By 1968, an array of 36 satellites named *Transit* provided navigation to the US military and commercial entities, including the Navy's missile submarines. *Transit's* geolocation technology delivered accuracy to a few meters, improving by two times the accuracy of the Earth's land maps (The Aerospace Corporation 2025). *Transit* operated for 28 years until it was replaced by the Global Positioning System (GPS).

Apollo space missions In 1969, after an enormous engineering effort by NASA, the Apollo 11 mission took three astronauts into lunar orbit. Neil Armstrong and Buzz Aldrin rode the lunar module to the surface and set foot on the moon. Completing this daunting mission allowed the United States to reclaim bragging rights of military, economic, and technological superiority over the Soviet Union (Loff 2015). Barbara Crawford Johnson (Achievement Award 1974) worked with NASA in support of the manned spaceflight programs, providing engineering for lunar landing, Skylab, and Apollo-Soyuz programs (ETHW n.d.). The Apollo 13 space mission launched in 1970, but had to be aborted two days later because of an on-board explosion affecting life support systems. A high stakes engineer-on-the-fly exercise took place over four days as NASA engineers found a way for the stranded three-man crew to stay alive and safely return to Earth. The lessons learned from the successful failure of this mission continue to guide safety for space travel today (Mohon 2020).

20.2.4 Banking

Automated Teller Machines (ATMs) A new machine invented in 1967 by John Shepherd-Barron allowed customers to retrieve cash from their Barclays Bank accounts after bank hours in London (Bátiz-Lazo 2015). Companies swiftly announced other ATM versions in the United Kingdom, Sweden, and Japan. In the United States, a patent was granted in 1973 to Donald Wetzel, Tom Barnes, and George Chastain for the Docutel automated teller machine (Gibson 2003). Security and anti-fraud innovations supporting ATM usage included carbon-14 markings on deposit slips and the requirement of a personal identification number.

20.2.5 Communications

Telephone software In 1970, Erna Schneider Hoover, a computer engineer at Bell Laboratories, patented a computerized telephone switching system for connecting calls, earning one of the first software patents ever issued (SWE Records 2000). She was inducted into the National Inventors Hall of Fame in 2008 (National Inventors Hall of Fame 2025).

20.2.6 In the Home

Microwave oven Amana Corporation introduced the home *Radarange* microwave oven in 1967, when miniaturization of components finally allowed a kitchen countertop-sized unit. While time-saving, it was still a luxury item for most middle-class families at a cost of $495. The frequency of the microwaves specifically agitates water molecules in food, causing your dinner to heat, but not the plate it is on. An entirely new food engineering segment was born. Many home cooks (and their teenagers) burned food while trying to adapt their favorite meals using this technology. By 1975, sales of microwave ovens outpaced gas ovens for the first time (Jorgensen 2017).

Home security In 1966, African American Marie Van Brittan Brown and her husband Albert Brown invented a closed-circuit camera door-monitor and home security system, paving the way for modern-era home security systems. They received a patent in 1969 (SWE Records 2000).

Textiles In a 1969 experiment in his research lab at W.L. Gore & Associates, Robert Gore stretched heated rods of polytetrafluoroethylene (PTFE) under specific conditions, creating expanded PTFE (ePTFE), later called GORE-TEX (Gore-Tex 2025). GORE-TEX gained immediate fame in jackets and outerwear for its strength and water resistance. The medical, industrial, and space industries also identified uses for the technology (Gore-Tex 2025). Giuliana Cavaglieri Tesoro (Achievement Award 1978) developed flame-resistant fibers and anti-static chemicals for synthetic fibers used in clothing and textiles (ETHW n.d.).

20.3 State of Women

Societal changes in the USA had a profound impact on women in the second half of the 1960s. With widening availability and use of birth control, women were entering the paid workforce in ever greater numbers. They expressed dissatisfaction regarding gender disparities in job choices, advancement, and pay. Women's pay was still 60 percent of the pay males received, despite equal pay legislation in 1963 (Walsh 2010a). Women remained barred from many states' jury duty and pregnancy could result in job termination.

The civil rights and antiwar movements influenced young women as they were bombarded with contradictory expectations and images about work and family. It raised their political and radical movement awareness (Walsh 2010a). "Most young women, at least in the middle class, expected to have access to the same careers and to receive the same compensation as men," wrote historians Maurice Isserman and Michael Kazin (2004) in *America Divided: The Civil War of the 1960s*.

With the founding of the National Organization for Women (NOW) in 1966, women took action to influence their rights in the workplace, in education, and in the community. NOW used demonstrations and legal challenges to advocate for changes to child care, pregnancy leave, reproductive rights, equal pay, and pension rights (Britannica 2024a). The initial focus from the founders of the movement was on enforcement of the new legislative gains happening in the courts. They sued the country's 1300 largest corporations for sex discrimination (Digital History 2021). NOW supporters were a mix of all ages, races, and genders. It was one of the first women's organizations to incorporate the issues facing African American women into its efforts (Schneir, 1994).

In January 1968, 87-year-old Jeannette Rankin led 5000 women in a march in Washington, DC, against the Vietnam War. The march was the capstone of Rankin's long career as a suffragist, pacifist, and the first woman elected to the US Congress in 1916 (History.com 2023). That same year, feminists protested at the Miss America contest in Atlantic City, arguing that the pageant was sexist (Walsh 2010a). Led by the New York Radical Women's group, protesters took issue with the focus on women's bodies versus their intellect.

Change was happening for women, albeit slowly. In 1965, Patsy Takemoto Mink, (Democrat-Hawaii), became the first woman of color and the first woman of Asian-Pacific Islander descent elected into the US House of Representatives (CAWP 2025). She served multiple terms. Senator Margaret Chase Smith (Republican - Maine), became the first woman to serve as the chair of the Senate Republican Conference in 1967 (CAWP 2025). Shirley Chisholm became the first African American woman elected to Congress in 1968, serving seven terms. She would seek the Democratic Party's nomination for President of the USA in 1972 (Freeman, 2005; CAWP, 2025).

The Boston Women's Health Book Collective published *Our Bodies, Ourselves* in 1971. A book for women, written by women, that contains information related to all aspects of a woman's health and sexuality. The personal experiences of women are quoted throughout the book, while the social and political context of women's health informs the content (Ginty 2004). Miriam Hawley, speaking at a woman's health seminar held at Boston's Emmanuel College that sparked the creation of the book, stated:

> We weren't encouraged to ask questions, but to depend on the so-called experts. Not having a say in our own health care frustrated and angered us. We didn't have the information we needed, so we decided to find it on our own (Ginty 2004).

The Equal Rights Amendment (ERA) movement would amplify the issues facing women and usher in changes affecting how society treated them.

20.4 State of SWE[1]

In 1968, SWE began recognizing the role men had in the acceptance of technical women in the workplace. It was a time of significant change and growth for SWE, propelled by conferences on women in engineering and the multitude of legislative rulings that sparked the emerging women's movement and continued civil rights agendas.

Membership statistics compiled in 1970 report SWE having 17 professional sections, 31 student sections, and just over 1100 members, including 371 student members. This was up by 350 women since the 1965 survey (SWE Records 2000).

In 1968, the Council of Section Representatives mandated the formation of a committee to evaluate the future financial needs of the Society, as it was operating at a deficit (Kindya 1990). A big push for additional corporate memberships provided the needed financial boost by 1971, and saved the society from going broke during the massive student influx of the mid-1970s.

The Men's Auxiliary of the Society of Women Engineers (MASWE), established in 1968, provided men a formal opportunity to take part in SWE and support women in engineering. It existed for nine years until men received full SWE membership rights.

In 1968, SWE created its second national award. The Rodney D. Chipp Memorial Award recognized outstanding support of women engineers and/or SWE by an individual male or corporation. Named for the late husband of Beatrice Hicks, SWE's first president, the inaugural recipient was Col. Clarence E. Davies, former secretary of the American Society of Mechanical Engineers, first secretary of the Engineers Council for Professional Development, and leader in the Engineers Joint Council (SWE Records 2000).

Another step forward within the technical society hierarchy occurred in 1969. Engineering honor society, Tau Beta Pi, admitted women as full members for the first time since the organization's founding in 1885. Previously, female engineering students were only eligible for the Tau Beta Pi Women's Badge. In 1970, the Baltimore-Washington Section's Naomi J. McAfee became the executive secretary of the American Society for Quality Control (ASQC). This was the first time a

[1] This section includes personal recollections about the time period and SWE from a survey of SWE Fellows and Late Career/Retiree Affinity group members conducted in February 2025.

woman served as an officer in ASQC. She would become SWE President in 1972–74 (Kindya 1990; SWE Records 2000).

In 1970, the Engineering Foundation and SWE joined forces on a conference relating to women in engineering, later referred to as the *Henniker Conferences.*[2] SWE President Olive Salembier led the planning and execution, with sponsorship from the Engineering Foundation and several interested corporations. The first forum in 1971 was *Henniker I*, focused on "Bridging the Gap between Society and Technology." It addressed reasons why women shy away from engineering careers, including sex stereotypes, societal attitudes, myths about the physical nature of engineering, and a range of misconceptions about women's attitudes toward work. The second part of the conference called for action plans aimed at establishing concrete programs for greater participation of women in government, industry and education. *Henniker II, Henniker III* and *Tidewater* conferences would follow in later years (Kindya 1990; SWE Records 2000).

Women engineering students during this time were typically part of a tiny group in their engineering and computer science classes—often three or fewer female students. They continued to use slide rules to calculate equations and complete homework. Computer mainframe usage was now part of curriculums. Male classmates generally accepted the women students, but the wives and girlfriends were not so welcoming!

> During my college years, 1967–1971, social unrest was all about the peace protests of the Vietnam War. The Women's Liberation movement was in high gear along with bra burning protests. Students dressed either clean cut or [as] dirty hippies, nothing in-between... Many engineering students during my junior & senior years were Vietnam veterans, very serious students with daytime jobs and raising families. They had already gone through tech school in service and [were] now making 110% on tests, really skewing the grade curve. I never wore pants (slacks or jeans) to classes but did take advantage of knee-high boots with my mini skirt length skirts & dresses. All the young men carried briefcases with their textbooks and slide rule inside while I carried a hard leather satchel with two large handles that could carry at least 4 thick textbooks and my big slide rule…and yes, I did develop back problems from carrying that heavy case. Suzanne Moore, SWE Fellow and SWE Life Member (personal correspondence 2025).

SWE student sections existed on campuses at major colleges and universities or began to be formed during this time. Jaclyn Spear, SWE Fellow, Life Member, and past SWE President (1994–1995) remembered, "I was one of fifteen women engineers in my class and was a founding member of the Cornell student section in 1971, along with my older sister who was 2 years ahead of me" (personal correspondence 2025).

Female engineers entering the workforce during this era were often the only technical women in their companies on the engineering team. Many gravitated to SWE to find support and role models from others in similar situations. Where SWE

[2] The *Henniker Conferences* were named after the meeting location of Henniker, New Hampshire, where they were held. The last conference, called *Tidewater*, was held at the Tidewater Inn located in Easton, Maryland.

professional sections did not exist, they created them or joined via the Members at Large designation.

> The one other woman engineer in my organization had been a member of SWE. She started a new SWE section in our area and encouraged me to join. That opened up visibility to top management for me and extra opportunities that I wouldn't have had otherwise. Connie Golden, SWE Fellow and Life member (personal correspondence 2025).

SWE was an important resource for exchanging information on the workplace environment, legal rights, and practicing skills like public speaking, necessary to gain promotions. It was also the place to get advice about handling gender bias, discrimination, microaggressions, and other sticky situations.

> For years, I was the typical first and/or only woman engineer wherever I worked… I helped form two [SWE] sections during those years since I personally got so much from knowing and interacting with other technical women. SWE has been a constant in my career and my life from university days, even as I changed both career paths and work organizations. It was an important part of my life. Kathleen Harer, SWE Fellow, Life Member, and Past SWE President (personal correspondence 2025).

20.5 Laws and Culture

1968 was a time of great unrest 1968 was one of the most tumultuous years in American history, marked by the civil rights movement, the Vietnam War, the assassinations of Dr. Martin Luther King, Jr. and Senator Robert F. Kennedy, and the antiwar movement (Wallenfeldt 2024). The tone of the Vietnam conflict changed with the Tet Offensive in January 1968. The graphic and extensive media coverage made a deep impression on Americans and ramped up antiwar protests. In August, thousands of antiwar and anti-establishment demonstrators clashed with police at the Democratic National Convention in Chicago, sending televised images of violence to the world and raising the possibility that America was imploding (Walsh 2010c).

The polarization of American life widened into politics with the election of Richard M. Nixon in 1969. Not even the Woodstock Festival, hosting a half-million young people espousing music, peace, and love, could bridge the divides facing Americans (Walsh 2010c). Violence culminated in 1970 when members of the Ohio National Guard killed four unarmed college students in protests at Kent State University; other college protests occurred throughout the USA.

Equality legislation Affirmative action legislation was implemented and challenged. Equality in pay and hiring won challenges at the Supreme Court. The Lesbian, Gay, Bisexual, Transgender, and Queer (LGBTQ) movement began.

The space race The Apollo mission successes and near misses paved the way for future space programs and rapid technology advances (and inspired one of this book's authors to pursue engineering).

Environmental movement The Sierra Club grew to over 100,000 members by 1970, becoming more combative than in its early period. The inaugural *Earth Day* was held on April 22, 1970, with 20 million Americans participating in one or more local events (Isserman and Kazin 2004). It was decentralized, allowing communities to highlight efforts and needs in their geographic location. What was not foreseen was the emergence of youthful environmentalists chanting anti-corporate rhetoric toward companies they felt were harming the planet.

Summary Table 20.1 provides examples of these changes.

Table 20.1 Culture and Legal Highlights for 1967–1971

Year	Event
1961–1969	**Affirmative Action:** President John F. Kennedy in 1961 issued Executive Order 10925, which required government contractors to *take affirmative action to ensure that applicants are employed, and that employees are treated during employment without regard to their race, creed, color, or national origin.* Affirmative Action started under the Civil Rights Act of 1964 to improve opportunities for African Americans while civil rights legislation was being completed. In 1967, President Lyndon Johnson signed Executive Order 11375, which specified that gender discrimination must also be countered. The criteria for affirmative action are race, disability, gender, ethnic origin, and age (MacLaury 1998). Businesses, educational institutions, and the state and federal government, had to demonstrate efforts to improve employment or educational opportunities, including hiring and promotions, for members of minority groups and for women. Cries of *reverse discrimination* and court challenges began almost immediately. Significant lawsuits were *Regents of the University of California v. Bakke* (1978) and *Adarand Constructors v. Pena* (1995), in which the Supreme Court ruled that federal affirmative action programs were unconstitutional unless they fulfilled a "compelling governmental interest." By 1996, Affirmative Action withered as California Civil Rights Initiative (*Proposition 209*) passed, prohibiting all government agencies from giving preferential treatment to individuals based on race or sex (Britannica 2025a).
1963–1967	**Equal Pay Act:** Equal pay for equal work was signed into law by President John F. Kennedy in a measure to end gender-based disparity. The Act requires that men and women who work in jobs that are substantively equal in terms of skill, effort, responsibility, and working conditions shall receive the same pay. In response to this ruling, some employers tried to convince the government that the ruling actually made the employment of women costlier. Among the reasons given to justify unequal pay were these: working women had a higher turnover rate because of family obligations; some state laws prohibited women from working at night; and other laws limited the actual number of hours women could work and the amount of weight women could lift. The government rejected these contentions in requiring equal pay. Equal Pay Laws had been in the books of two states (Michigan and Montana) since 1919, but as of 1970, there were fifteen states that had no such laws (Fugiero 2024; SWE Records 2000).
1967	**Supreme Court Integration:** Thurgood Marshall was appointed by President Lyndon B. Johnson as the first black justice of the Supreme Court (Walsh 2010b).
1968–1973	**Sex-segregated Want Ads:** The Equal Employment Opportunities Commission rules that sex-segregated help-wanted newspaper ads are illegal. The US Supreme Court upheld the ruling in 1973 (SWE Records 2000).

(continued)

Table 20.1 (continued)

Year	Event
1968	**Dr. Martin Luther King, Jr. assassination:** On April 4, 1968, Dr. Martin Luther King, Jr. was assassinated in Memphis, Tennessee, where he had gone to lead a march by striking sanitation workers. Dr. King had recently begun portraying inequality in economic as well as racial terms. In response to King's death, more than 100 American inner cities exploded in rioting, looting, and violence (Wallenfeldt 2024).
1968	**Senator Robert F. Kennedy assassination:** An assassin targeted Kennedy shortly after his victory in the California Democratic presidential primary on June 5, 1968 (Walsh 2010c). He died the following day. News of his death stunned the world, coming so quickly after Dr. Martin Luther King's demise and because his brother, President John F. Kennedy, had been assassinated in 1963.
1969	**One Giant Leap For Mankind:** On July 20, 1969, Astronauts Neil Armstrong and Buzz Aldrin were the first humans to walk on the surface of the moon. This was the culmination of the ambitious goal set in 1961 by President John F. Kennedy (JFK Library 1961). Armstrong later praised the "hundreds of thousands" of people behind the project. "Every guy that's setting up the tests, cranking the torque wrench, and so on, is saying, man or woman, 'If anything goes wrong here, it's not going to be my fault'" (NASA 2019). The 2016 book *Hidden Figures* by Margot Lee Shetterly tells the story of the women "computers" working tirelessly behind the scenes on complex orbital calculations. Katharine Johnson, calculated rocket trajectories for the Mercury and Apollo missions. When NASA used electronic computers for the first time in 1962 to calculate John Glenn's orbit around Earth, officials called on Johnson to verify the computer's numbers; Glenn had asked for her specifically and had refused to fly unless and until Johnson verified the calculations (NASA 2015).
1969	**Stonewall Riots:** Violent uprisings between police and gay rights' activists that began outside the Stonewall Inn, a gay bar in Greenwich Village, New York City. In most states by the time, the solicitation of same-sex relations was an illegal act. Gay bars were places of refuge where gay men and lesbians could socialize in relative safety from public harassment. As the riots progressed, an international gay rights movement was born (Britannica 2024b).
1970	**Environmental Protection Agency (EPA):** The EPA was created by President Richard Nixon in 1970 to protect human health and the environment (EPA 2024). It was the direct result of the environmental movement of the late 1960s and industrial accidents that contaminated air and watersheds.
1971	**Phillips v. Martin Marietta, 400 U.S. 542**. The Supreme Court ruled an employer violates Title VII of the Civil Rights Act of 1964 when it refuses to hire women with young children while hiring men who are similarly situated (Justia US Supreme Court 1971).
1971	**Equal Rights Amendment:** The US House of Representatives passed the Equal Rights Amendment in a vote of 354 to 24 (NOW 2025).

References

Abby. 2024. Computer Mouse: Complete History. November 21. https://history-computer.com/technology/computer-mouse-complete-history/. Accessed February 2, 2025.

The Aerospace Corporation. 2025. A Brief History of GPS. https://aerospace.org/article/brief-history-gps. Accessed February 4, 2025.

Bátiz-Lazo, Bernardo. 2015. A Brief History of the ATM. March 26. https://www.theatlantic.com/technology/archive/2015/03/a-brief-history-of-the-atm/388547/. accessed February 3, 2025.

Blankenbaker, John. 2010. The first personal computer: Kenbak-1. February 26. http://www.kenbak-1.net/. accessed February 3, 2025.

Britannica, T. Editors of Encyclopaedia. 2024a. National Organization for Women. *Encyclopedia Britannica.* December 6. https://www.britannica.com/topic/National-Organization-for-Women. Accessed February 7, 2025.

Britannica, T. Editors of Encyclopaedia. 2024b. Stonewall Riots. *Encyclopedia Britannica.* December 20. https://www.britannica.com/event/Stonewall-riots. Accessed January 23, 2025.

Britannica, T. Editors of Encyclopaedia. 2025a. Affirmative Action. *Encyclopedia Britannica.* January 14. https://www.britannica.com/topic/affirmative-action. Accessed January 23, 2025.

Britannica, T. Editors of Encyclopaedia. 2025b. Dot Matrix Printer. *Encyclopedia Britannica.* January 15. https://kids.britannica.com/students/article/Dot-matrix-printer/321714. Accessed February 2, 2025.

CAWP. 2025. Milestones for Women in American Politics. https://cawp.rutgers.edu/facts/milestones-women-american-politics. Accessed February 7, 2025.

Digital History. 2021. Feminism Reborn. https://www.digitalhistory.uh.edu/disp_textbook.cfm?smtID=2&psid=3341. Accessed February 7, 2025.

EPA. 2024. Origins of EPA. May 31. https://www.epa.gov/history/origins-epa, Accessed February 18, 2025.

Engineering and Technology History Wiki (ETHW). n.d. SWE Awards. https://ethw.org/SWE_Awards. Accessed March 22, 2025.

Ford, Paul. 2014. The Great Works of Software. April 30. https://medium.com/message/the-great-works-of-software-705b87339971. Accessed February 2, 2025.

Freeman, Jo. 2005. Shirley Chisholm's 1972 Presidential Campaign. University of Illinois at Chicago Women's History Project. February. http://www.uic.edu/orgs/cwluherstory/jofreeman/polhistory/chisholm.htm. Accessed March 3, 2025.

Fugiero. 2024. M. Equal Pay Act of 1963. *Encyclopedia Britannica.* October 18. https://www.britannica.com/topic/Equal-Pay-Ac . Accessed January 23, 2025.

Gibson, Kevin. 2003. ATM Marketplace. Rise and Fall of the Docutel part II. March 17. https://www.atmmarketplace.com/articles/rise-and-fall-of-docutel-part-ii-of-ii/. Accessed February 4, 2025.

Ginty, Molly M. 2004. Our Bodies, Ourselves Turns 35 Today. May 4. https://web.archive.org/web/20090903164405/ http://www.womensenews.org/article.cfm/dyn/aid/1820/context/archive. Accessed February 2, 2025.

Gore-Tex. 2025. History. https://www.gore-tex.com/about/history. Accessed February 7, 2025.

Hart, Michael S. 1992. The History and Philosophy of Project Gutenberg. August. https://www.gutenberg.org/about/background/history_and_philosophy.html. Accessed February 2, 2025.

Hauben, Michael. 2025. ARPANET The First Internet. https://www.livinginternet.com/i/ii_arpanet.htm. Accessed February 3, 2025

Hemmendinger, D. Pascal. 2024. Encyclopedia Britannica. October 18. https://www.britannica.com/technology/Pascal-computer-language. Accessed February 7, 2025.

History.com, Editors. 2023. The Jeannette Rankin Brigade: 5,000 women march against Vietnam War. A&E Television Networks. September 15. https://www.history.com/this-day-in-history/jeannette-rankin-brigade-vietnam-war. Accessed February 7, 2025.

IBM. 2025a. Dynamic random-access memory (DRAM). https://www.ibm.com/history/dram?mhsrc=ibmsearch_a&mhq=dram. Accessed February 4, 2025.

IBM. 2025b. Floppy Disk Storage. IBM. https://www.ibm.com/history/floppy-disk. Accessed February 7, 2025.

IBM. 2025c. Robert Dennard. https://www.ibm.com/history/bob-dennard. Accessed February 4, 2025.

Intel. 2011.Quotes: Intel 4004 processor celebrates 40th anniversary. May 4. https://download.intel.com/newsroom/kits/40thanniversary/pdfs/40thCPU_Quotes.pdf. Accessed February 2, 2025.

Isserman, Maurice and Kazin, Michael. 2004. *America Divided: The Civil War of the 1960s.* Second Edition. Oxford University Press. ISBN 0-19-516046-0

JFK Library. 1961. Historic Speeches. May 25. https://www.jfklibrary.org/learn/about-jfk/historic-speeches/address-to-joint-session-of-congress-may-25-1961. Accessed February 17, 2025.

Jorgensen, Timothy J. 2017. Hot Food, Fast: The Home Microwave Oven. March 16. https://www.smithsonianmag.com/innovation/hot-food-fast-home-microwave-oven-turns-50-180962545/. Accessed February 3, 2025.

Justia US Supreme Court. 1971. *Phillips v. Martin Marietta Corp.*, 400 U.S. 542. https://supreme.justia.com/cases/federal/us/400/542/. Accessed January 23, 2025.

Kirschenbaum, Matthew. 2013. What was the first novel ever written on a word processor? Slate. March 1. https://slate.com/culture/2013/03/len-deightons-bomber-the-first-book-ever-written-on-a-word-processor.html. Accessed February 3, 2025.

Kindya, Marta. 1990 Four Decades of the Society of Women Engineers. Society of Women Engineers Records (LR001539). Box 180, Folder 9, Walter P. Reuther Library, Wayne State University. Accessed January 23, 2025.

Lewis, Peter H. 1985. Peripherals: Getting the most out of a Dot-Matrix Printer. December 17. https://www.nytimes.com/1985/12/17/science/peripherals-getting-the-most-out-of-a-dot-matrix-printer.html. Accessed February 2, 2025.

Loff, Sarah A. 2015. NASA. Apollo 11 Mission Overview. April 17. https://www.nasa.gov/missions/apollo-11-mission-overview/. Accessed February 4, 2025.

MacLaury, Judson. 1998. History of the Office of Federal Contract Compliance Programs. US Department of Labor. https://www.dol.gov/agencies/ofccp/about/history. Accessed January 29, 2025.

Mohon, Lee. 2020. NASA. Apollo 13: The Successful Failure. April 06. https://www.nasa.gov/missions/apollo/apollo-13-the-successful-failure/. Accessed February 4, 2025.

NASA. 2019. One Giant Leap for Mankind. July 20. https://www.nasa.gov/history/july-20-1969-one-giant-leap-for-mankind/. Accessed January 23, 2025.

NASA. 2015. Katherine Johnson: The Girl Who Loved to Count. November 15. https://www.nasa.gov/centers-and-facilities/langley/katherine-johnson-the-girl-who-loved-to-count/. Accessed January 23, 2025.

National Inventors Hall of Fame. 2025. Erna Schneider Hoover. https://www.invent.org/inductees/erna-schneider-hoover. Accessed February 3, 2025.

NOW. 2025. Chronology of the Equal Rights Amendment, 1923-1996. https://now.org/resource/chronology-of-the-equal-rights-amendment-1923-1996/. Accessed March 22, 2025.

Peddie, Jon. 2021. The Birth of the Personal Computer: Everything You Know is Wrong. November 15. https://www.jonpeddie.com/news/the-birth-of-the-personal-computer-everything-you-know-is-wrong/. Accessed February 2, 2025.

Raskin, Jef. 1974. FLOW: A Teaching Language for Computer Programming in the Humanities. Computers and the Humanities. July. Accessed February 3, 2025.

Schneir, Miriam. 1994. Feminism in Our Time: The Essential Writings, World War II to the Present. Vintage Books.

Society of Women Engineers (SWE) Records (LR001539). 2000. SWE 50 Years 1950-2000. Box 180, Folder 9, Walter P. Reuther Library, Archives of Labor and Urban Affairs, Wayne State University. Accessed January 10, 2025.

Tomlinson, Ray. 2001. The first network email. August 15. https://web.archive.org/web/20060506003539/ http://openmap.bbn.com/~tomlinso/ray/firstemailframe.html. Accessed February 3, 2025.

Wallenfeldt, J. 2024. Assassination of Martin Luther King, Jr. Encyclopedia Britannica. October 23. https://www.britannica.com/event/assassination-of-Martin-Luther-King-Jr. Accessed January 23, 2025.

Walsh, Kenneth T. 2010a. The 1960s: A Decade of Change for Women. US News & World Report. March 12. https://www.usnews.com/news/articles/2010/03/12/the-1960s-a-decade-of-change-for-women. Accessed February 6, 2025.

Walsh, Kenneth T. 2010b. Gains in the 1960s Made Obama's Election Possible. US News & World Report. March 10. https://www.usnews.com/news/articles/2010/03/10/gains-in-the-1960s-made-obamas-election-possible. Accessed February 6, 2025.

Walsh, Kenneth. 2010c. The 1960s: Polarization, Cynicism and the Youth rebellion. US News & World Report. March 12. https://www.usnews.com/news/articles/2010/03/12/the-1960s-polarization-cynicism-and-the-youth-rebellionredirect. Accessed February 6, 2025.

Webster, Edward C. 2000. *Print Unchained: Fifty Years of Digital Printing: A Saga of Invention and Enterprise.* West Dover, VT: DRA of Vermont. ISBN 0-9702617-0-5

Chapter 21
Marguerite Rogers, PhD (1915–1989) Achievement Award Recipient 1967

For her outstanding contributions to the field of air-delivered tactical weapons

Jill S. Tietjen

21.1 Introduction

Dr. Marguerite Rogers was considered the foremost expert on air-delivered tactical weapons in the entire Department of Defense. Rogers received all three of her degrees from what was then called Rice Institute (today Rice University) and embarked on an academic career before working for the Navy and conducting research in ordnance during World War II. She continued her work in ordnance while holding several positions in academia and spent most of her career at the Naval Ordnance Test Station, China Lake, California. Rogers contributed significantly to free-fall weapons, received many awards, and left a significant legacy as a role model and trailblazer. In 2023, the Dr. Marguerite "Peggy" Rogers Laboratory opened at the Naval Air Warfare Center Weapons Division at China Lake, California.

21.2 Early Life and Education

Marguerite "Peggy" Moilliet (later Rogers) was born in Minatitlan, Vera Cruz, Mexico where her British father was a physician specializing in tropical medicine with Eagle Oil Company. Her father died of malaria when she was five, and the family–the one surviving older brother (the other older brother died in infancy of whooping cough), Peggy, and her mother–moved to Seymour, Texas, where the

J. S. Tietjen (✉)
Technically Speaking, Inc., Greenwood Village, CO, USA

C. D. Craig et al. (eds.), *Women Engineering Legends 1952-1976: Society of Women Engineers Achievement Award Recipients*,
https://doi.org/10.1007/978-3-032-00224-2_21

family had relatives. Peggy's mother became a schoolteacher (SWE Records; Rogers, n.d.)

Rogers manifested an interest in science early and she was encouraged by her mother and a high school teacher to continue in the field. She graduated as valedictorian of her high school class in Seymour, Texas. At the time she graduated, she had earned the highest scholastic record ever made at that school. She went to Rice Institute (today Rice University), because it was free to Texas students. In addition, she had a scholarship to pay for books and supplies and her Uncle Jock sent her money to help with living expenses. She met Fred Terry Rogers, Jr., at Rice and they married in 1937 after he graduated, but when she was still an undergraduate. When they got married, the school revoked her scholarship, but the chairman of the Physics Department employed her in various jobs to compensate. She earned all three of her degrees from Rice, BA with honors in 1937 (Phi Beta Kappa), MA in physics in 1938, and PhD in physics in 1940. Her PhD research was in nuclear physics—she studied the properties of polonium. She was the first woman to graduate from Rice with a PhD in physics. Peggy and her husband had five children (Fred Terry Rogers Papers 1929–1956; SWE Records; Rogers n.d.)

21.3 Career

After her graduation, Rogers served on the faculty at the University of Houston and advanced to the rank of assistant professor. She then left the University of Houston and was employed at the Naval Avionics Facility, Indianapolis (NAFI), from 1943 to 1946. At NAFI, Rogers was Head of the Optics Section of the Research Department. Here, she contributed significantly to basic and applied research work relating to gun sights, rocket sights, and torpedo directors. For her work at NAFI, she received a citation from the Navy for meritorious service to naval ordnance development during the war. Her supervisor in Indianapolis was Dr. L.T.E. Thompson who would later recruit both Rogers and her husband to work for the Navy in California (Ogilvie & Harvey 2000; SWE Records; Rogers n.d.).

Rogers left NAFI and returned to an academic career serving as a Research Associate at the University of North Carolina at Chapel Hill. She published papers relating to the physical properties of steel and developed a differential equation for the overall behavior of a servo-operated rocket potentiometer and researched other topics as well. She worked briefly at the Oak Ridge National Laboratory where her focus was solid-state physics research while she and her husband waited for civilian quarters to be built at China Lake, California where she had been invited to work by Dr. L.T.E. Thompson (SWE Records).

In 1949, Rogers (Fig. 21.1) began her employment with the Naval Ordnance Test Station (NOTS). During the years 1949–1953, she conducted analytical studies essential for the ballistic equations and fire-control errors relating to fire-control systems and rockets as well as studies relating to various machine guns (SWE Records; Rogers n.d.).

Fig. 21.1 Marguerite (Peggy) Rogers. (Photograph courtesy of Society of Women Engineers Records, Walter P. Reuther Library, Archives of Labor and Urban Affairs, Wayne State University)

Rogers returned to academia from 1953 to 1956 as Professor of Physics and Head of the Sciences Division at Columbia College in South Carolina. Her research during this time, performed in conjunction with the University of South Carolina, related to terminal ballistics of textile armor assemblies, especially considerations related to the dissipation of projectile energy. This research was performed under contract to the US Army Quartermaster Corps and later by the Army Office of Ordnance Research. Additional research studies that she conducted related to the penetration of bullets into targets under contract to the Office of Naval Research. When her husband died suddenly of a heart attack in 1956, Rogers and her family entourage (five children and her mother—who took care of the "home front") moved to England to be closer to her brother. Here, she spent a year (1956–1957) at the Royal Technical College in Salford, England as a lecturer in physics. The move to England did not go well and the family returned to the USA where Rogers took a position back at China Lake (SWE Records; Rogers n.d.).

Upon her return, her career trajectory took off. The rest of her career was spent at China Lake where she would eventually earn the nickname the "First Lady of China Lake." In 1957, when she returned, she served as an electronics specialist at NOTS. Initially, she investigated new navigational and fire-control systems. In 1958, she was selected to head the Heavy Attack Systems Analysis Branch in the Aviation Ordnance Department, directing the analysis for a variety of weapon systems. She received a Superior Accomplishment Award and a Sustained Superior Performance Award while in this position. By 1961, she was directing analysis work in avionics, target recognition, space vehicles, and free-fall ordnance and warheads (SWE Records; Rogers n.d.).

In 1962, she became head of the Air-to-Surface Weapons Division. As director of the Navy's free-fall program, she oversaw approximately 15 major weapons and warheads in various stages of development, plus the applied research program leading to the weapons of the future. The weapons that she oversaw fell into four categories: retarded weapons (Snakeye I and II), cluster weapons (Sadeye, Gladeye, and Rockeye I and II), chemical weapons (Weteye, Bigeye, Padeye and Misteye), and special-purpose weapons (Fireye, Deneye, Briteye, and Walleye Warhead) (SWE Records).

Rogers became the foremost expert on air-delivered tactical weapons in the entire Department of Defense. In 1965, she received the L.T.E. Thompson award, the highest award that the NOTS can present for "Her outstanding technical leadership and effectiveness in managing and directing the development of a large portion of this country's conventional warfare arsenal, the Free-Fall Weapons Systems Program." In 1966, she became Head of the Weapons Systems Analysis Division, directing planning and advanced design of certain guided weapons and the free-fall program. Although she was not a software specialist, she led the organization that maintained the software for airplanes, modified it to employ new weapons, and thoroughly tested it before delivery was made to the fleet. She was the first woman department head at NOTS and the first woman to hold a public law position (SWE Records; Rogers n.d.).

She was emphatic in taking her career responsibilities seriously—"those young folks on the ships and airplanes are our children, and we owe it to them to provide the best equipment we can." She also said "It's very satisfying to me having contributed something to the nation. Working for industry could never have provided that opportunity. Public service either means a great deal to you, or it's not very important. I have great respect for all public servants, whether they work for the federal government, or for the police and fire departments and so on. To me, the national defense was worth working for" (Rogers n.d.).

After her retirement from NOTS, she became the Department Head for Computer Sciences Corporation. Rogers died in 1989 after fighting a serious health problem for decades. She had been diagnosed with a blood disorder, *polycythemia rubivera*, in the early 1960s. Rogers believed this illness was a result of working with polonium during her PhD research at Rice, at a time when there was no radiation safety protection of any kind (SWE Records; Rogers n.d.).

21.4 Organizations

Rogers was a member of Phi Beta Kappa, Sigma Xi, the American Physical Society, and the American Association for the Advancement of Science. She served on the Tri-Service Coordinating Committee, Weapons Development Department for more than 10 years. She served as Vice President for the American Ordnance Association, China Lake Chapter; and Treasurer of the Scientific Research Society of America China Lake Chapter (SWE Records; Rogers n.d.)

She was instrumental in founding the Ridgecrest (California) Chapter of 4-H Clubs and acted as advisor for the club. She was active in the Fire Mountain Foundation and the Maturango Museum. She was involved with St. Michael's Episcopal Church and served as a Sunday School Teacher. She served as President of the Women's Guild, and the United Church Women, Indian Wells Chapter. She served as Chair of the China Lake Work Experience Program Student Committee. Rogers had been an equestrian since returning to California in 1957 and raised, bred, and showed Peruvian Paso horses. She was a member of the Valley Riders. Around the community of Ridgecrest, California, she was known as Dr. Peggy (SWE Records; Rogers n.d.).

21.5 Honors and Awards

Rogers received SWE's Achievement Award in 1967 (Fig. 21.2), with a citation that read *For her outstanding contributions to the field of air-delivered tactical weapons.* Receiving the Achievement Award was important to Rogers, as she related at a panel on the occasion of the 1968 SWE Convention (SWE Records):

> I know that receiving the Achievement Award has been a very important item to me—not only personally, because I was tremendously flattered and thrilled to get it, not only because

Fig. 21.2 Marguerite M. Rogers (center) accepts the Achievement Award from Alice Morgan (right) at the Society of Women Engineers National Convention in Washington, DC, June 21–25, 1967. (Photograph courtesy of Society of Women Engineers Records, Walter P. Reuther Library, Archives of Labor and Urban Affairs, Wayne State University)

> I enjoyed meeting all of you people—but it gives one a backing. Being a part of this group gives one a backing that is convincing and must not be overlooked. Now, since I am an engineer by fiat, I gained a luster of practicality and forthright youthfulness that I never had before. This is, however, something that—the moral support that we give each other in an important field should not to be overlooked as an exceedingly important aspect of the Society of Women Engineers. And I feel very proud to have been an Achievement Award winner, and I have found that it carries a lot of weight among my colleagues and the people that I deal with.
>
> In the Achievement Award, I had the feeling that I had been selected by a group of my peers who were competent to select me. It therefore gave me a sense of capability and a feeling of stature that I think cannot be underestimated. For me, it was a tremendous lift to have that feeling, because I think all of us sometimes wonder if what we're doing is really worthwhile. Is it really valid? And when I was selected as this, I had the feeling it was a credential. It was a vouching to myself, as well as the rest of the world. And I found that was a tremendous experience to me (SWE Records).

Also in 1967, Rogers received the American Ordnance Association (later the American Defense Preparedness Association) Harvey C. Knowles Award (the second person to receive the award) which is presented annually to an American citizen for a major technical contribution to armament progress. Her citation read, "A distinguished authority in the analytical planning and designing of air-delivered weapons. Rogers has demonstrated the highest ideals of patriotic citizenship, scientific proficiency, and technical competence." She had previously received the Kiwanis Citizenship Award in 1965 (Army Research and Development 1977; SWE Records; Rogers n.d.).

Roger's many federal awards include the Superior Civilian Service Award for her "strong leadership and management in the development of the free-fall weapons. .." in 1966, and the Federal Women's Award in 1975 (the first Navy nominee to receive the award). In 1980, she received the Navy Distinguished Civilian Service Award, the Navy's highest civilian Award. The following year, she received the Department of Defense Distinguished Civilian Service Award, only six of which are given each year (NWC Rocketeer 1981; SWE Records; Rogers n.d.).

Rogers was a Fellow of the Texas Academy of Science (1946). She received the China Lake Business and Professional Women's Club award for "Woman of the Year" in 1965 as well as the Women of Achievement award from the California State and Professional Women's Club. In 1969, she received the Engineering Merit Award of the Institute for the Advancement of Engineering (SWE Records; Rogers n.d.).

21.6 Legacy

A significant role model throughout her career, Rogers had faced severe discrimination in her younger days and took equality of opportunity very seriously. She was considered the first role model for women in science and engineering at China Lake.

Rogers was the first female department head at NOTS and the first woman in the Navy to hold a public law position. In 2023, the Dr. Marguerite "Peggy" Rogers Laboratory was opened at the Naval Air Warfare Center Weapons Division at China Lake, California. It is a major research, development, test and evaluation facility (NAVAIR News 2023; SWE Records; Rogers n.d.).

References

Army Research and Development. 1977. ARRADCOM Man Wins ADPA's Harvey Knowles Award, August-September: 28. https://asc.army.mil/docs/pubs/alt/archives/1977/Aug-Sep_1977.PDF.

The Fred Terry Rogers Papers. 1929–1956, MS 29, Woodson Research Center, Fondren Library, Rice University. https://archives.library.rice.edu/repositories/2/resources/28.

NAVAIR News. 2023. NAWCWD cuts ribbon on first laboratory earthquake recovery project, October 25. https://www.navair.navy.mil/news/NAWCWD-cuts-ribbon-first-laboratory-earthquake-recovery-project/Wed-10252023-1139.

NWC Rocketeer. 1981. DoD Distinguished Civilian Service Award presented to Dr. Marguerite Rogers, July 31: 1&5. http://www.chinalakealumni.org/Downloads/Rocketeer/1981/Rktr07.31.1981.pdf.

Ogilvie, Marilyn and Joy Harvey, eds. 2000. *The Biographical Dictionary of Women in Science: Pioneering Lives from Ancient Times to the Mid-20th Century*. New York: New York: Routledge.

Rogers, Alexander (Sandy). n.d. Unpublished.

Society of Women Engineers (SWE) Records (LR001539). Box 192, Folder 9, Walter P. Reuther Library, Archives of Labor and Urban Affairs, Wayne State University.

Chapter 22
Isabella Karle, PhD (1921–2017) Achievement Award Recipient 1968

In recognition of her significant contributions to the development of unique procedures for crystal structure analysis

Jill S. Tietjen

22.1 Introduction

Growing up in Detroit and not speaking English until she attended elementary school, Isabella Lugoski Karle determined in high school that she loved chemistry and would pursue a career in the field. During her only semester at Wayne State University, a professor there educated her about career opportunities and the need for graduate education. Thus, she decided she would go to graduate school. Karle got all three of her degrees at the University of Michigan and met her husband there. After getting her PhD, she worked on the Manhattan Project until World War II was over. She became the first female faculty member in the chemistry department at the University of Michigan. The Karles were then hired by the Naval Research Laboratory (NRL). Their work at NRL proceeded as had much of their prior work—Jerome undertook the theoretical efforts and Isabella performed the experiments. Spending the rest of her career at NRL, Karle revolutionized crystallography, laying the foundation for three-dimensional structures of molecules and the ability to synthesize those molecules. Her techniques, which have now resulted in the identification of thousands of molecular structures, have enabled the development of new pharmaceutical products and other synthesized materials.

J. S. Tietjen (✉)
Technically Speaking, Inc., Greenwood Village, CO, USA

C. D. Craig et al. (eds.), *Women Engineering Legends 1952-1976: Society of Women Engineers Achievement Award Recipients*,
https://doi.org/10.1007/978-3-032-00224-2_22

22.2 Early Life and Education

Isabella Lugowski was born in Detroit in 1921 to parents who had immigrated from Poland. Her father Zygmunt A. Lugoski was a house painter and her mother Elizabeth Graczyk was a seamstress. Her mother and a friend decided to go into the restaurant business—Karle was their accountant, even before she started elementary school. They did not speak English at home; that she learned when she started school. During her high school years, she decided to study chemistry. She got some pushback from her father, who wanted her to be a lawyer, but her parents were very supportive of her going to college (Atomic Heritage Foundation 2015; Hargittai 2015; Proffitt 1999).

> I didn't speak English until I went to elementary school. The home language was Polish. My mother realized fairly early that I like numbers. I soon became the accountant, so to speak. Fresh meat was delivered every day [to the restaurant] and the butcher left the bill that had to be paid once a week. So once a week, I added up all the numbers of the money that was owed him. Then my mother decided that if I could do that, I could learn to read and write, too. My mother, although she never had attended any schools, knew both Polish and English so she taught me to read in Polish. So before I went to school, I could read and write in Polish, which was handy because the alphabet is the same except for three letters. I knew how to sound out words, I knew the alphabet, I knew how to write them down. (Atomic Heritage Foundation 2015)

> But by the time I got to be six years old, it was mandatory that I go to the elementary school in which only English was spoken. That is, by the teacher. I think every student spoke a different European language. . . But we all spoke English after the first year. I went through school rather quickly in that I didn't go to second grade. I didn't go through fifth grade. There was another grade I didn't go through so that I was young when I got to high school. There was no hesitation to put students ahead if they could manage the work. Depression days had come in the late 1920s. There wasn't enough money to pay teachers anyway or to have large schools. (Atomic Heritage Foundation 2015)

> I didn't know what science meant until I was in high school. I read that if I had wanted to go to university, I would have to have some coursework in high school in science. My homeroom teacher told me, "Well, you could have chemistry or you could have physics or you could have biology. One will be sufficient. So I picked one out of the air, which was chemistry. That was a fortunate choice because I thought that the teacher, who was a female chemistry teacher – unusual in those days – was a very good teacher. And she was the one who sparked a very intense interest in which I was learning in her class (Atomic Heritage Foundation 2015).

> Both of my parents were very hopeful that I would go onto the university. Nobody in the family ever had. In fact, nobody in the family had ever finished high school. My parents sort of shrugged their shoulders when I told them I was going to study chemistry at the university (Atomic Heritage Foundation, 2005).

She graduated from high school in the middle of the school year and started at Wayne State University. She went to classes and lived at home. That experience was formative as her chemistry professor, assistant professor Dr. Joseph Jasper, at Wayne State was the first person to tell her she should consider going to graduate school to get a master's degree and a PhD—and that became her goal (Atomic Heritage Foundation 2015; Shearer & Shearer 1997).

During her semester at Wayne State, a former high school teacher came to the house and told her the following day there would be a statewide examination for a scholarship to the University of Michigan. The teacher would pick her up and take her to the examination. Karle was number four out of all the students who took the examination and got a 4-year scholarship to go to the University of Michigan. She also got another fellowship and enrolled in the chemistry program at the University of Michigan the following fall (Atomic Heritage Foundation 2015; Shearer & Shearer 1997).

Karle graduated with highest honors in 1941 and wanted to continue her studies as a graduate student. But she discovered that all graduate teaching assistantships in the chemistry department were awarded only to men. The American Association of University Women came to her rescue with a fellowship that allowed her to start her graduate work. The following year, she was awarded the Horace H. Rackham fellowship. Karle stayed at the University of Michigan, receiving her MS in 1942 and marrying fellow graduate student Jerome Karle that same year. She received her PhD in 1944. For her PhD, she worked on electron diffraction by gaseous molecules with Dr. Lawrence Brockway (see Fig. 22.1) (Shearer & Shearer 1997).

Fig. 22.1 Isabelle Karle with original electron diffraction apparatus. (Courtesy of AIP Emilio Segrè Visual Archives)

22.3 Career

After obtaining their PhDs from the University of Michigan, both Karles worked on the Manhattan Project at the University of Chicago. They were both involved in finding procedures to make the fission fuel plutonium without impurities. Isabella succeeded in making pure plutonium chloride. When the war was over, they worked for 2 years at the University of Michigan despite the anti-nepotism rules in place at the time. Isabella Karle was the first female member of the chemistry faculty at Michigan (NSTMF 2025).

In 1946, they were both hired at the Naval Research Laboratory (NRL) and even allowed to work together. Karle (Fig. 22.2) characterizes the NRL "we were immediately thrust into an atmosphere of exciting research and surrounded by exceptional people." Their first work was on electron diffraction (NRL 1994).

> Jerome designed an electron diffraction apparatus, which was very large and complicated and I assembled the pieces, aligned and fine-tuned all the components, and eventually collected and analyzed data. During the process of building the apparatus, when the on-site foundry and machine shops were supposed to cast and fabricate the large brass segments, [the head of Support Services] suggested and authorized that two of everything be fabricated... We had a successful first instrument much more quickly because if one piece didn't work the other one did. Eventually, we had two apparatuses. So, as a result, we had papers with experimental data already analyzed and in print, barely two years from the time we first started designing the equipment (NRL 1994).

Fig. 22.2 Isabella Karle. (Courtesy of AIP Emilio Segrè Visual Archives)

Their work at NRL proceeded as had much of their prior work—Jerome undertook the theoretical efforts and Isabella performed the experiments (Shearer & Shearer 1997).Their next efforts were directed toward X-ray diffraction and determining the three-dimensional structure of molecules (see Fig. 22.3) (NRL 1994).

> Our subsequent work concerned X-ray diffraction and the elucidation of the three-dimensional structure of molecules. A key problem existed; in fact, half of the necessary information seemed to be lost in the process of the diffraction procedure. It was called the phase problem. If we could evaluate the phases as well as the intensities, then we could work backwards and determine the structures of these molecules. (NRL 1994).
>
> Well, Jerome Karle, with a colleague, Herbert Hauptman, who worked in our group at the time, solved the problem of obtaining the phases theoretically. The results were presented in terms of an infinite set of inequalities, probabilistic functions, and transcendental functions. It was I who bridged the gap between the theory and experiment by translating these mathematical results into practical procedure. . . (NRL 1994).
>
> There was a great amount of trepidation when we gambled our research budget to have IBM do a final refinement on my results. You see, at that time, this was the mid-1950s—IBM was the only company that had a computer that was capable of doing the problem that we needed to be done. . . My results worked and that was the beginning of doing several hundred (I've not counted them) structures over the years, in various areas of science. The successes we've had, in this sort of activity, have spawned numerous advance-study workshops that took place around the world for the next 20–25 years. (NRL 1994).
>
> The notes and instructions that I provided for the workshops became teaching materials in various universities and the basis for a number of computer programs. These courses were instrumental in expanding the number of crystal structures solved around the world, from a handful each year to much more than 10,000 a year, at present. (NRL 1994).

Fig. 22.3 Isabella Karle holding a construction of a molecule. (Gift of Dr. Karle, courtesy AIP Emilio Segrè Visual Archives)

One of the first successes of solving crystal structures was determining the structure of the venom of a South American frog. The secretions from the frog skin were used by indigenous people as poison on arrow tips. Understanding the mechanism used by the venom to block specific nerve transmission proved to be very useful for the study of nerve transmission for medical purposes. Based on her work, a synthetic form of the venom could be manufactured (Shearer & Shearer 1997).

The Karles (see Fig. 22.4) retired from the Naval Research Laboratory after more than sixty years of service (Hargittai 2015). Over the course of her career, Karle had published nearly 300 scientific papers (Shearer & Shearer 1997). She said, "I've had a satisfying and fascinating career" (Shearer & Shearer 1997).

22.4 Honors and Awards

Over the course of her distinguished career, Karle received many awards and honors. Her first award was SWE's 1968 Achievement Award of which she said, "I was rather surprised and pleased that the women engineers were the ones who gave me my first award" (Atomic Heritage Foundation 2015). She was elected to the National Academy of Sciences in 1978. Jerome Karle was the co-recipient of the 1985 Nobel Prize in Chemistry, an award that he shared with Herbert Hauptman. It has been said that the sadness Jerome felt when he learned that Isabella would not share the Nobel Prize never left him (Hargittai 2015).

Fig. 22.4 Isabella and Jerome Karle at the Naval Research Laboratory. (Courtesy of Wikipedia)

Although she believed it would have been wonderful to receive the Nobel Prize with Jerome, Isabella Karle received much recognition for her work. She received the Aminoff Prize from the Royal Swedish Academy of Science in 1988 which was established for pioneers in crystallography. In 1993, she became the first woman to receive the Bower Award and Prize for Achievement in Science from the Franklin Institute "for determining three-dimensional structure of molecules with X-ray diffraction" (Hargittai 2015). Karle received the National Medal of Science from President Bill Clinton in 1995 "For the development and application of a method for determining essentially equal-atom crystal and molecular structures by X-ray analysis, thereby having a profound effect on the practice of organic and biological chemistry" (NSTMF 2025). She also received several honorary degrees, awards from the US Navy, and other honors (Bailey 1994; Shearer & Shearer 1997).

Karle was also active in the profession, serving as President of the American Crystallographic Association. She was a member of the American Physical Society and the American Chemical Society (Bailey 1994).

22.5 Personal

She met her husband, Jerome Karle, at the University of Michigan. Here is how Jerome described their initial meeting (Hargittai 2015):

> I met her the first day that I went to school at the University of Michigan. I was one of those people who, when we had to set up an apparatus, for example, would go and do it, rather than wait until the lab period started. The places in the teaching laboratory were assigned according to alphabetical order. Her last name started with an L [Lugoski] and mine started with a K. When Isabella arrived, she saw me and the entire apparatus set up to carry out an experiment. I do not remember what she said, but she was surprised. That is how I met Isabella. The first year I did not see her very often, except in class. We went for an occasional walk in the evening. Once she was not feeling well and I brought material that she had missed in class to her lodging. The next year we started to have lunch together and by the end of that school year we were married (Hargittai 2015).

They had three daughters, all of whom pursued scientific careers. Karle considered herself very fortunate to have specialized in the field of crystallography (Atomic Heritage Foundation 2005; Hargittai 2015).

> Crystallography wasn't something that you had to watch all the time. You could take it home with you, you could think about it while minding the babies. Most of the projects in crystallography, for example, start with an idea or a substance, and there is an end when you get the crystal structure. This is an isolated thing, and in order to have a research project, you want a number of related things that would go together, but it was possible to do it stepwise. In other projects there may be so much more interaction (Hargittai 2015).

But, nevertheless, she did have help with her children (Hargittai 2015).

> We were fortunate that after World War II, there were a lot of women of the grandmother age whose children didn't want to live on the farm anymore. The Virginia Mountains are about sixty miles from here. Many of the children came to Washington during the war and

> didn't want to return after it ended. So the elder ladies came and we could have a housekeeper/babysitter during the week who would then be visiting with her children in the city during the weekends. That worked out very well until our children were well enough grown and we didn't need help anymore (Hargittai 2015).
>
> Having children required organization, but I never felt them to be an obstacle for my career. It never interfered with our professional lives. When they got to be a little older, and by a little older I mean at least seven years old, we took our children with us for our summer travels to various meetings in Europe (Hargittai 2015).

She remained active, engaging in activities such as ice skating, swimming, hiking, and yard work, throughout her life (Shearer & Shearer 1997).

22.6 Legacy

The National Science and Technology Medals' Foundation explains her legacy this way (NSTMF 2025):

> It was at NRL that Karle used a novel X-ray scattering technique to determine the structure of crystals, a technique that is used to study the biological, chemical, metallurgical and physical characteristics, allowing processes to be designed to duplicate the molecules being studied.
>
> Karle's method drastically improved the speed and accuracy of chemical and biomedical analysis and remains the basis of all advanced X-ray crystallography, including computerized programs, used around the world today. The technique has played a major role in the development of new pharmaceutical products and other synthesized materials.

References

Atomic Heritage Foundation. 2005. Voices of the Manhattan Project. Isabella Karle's Oral History Interview.https://ahf.nuclearmuseum.org/voices/oral-histories/isabella-karles-interview-2005/.

Atomic Heritage Foundation. 2015. Voices of the Manhattan Project. Isabella Karle's Oral History Interview.https://ahf.nuclearmuseum.org/voices/oral-histories/isabella-karles-interview-2015/.

Bailey, Martha J. 1994. *American Women in Science: A Biographical Dictionary*. Santa Barbara, California: ABC-CLIO.

Hargittai, Magdolna. 2015. *Women Scientists: Reflections, Challenges, and Breaking Boundaries*. Oxford, England: Oxford University Press.

National Science & Technology Medals Foundation (NSTMF). Isabella L. Karle. https://nationalmedals.org/laureate/isabella-l-karle/. Accessed January 23, 2025.

Office of Naval Research (NRL). 1994. Naval Research Reviews, Two. Volume XLVI. https://apps.dtic.mil/sti/tr/pdf/ADA281435.pdf.

Proffitt, Pamela, ed. 1999. *Notable Women Scientists*. Farmington Hills, Michigan: Gale Group.

Shearer, Benjamin F. and Barbara S. Shearer, Editors. 1997. *Notable Women in the Physical Sciences: A Biographical Dictionary*. Westport Connecticut: Greenwood Press.

Chapter 23
Alice Mary Stoll (1917–2014) Achievement Award Recipient 1969

In recognition of her significant contributions in the field of the development of fire-resistant fibers and fabrics, based on her pioneering studies of heat transfer by flame contact

Jill S. Tietjen

23.1 Introduction

Alice Stoll's life work has saved thousands of lives through fire-resistant fabrics and the foundational research that led to the understanding of how burns occur as a function of time and temperature. Born and educated in New York City, Stoll's seminal contributions occurred at the Naval Air Development Center in Pennsylvania. There she developed two different Stoll Curves. One demonstrated human tolerance to centrifugal acceleration. Knowledge and application of this Stoll Curve would later enable designers of aircraft and space vehicles to determine when pilots needed to be protected against high g-force conditions. Her second Stoll Curve is still foundational today to the understanding of how to protect humans from flames. It led to the development by DuPont of Nomex®, a fabric that is used in everything from military uniforms to aircraft upholstery to hospital bed linens to ironing board covers. Stoll received many awards during her lifetime and was inducted posthumously into the National Inventors Hall of Fame.

J. S. Tietjen (✉)
Technically Speaking, Inc., Greenwood Village, CO, USA

C. D. Craig et al. (eds.), *Women Engineering Legends 1952-1976: Society of Women Engineers Achievement Award Recipients*,
https://doi.org/10.1007/978-3-032-00224-2_23

23.2 Early Life and Education

Alice Stoll (Fig. 23.1) was born on Long Island, New York, in 1917. After completing high school, she attended Hunter College earning her BA in chemistry and physics in 1938. Stoll worked for several years and served in the US Navy before returning to graduate school. In 1948, she received her master's degree in physiology and biophysics from Cornell University Medical College (New York City) (NIHF 2025; Proffitt 1999; SWE Records).

> I was born in Long Island, raised in Staten Island attended only parochial schools, graduating from St. Peter High School. [I got interested in physics because] I guess it was a teacher I had in high school, she was great. She was brilliant. And she taught the subject like made daylight out of it. And I just loved it. That was it. It was hard to get into the field. Hunter College even then was 6,000 women and they didn't have a physics program. They had a minor program but not a physics major. So I majored in chemistry and took a minor in physics and graduated in 1938. But I still love physics best (SPCWHS n.d.).

23.3 Career

After earning her bachelor's degree, Stoll was employed as a research assistant at the Cornell University Medical College in New York City until 1943. Her areas of emphasis included allergy, basal metabolism, and infrared spectrophotometry (SWE Records).

Fig. 23.1 Alice M. Stoll portrait (Photograph courtesy of Society of Women Engineers Records, Walter P. Reuther Library, Archives of Labor and Urban Affairs, Wayne State University)

If it were not for Hunter College, neither one of us [Alice Stoll and lifelong collaborator and friend Maria Chianta] could have attended college at all. It was the height, or should I say, depth of the "Great Depression" which today is but a few lines in the history books. There was not a job to be had in '34, and in '38 we were either "over qualified" or "inexperienced." In any event, after five months of pounding the pavement I and a few other lucky science graduates were fortunate enough to be accepted as volunteers in The New York Hospital (SPCWHS n.d.)

During this time, Maria and I both advanced to paid positions . . .I as a Chemist in the Allergy Research Lab at New York Hospital. . . . I moved into the Infra-red Spectrophotometry at the Cornell University Medical School (SPCWHS n.d.)

Stoll was on active duty in the US Navy during World War II. She enlisted in the US Naval Reserves in July of 1943. She became a Pharmacist's Mate second Class in September of 1943, graduated from Officer Candidate School in March 1944, and then served as an ensign in the Hospital Corps from May 1944. She was promoted to Lieutenant Junior Grade before she was released to inactive duty in April 1946. Stoll continued to serve in the US Naval Reserves until August 1966, when she retired with the rank of Commander. (Fig. 23.2) Her primary activity within the reserves, both when she was enlisted personnel and serving as an officer, was research in parasitology (SWE Records).

However, World War II was accelerating and expanding and women were being accepted into the military. So in July 1943, we [Alice and Maria] joined the Navy to "do our bit." For boot training we were sent – where else?! – to the Bronx Campus of Hunter College! (SPCWHS n.d.).

After basic training, . . . we rotated through the laboratories of the Naval Hospital in Bethesda, Maryland and soon were stationed, . . . I doing Parasitology Research in the Naval Medical Research Institute, still in Bethesda. We served about three years on active duty . . . I still at Bethesda in charge of the Parasitology Research Lab at NMRI. . . . I loved

Fig. 23.2 Alice Stoll in her Navy uniform. (Courtesy of Southeastern Pennsylvania Cold War Historical Society)

> the Navy and I stayed on in the Reserves another twenty years, retiring in 1966 with the rank of Commander (SPCWHS n.d.).

After her release from the Navy, Stoll returned to Cornell University in New York City to work on her master's degree, staying on as a research and teaching assistant and research associate. She opted not to go for her PhD. She was employed in the Department of Physiology, where she worked in environmental physiology, instrumentation, and research methods. She also worked as a consultant for laboratories, including the Aerospace Medicine Lab at Ladd Air Force Base in Alaska, during 1952 and 1953. From 1946 to 1966, she also spent time on her duties within the Reserves (AIP Publishing n.d.; Proffitt 1999; SWE Records)

> After the war . . . I went to graduate school at Cornell Medical – New York Hospital. . . . I got the first degree awarded in the new science of Biophysics (later becoming a charter member of the Biophysical Society.) . . . I felt satisfied when I got the biophysics degree from Cornell. I did research in temperature regulation and heat transfer at Cornell Med (SPCWHS n.d.).

> No. I don't have a PhD. I went in the Navy when I was 25 and I was 28 when I got out. By the time I got my master's I was 30 and I said enough of this stuff. I've got to get some work done in this life. So they tried to get me to go get my PhD, but in those days a PhD was not three or four years, it was ten years. I'll be forty years old by the time I finish this. That's ridiculous. So I didn't do that. I just went to work (SPCWHS n.d.).

While she was at Cornell, Stoll, along with two others, filed for a patent that was awarded August 10, 1954. The US Patent number 2,685,795 was for a Pan-radiometer. The pan-radiometer, as described in the patent, was an instrument for measuring environmental radiation. The instrument was sold commercially as a Thermistor Radiometer. A thermistor measures both skin temperature and environmental radiant temperature. She stayed at Cornell until 1953 (AIP Publishing n.d.; SWE Records).

Stoll joined the US Naval Air Development Center (NADC) in 1953 and stayed there until her retirement in 1980. Her positions included physiologist, special technical assistant to the research director, Head of Physiology Department, Head of Thermal Division, and Head of Biophysics and Bioastronautics for the crew systems department until her retirement in 1980 (Bailey 1994; Proffitt 1999; SWE Records).

> I was doing research at Cornell Medical College, New York City. I was in the physiology and biophysics division there. We were studying pain, thermal exchange with the environment and burns and things like that. My boss [Dr. James Hardy] was offered the directorship at the Aviation Medical Acceleration Laboratory at NADC. He accepted that and took with him seven people from Cornell. I was one of those people (SPCWHS n.d.).

> In 1953 we [Alice and Maria] accepted positions with the Navy Department in the Aviation Medical Acceleration Lab, Johnsville, Pennsylvania. Here we both enjoyed long careers in various fields . . . acceleration in preparation for space exploration, astronaut training on the centrifuge, nuclear, biological and chemical warfare defense, (Figures 23.3 and 23.4) protective clothing and all kinds of heat transfer from nuclear level thermal radiation to flame contact and rocket exhaust exposures. On the practical side we developed the fire resistant flight clothing in current use in the military, as well as infant wear and racing car drivers suits. We provided technical consultation and direction for the establishment of the

Fig. 23.3 Alice Stoll with mannequin at Nuclear Radiation Test Facility. (Courtesy of Southeastern Pennsylvania Cold War Historical Society)

Fig. 23.4 Alice Stoll at Nuclear Radiation Test Facility. (Courtesy of Southeastern Pennsylvania Cold War Historical Society)

> Flammable Fabrics Safety Lab at the US Bureau of Standards, etc. Our field studies have tested fire suppression systems, burn hazards of ejection seat deployment and tank explosions as well as protective clothing for flight attendants in civilian aircraft, conducted for FAA (SPCWHS n.d.).

As a physiologist, she conducted research on acceleration and cardiovascular effects. Stoll examined data from previous research that had investigated the effects of centrifugal acceleration on chimpanzees and humans, knowledge necessary for applications in aviation and spaceflight. She used these data to develop a Stoll Curve demonstrating human tolerance to centrifugal acceleration. Knowledge and application of the Stoll Curve would later enable designers of aircraft and space vehicles to determine when pilots needed to be protected against high g-force conditions (Bailey 1994; NIHF 2025; SWE Records).

As Stoll recounts in an oral history, the work on acceleration from her arrival at NADC in 1953 until 1956, and the development of the Stoll Curve for human tolerance of acceleration, stemmed from a happy accident (SPCWHS n.d.).

> Actually when I went to NADC, I was in the thermal research area to begin with. But we hadn't yet set up a laboratory at NADC. And so I really had nothing to do. So Doctor Beckman, who is the author of the experimental data and who was head of physiology at that time, told me that he had this big bin full of acceleration data and I thought acceleration lab, I should do something about that. And I asked would it be all right if I looked at the data and made an analysis of whatever. And he said, "Oh, that would be great." So that's what I did. And although his article had been published a couple of years before, I took a different view on it. And analyzed the three hundred pieces of evidence from the centrifuge work and came up with an article (SPCWHS n.d.).

She became the first woman to ride in the centrifuge (SPCWHS n.d.)

> I wanted to ride that centrifuge. And, of course, women, no your breasts are going to get squashed. That's the first thing they worry about. And your heart's not going to take it and so on. There happened to be a federal law that said you can't subject anybody to any experiment that you wouldn't subject yourself to. So I says there's a law. And I'm doing this with human subjects here so I want to ride the centrifuge. And the law says I can't subject blah blah blah blah blah. Well, okay, what could they do? They let me ride the centrifuge. And I rode it to gray out. That was about 7.5 g. The same as a man. And the heart rate [was normal]. Everything was just fine. So after that they took all the secretaries and everybody and they had a whole study of women on the centrifuge. It was great. (SPCWHS n.d.).

Her subsequent research examined thermal tissue damage and pain sensation. At one point, she also carried out significant administrative duties when she acted as the research director during his absence for military service. As Head of the Thermal Division, she conducted studies in thermal radiation and flame contact effects, developed fire-resistant materials, and devised methods of study. She continued this type of research in her subsequent position, which also included examining spacecraft atmospheres. Some of her work stemmed from the fact that the armed services were developing supersonic planes and wished to understand the impacts those planes would have on their crews. Thus, her work included understanding the effects on crews of extremely cold temperatures at high altitudes, breaking the sound barrier, and the danger of fire in a closed environment (Bailey 1994; SWE Records).

Stoll revolutionized the approach to protection of military and civilian personnel against skin burns due to flame contact and thermal radiation. She invented an apparatus and the associated method to analyze heat transfer by flame contact, for which research had not been previously conducted. Knowledge did not exist on the production and prevention of thermal burns. Because of her research and apparatus, it became possible to rate different materials as to their ability to protect humans from thermal pain and blisters. She also discovered, due to an accidental burn of a test participant, that clothing not only acts to protect skin but can also catch on fire and burn skin. This led her to the study of burning properties of fabrics when the concentration of oxygen was very high—such as would be found in an aerospace environment. She conducted much of this research jointly with chemist Maria Chianta (NIHF 2025; Proffitt 1999; SWE Records).

> I think it was 1957 that we were asked to find a substitute for nylon. Nylon was used in the flight suits of the aviators. And the trouble with it was that it was flammable. When they were in a fire, the suit would melt to the skin. When they pulled the burnt fabric off, the skin came with it. We were asked to find a substitute that would not melt and cause this affliction (SPCWHS n.d.).
>
> You have to start with the basics. You have to understand what creates the burn. What the temperatures are and the exchange of heat that causes a burn. We went into that with the instruments that we used in the thermal lab – radiometers – and we created various heat sources – infrared, and the flames, and that kind of thing. And at the same time, we studied the actual transfer of heat from the environment to the skin. In the process, we had to devise the flame, the fire pit (SPCWHS n.d.).
>
> We drew up our specifications for the fire pit and had it built by an outside contractor. We were simulating a flight man getting out of a plane that landed on a carrier deck. He would have to run through flames (totally enveloping flames) across the deck. So we timed that and the average time was two and a half seconds. But we made the criterion three seconds to give a margin of safety. And then when we built the pit, we also had to devise a means of giving it a square wave exposure to the flames because all of the basic research is based on square wave impulse. It's the only way you can control how much heat transfer is done in a certain amount of time (SPCWHS n.d.).
>
> We started looking at many fabrics, subjected them to the flame contacts, infrared heat and so on and in the course of this exercise, we learned about a filament that DuPont was working with. It was called HT-1, which was high temperature 1. This filament was being created to make a filter that would fit over the chimneys of the plants that were producing very hot smoke and flames. It was to be flame proof and catch the burning cinders so they didn't get into the environment (SPCWHS n.d.).
>
> We contacted the people in charge of that and told them what we wanted to do. Well this material was a filament and it was very thick – totally unsuitable for clothing. But we figured, if we work on it – it doesn't melt – which is a big advantage – and DuPont can make it thinner. And so for the next ten or fifteen years, DuPont and our lab cooperated in producing fabric out of this filament. They also got the material to be spun material that you could make into more common fabric. But this was a long, long row to hoe (SPCWHS n.d.).

DuPont's path to the development of Nomex® was long, but satisfactory. Their development process spanned from HT-1 to HT-4. At that point, it was suitable for clothing and then they had to get a name for the new fabric. (SPCWHS) DuPont first sold Nomex® in 1967 (DuPont 2025).

> I didn't know this but companies convene a panel to find a name and at DuPont they use a six [sic] letter name. Because it had the no melting quality, they were going to call it Nomel. Then, of course, one of the criteria is that you have to spell it backwards to make sure it doesn't spell anything. That one spelled lemon. They changed it to Nomex (SPCWHS n.d.).

Her efforts demonstrated that fabric constructions of inherently fire-resistant fibers were not only feasible but also far superior to fabrics treated with fire-retardants.

Her efforts resulted in the acceptance of a material (Nomex®) for standard-issue fire-retardant clothing to all branches of the military. Hospitals use Nomex® for bed linens and sleepwear for paraplegics. It is used in protective clothing for race car drivers and firemen, and in any number of everyday uses, from thermal underwear to ironing board covers to aircraft upholstery. It is used by industrial workers, especially those who could be exposed to hazards, including flash fires and electrical arcs. Her work also led to efforts on the part of several textile manufacturers to develop better fire-resistant fibers and fabrics (Bragg 2024; NIHF 2024; Proffitt 1999; SWE Records).

Stoll said she became an engineer through the back door as she told *Mechanical Engineering* magazine in 1973 (Bragg 2024):

> I entered the engineering field via the backdoor in the days when bioengineering had not yet been invented and interdisciplinary activities scarcely existed. Medical and biological researchers in those days bought instrumentation off the shelf or made their own. Since I greatly enjoy both designing and constructing hardware to provide solutions to laboratory problems, as time went on I became more and more engrossed in this kind of work. Finally, sometime around 1960, the engineering community decided that 'by their works ye shall know them' and gathered me into their fold.

23.4 Publications and Significant Research

A timeline of Stoll's papers, research, and results includes (SWE Records):

- Fundamental biophysical data published in 1959, which related time and temperature to the production of burns in humans and presented an analysis of the relationship between pain and tissue injury.
- The factors associated with the protection possible through Nomex® flight clothing was published in 1962.
- Three papers in 1964 described the apparatus and method (US Patent 3,148,531—Fig. 23.5) for analyzing flame contact heat transfer between thin layers of materials in two-layer systems, the experimental validation of the mathematical equation involved, and the application of the apparatus and method as described in the patent to the determination of thermal constants in materials in thin layers.

Additional work that she accomplished led to the development of the Stoll Curve. These included: (SWE Records).

- Papers from 1965 to 1967 discussing the significance of the observed heat transfer data with respect to protection from flame contact as compared with radiation (generally, two-layer systems that were effective with regard to flame contact had little or no use in protection from thermal radiation).
- The paper (1969) that laid out the derivation and use of what is now known as the Stoll Curve—a practical rating scale for evaluating thermal protection. Correlations were developed from empirical observations and compared to knowledge from heat transfer studies. The absorbed energy that results in pain

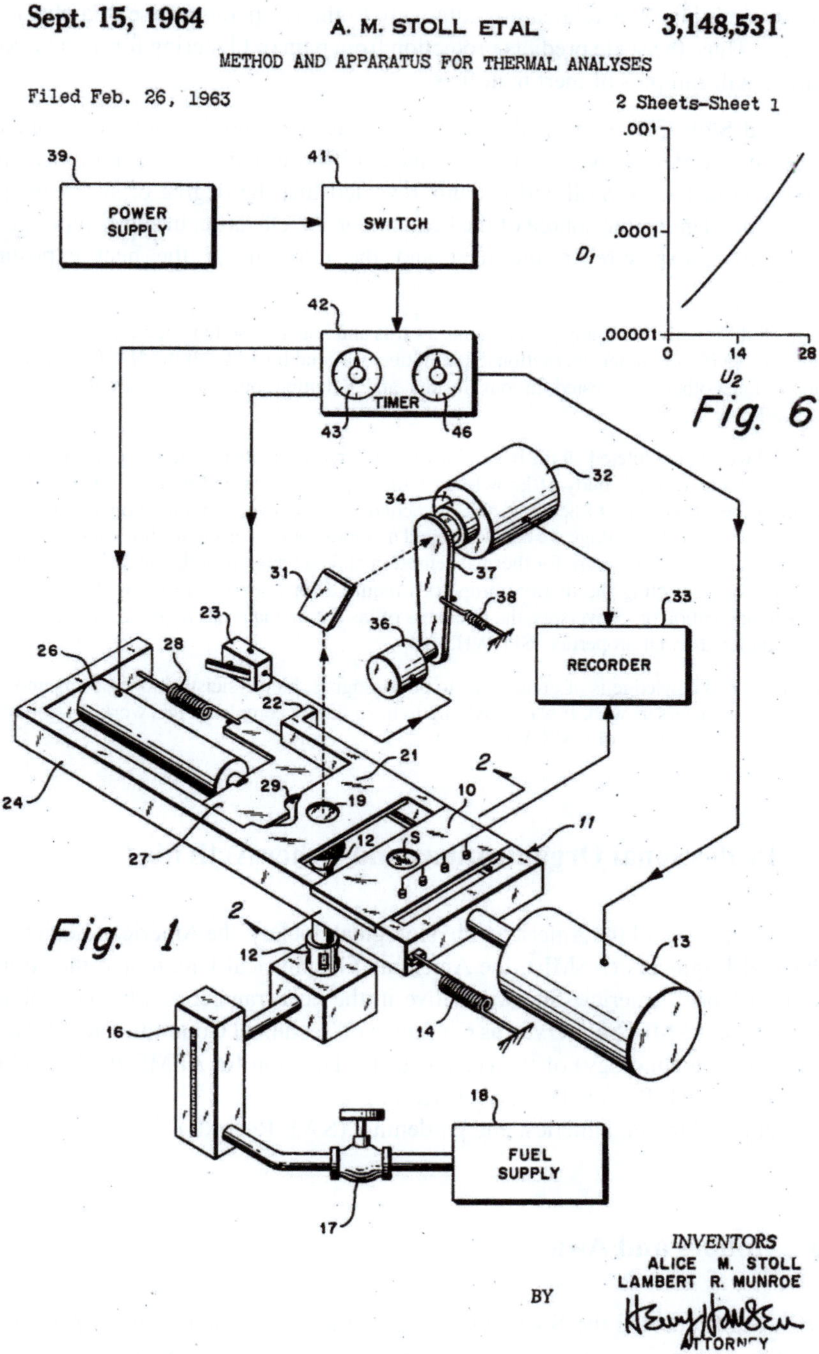

Fig. 23.5 Page 1 from US Patent Number 3,148,531—Alice Stoll "Method and Apparatus for Thermal Analyses"

and tissue injury was related quantitatively to the temperature rise in a physical system. Thus, the scale predicts protection from pain or blistering for human skin using small samples of inert materials.

The second Stoll Curve charts the heat levels and the time duration necessary to produce second-degree burns and pain under differing exposure conditions. The research conducted by Stoll and Chianta revealed that the degree of skin damage was not dependent on the source of the heat, but on the elevation in skin temperature resulting from exposure to the heat and the duration of the heat exposure (NIHF, 2025).

> From all these activities many publications, awards and honors resulted, including appointments to NATO and other international committees, official travel to all the NATO nations and various others interested in our results and adapting our methods to their use (SPCWHS n.d.).
>
> By 1980 we had completed all the basic research required to create an internally consistent, experimentally verified, body of knowledge with respect to pain and burn in human skin produced via all avenues of heat transfer, viz., conduction, convection, radiation and flame contact. From this knowledge we had developed methods for measuring the thickness of the skin and the skin temperature for thermal sensation and damage given the intensity of heat applied, and predicting the thermal properties required for materials to be used for skin damage prevention or, conversely, the thickness of insulation materials required for materials of known thermal properties (SPCWHS n.d.).
>
> This body of knowledge is of great value to physiologists, biophysicists, bioengineers and materials engineers as well. It was satisfying to have finally completed this work – a good time to retire and so we did (SPCWHS n.d.).

23.5 Professional Organizations and Other Activities

Stoll was a member of the American Physiological Society, the American Society of Mechanical Engineers (ASME), the American Geophysical Union, and the Arctic Institute of North America. She was active in the heat transfer and biotechnology sections of the ASME. She served as Chairman of Technical Committee K-17 (Heat Transfer in Biotechnology) of the Heat Transfer Division for ASME from 1965 to 1968 (Bailey 1994: Proffitt 1999; SWE Records).

Stoll enjoyed travel, athletics, and gardening. (SWE Records).

23.6 Honors and Awards

In 1969, Stoll received the Society of Women Engineers' Achievement Award *In recognition of her significant contributions in the field of the development of fire-resistant fibers and fabrics, based on her pioneering studies of heat transfer by flame contact.* In 1972, she received the Paul Bert Award from the Aerospace Medical Association. This award recognizes outstanding research contributions in aerospace physiology (AsPS 2025; ETHW 2025; Proffitt 1999).

Stoll was a Fellow of the American Association for the Advancement of Science and the Aerospace Medical Association. She was an Associate Fellow and charter member of the Biophysical Society (Bailey 1994: Proffitt 1999; SWE Records).

She received many awards from the Naval Air Development Center including incentive awards for publications of technical papers in acceleration research, flame contact studies, thermal protection capacity of aviator's textiles, a thermistor temperature-gradient measuring device, atmospheric oxygen enrichment effects on burning rates, skin temperature measurements, thermal protection principles, chapter on heat transfer in biotechnology, and invention disclosure and patent of instrument for heat transfer analysis. Stoll was named the Civil Servant of the Year 1965 by the Federal Business Association and the Federal Personnel Council of Philadelphia for significant achievements in thermal research. She also received a letter of commendation in the *Navy Officer's Jacket* for efforts within the Office of Naval Research (SWE Records).

Stoll was inducted into the National Inventors Hall of Fame in 2024 in recognition of her development of fire-resistant fibers and fabrics (NIHF 2024).

23.7 Legacy

Alice Stoll has not one, but two curves named after her as the "Stoll Curve." One Stoll Curve relates to the force of acceleration and g-forces on the body. This Stoll Curve was important fundamental data needed to safely put humankind in space and to protect fighter pilots. The other Stoll Curve relates heat exposure over time and predicts burn levels. The research conducted to develop the second Stoll Curve and the associated work performed in collaboration with DuPont led to the fabric Nomex® today used by firefighters, race car drivers, industrial workers, hospital linens, infants' clothing, and more all around the world.

The Stoll Curve for heat flux is still applied today. It is used in computer programs for applications in arc flash and flash fire laboratories. The calculations help determine the ability of fabrics to serve as personal protective equipment for industrial workers. (Tyndale 2023).

In 1998, The Maria A. Chianta and Alice M. Stoll Chair in Chemistry/Physics was established at Hunter College.

Acknowledgement With deep appreciation to Eleanor O'Rangers and the Southeastern Pennsylvania Cold War Historical Society for sharing materials on "Alice" with the author and for nominating Alice Stoll to the National Inventors Hall of Fame.

References

Aerospace Physiology Society (AsPS). AsPS Annual Awards—Paul Bert Award. https://www.aerospacephysiologysociety.org/awards. Accessed February 1, 2025.

AIP Publishing. Review of Scientific Instruments. n.d. https://pubs.aip.org/aip/rsi/article-abstract/25/2/184/298380/A-Wide-Range-Thermistor-Radiometer-for-the?redirectedFrom=fulltext. Accessed January 29, 2025.

Bailey, Martha. 1994. *American Women in Science: A Biographical Dictionary*. Santa Barbara, California: ABC-CLIO.

Bragg, Marsha Lynn. 2024. Research Physiologist Named to National Inventors Hall of Fame. *SWE Magazine* (Society of Women Engineers). https://swe.org/magazine/alice-stoll/. Spring.

DuPont. Nomex® fiber Technical guide. https://www.dupont.com/content/dam/dupont/amer/us/en/personal-protection/public/documents/en/Nomex(R)%20Fiber%20Technical%20Guide.pdf. Accessed February 4, 2025.

National Inventors Hall of Fame (NIHF). 2025. *Alice Stoll*. https://www.invent.org/inductees/alice-stoll. Accessed January 27, 2025.

National Inventors Hall of Fame (NIHF). 2024. 2024 NIHF Inductee Alice Stoll: A Leader in Fire Safety. https://www.invent.org/blog/inventors/alice-stoll. March 14.

Proffitt, Pamela, editor. 1999. *Notable Women Scientists*. Farmington Hills, Michigan: Gale Group.

Engineering and Technology History Wiki (ETHW). n.d. SWE Awards. https://ethw.org/SWE_Awards. Accessed February 1, 2025.

Society of Women Engineers (SWE) Records (LR001539). Box 193, Folder 1, Walter P. Reuther Library, Archives of Labor and Urban Affairs, Wayne State University.

Southeastern Pennsylvania Cold War Historical Society (SPCWHS). n.d. Unpublished.

Tynedale USA. 2023. What is a Stoll Curve? https://tyndaleusa.com/blog/2023/02/01/what-is-a-stoll-curve/. February 1.

Chapter 24
Irmgard Flügge-Lotz, PhD (1903–1974) Achievement Award Recipient 1970

For her significant contributions to the field of fluid mechanics, in particular, wing theory and boundary layer theory

Cecilia D. Craig

24.1 Introduction

Born in Germany in 1903, Flügge-Lotz learned her love of mathematics and science from her parents. She studied in Hannover and earned undergraduate and graduate degrees in engineering, graduating in 1927 with a PhD in engineering from Hannover Technical University (Vogt 2005).

Though Flügge-Lotz trained as an engineer, she worked as a research scientist in the 1930s. She developed the Lotz method, a process for calculating the lift of a three-dimensional wing. This became widely used, as did other key research, including the theory of on–off control systems. Her early career was set against a backdrop of Nazi Germany and a time when women had to fight to gain research positions and the titles that went with those (Vogt 2005). She married a civil engineer, Wilhelm Flügge, in 1938, an avid anti-Nazi scientist (Stanford News 1969).

After WWII, for a brief time in Paris, they both worked for a French research institute and ultimately took offers to join the engineering staff at Stanford University in 1948. She initially joined Stanford University as a lecturer and did not become a full Professor of Aeronautical Engineering and Engineering Mechanics until 1960 (Flügge-Lotz 1972). In 1968, she was awarded the designation of Professor Emerita.

In 1970, the Society of Women Engineers recognized her with the Achievement Award, and the American Institute of Aeronautics and Astronautics (AIAA) elected her a Fellow that same year. In 1971, she was the AIAA Theodore von Kármán

C. D. Craig (✉)
San Jose, CA, USA
e-mail: Ceal@Research.Druai.com

C. D. Craig et al. (eds.), *Women Engineering Legends 1952-1976: Society of Women Engineers Achievement Award Recipients*,
https://doi.org/10.1007/978-3-032-00224-2_24

Fig. 24.1 Portrait of Dr. Irmgard Flügge-Lotz. (Courtesy of Stanford University)

lecturer, the first woman so honored. Flügge-Lotz (Fig. 24.1) published over 50 papers and two books.

24.2 Early Life and Education

Born in 1903 in Hameln, Western Germany, Irmgard Lotz's parents were influential in her life, nurturing her technical interests. Irmgard's father was a mathematician. Her mother's family was in construction (Vogt 2005) and she had one sister (Ogilvie and Harvey 2000). With her uncle, she visited construction sites. Her mother encouraged her to study technical subjects (Stanford News 1969).

As a child, she went to local Sunday matinées where, at half-price admission, technical movies were offered showing technical problems and solutions (Stanford News 1969). Her father would take her to airshows in the early days of aviation in Germany. These experiences, plus her mother's family's construction activities, sparked her interest in engineering (Flügge-Lotz 1972). Her father was often ill after WWI, and Flügge-Lotz tutored students to bring money into the home (Proffitt 1999). Thus, she learned early in life that a woman "had to stand alone and independent" (Flügge-Lotz 1972, p 1).

Flügge-Lotz majored in physics and mathematics at Hannover's gymnasium (one of the best in Germany at the time) and graduated in 1923. She was one woman in a class of about 1000 engineering students (Ogilvie and Harvey 2000). Then she earned a Diplom-Ingenieur or Dipl. Ing. (engineering) in 1927 at Hannover Technical University. Upon graduating, she became an assistant to a professor in applied mathematics and projective geometry at Hannover (Vogt 2005). She tutored high school students to pay for her doctoral studies tuition (Stanford News 1969).

She began dissertation research to solve the problem of manufacturing nails at a high rate of speed to avoid overheating. Thus, her doctoral thesis concerned the mathematical theory of heat conduction in circular cylinders (Johnsson et al 2023). This research in applied mechanics on heating processes earned her a Doktor-Ingenieur or Dr. Ing. (Doctor of Engineering) degree in 1929. (SWE Records n.d.; Vogt 2005).

The few women in the universities at that time in Germany were Dr. [Flügge-] Lotz shared, "eager to work out our plans for ourselves....my male fellow students did not consider me a rarity, but just another student" (Flügge-Lotz 1972, p 2). Some professors expressed concerns that she might not find work if she aimed for a man's career. She was willing to take the risk. She selected engineering mechanics as her field of study because, as a newer field, fewer issues arose from long-held opinions (Flügge-Lotz 1972).

24.3 Early Career and WWII

Dr. Flügge-Lotz started at the Aerodyamische Versuchsanstalt (AVA), or Aerodynamics Research Institute, in Göttingen and worked there from 1928 to 1938, beginning as a research scientist and ending as department head for theoretical aerodynamics. Initially, she did clerical work for half the time and research for the other half. This changed after Ludwig Prandtl and Albert Betz, aerodynamics experts at the AVA noticed her work (NAE n.d.).

"When Irmgard Lotz started working at the Kaiser Wilhelm Institute for Fluid Mechanics under the direction of Ludwig Prandtl, she was the only woman among twenty-five assistants" (Vogt 2005). She was an assistant at the Aeronautical Research Laboratory from 1929 to 1934 until she became head of the aerodynamics theory department, though not officially. She lectured for weekly events at the University of Göttingen for five years on topics of "complex solutions of potential fluids," wind tunnel measurements, ray calculations, and "new arithmetical calculating machines" (Vogt 2005, p 201). Prandtl attempted to appoint her to a special research position, but the Nazi regime rejected it, likely because she was a woman (Vogt 2005).

Before working for Prandtl, while working at AVA in Göttingen, a top-level institution for aeronautical research, she created a process for calculating the lift of a three-dimensional wing. Flügge-Lotz worked on fluid and wing flow dynamics at the Kaiser Institute. Rapidly, she became well-known in fluid mechanics. After joining Prandtl's group, which had been "unsuccessfully working on solving a differential equation for his lifting-line theory for the spanwise lift distribution of an airplane wing" (Johnsson et al. 2023, p 4695), she solved the equation for the general case (Proffitt 1999). Then, she developed an easier method to find solutions using "simple tools (slide-rule and desk computer)" (Flügge-Lotz 1972., p 4). This became known as the *Lotz method* "for calculating spanwise (wingtip to wingtip) distribution of a wing's lifting force" (Flüggee-Lotz 1931; Vogt 2005, p 200).

At the Institutes of the Kaiser Wilhelm Society, between 1912 (the year it was founded) and 1945 (the end of the Nazi regime), Vogt identified at least "225 women scientists with doctorates" (191) in records. Mathematician Irmgard Lotz was one of four women who received doctorates in technical colleges (Vogt 2005).

Scientific women in Germany did find a community of practice in the military institutes, where they were essentially hidden, versus in universities. These special institutes provided opportunities for women scientists, mathematicians, and engineers, coupled with some institute leaders who paid the women the same and gave them opportunities, more than what happened at universities in Germany (and the US to an extent) (Vogt 2005).

Irmgard Lotz married Wilhelm Flügge in 1938. He was a civil engineer and an avid anti-Nazi, which kept him and Flügge-Lotz from university appointments. Working in research was considered acceptable, and they worked for that kind of group until after WWII ended (Stanford News 1969). She hyphenated her name since she had already published papers under her maiden name, Lotz. Married to another outstanding scientist, "marriage considerably impeded [her] career advancement" (Vogt 2005, p 202). Well-known in the fluids community, she had unofficial status at times, but the Nazis prohibited married women from working. "The Nazis felt that non-Nazi teachers might contaminate the youth" (Stanford News 1969), obstructing any university appointments. They moved from Göttingen after discrimination and harassment escalated and moved to the Deutsche Versuchsansalt fur Luftfahrt or German Aeronautical Research Establishment (DVL) (Stanford News 1969).

The Flügges moved to Berlin in 1939, where Wilhelm began working for the Deutsche Versuchsanstalt fur Luftfahrt (DVL). Soon after arriving, DVL offered Irmgard a job as a consultant on aerodynamics and fluid dynamics. Their work here was not directly involved with war efforts (Stanford News 1969). She did significant research on on–off controls during this time. These low-cost approaches were quickly and widely adopted (Proffitt 1999). She continued this research and application development later at Stanford University.

With the close of WWII, German scientists were pushed into various Allied centers. Going back to doing research in Germany was not viable (research that a military might use was limited in Germany until 1955). Thus, they began to look to new horizons (Vogt 2005).

24.4 Mid-Career

Flügge-Lotz and her husband moved to work in Paris from 1946 to 1948 where she was a research group chief for the French National Office for Aeronautical Research (ONERA). She gathered colleagues to work for ONERA. "The French allowed me to choose my associates. It was a good feeling that these men with their families were confident to work with me in a foreign country" (Flügge-Lotz 1972, p 5).

During this period at ONERA, she continued research on automatic control theory (Proffitt 1999).

In 1948, the couple accepted positions at Stanford University and moved to California. Stephen Timoshenko, at Stanford University, facilitated their visit to California. As a result, Wilhelm was offered a full professor position and Irmgard a "lecturer and research supervisor" role (Vogt 2005, p 204). As she was a woman and they were both teaching in the same department, Stanford policy dictated they could not both be professors, even though she was more highly regarded in theoretical aerodynamics. Students and colleagues questioned this lower-level position, though she was doing the work of a full professor (Vogt 2005). She developed a "year-long sequence of courses in mathematical hydro- and aerodynamics for first-year graduate students" (NAE n.d.). Nonetheless, she quickly had PhD students and, in 1949, began teaching boundary layer theory and published papers at Stanford University describing how to use computers for finite difference methods (Vogt 2005).

Flügge-Lotz developed a theory for discontinuous or on–off control systems in her first years at Stanford. These on–off control systems are often used in furnaces or other heating and air conditioning systems; when the temperature reaches a certain point, the system turns off or on. Her textbook, *Discontinuous Automatic Control*, published in 1953, was a foundational step for control systems (Johnsson et al. 2023).

Continuing work on control systems, Flügge-Lotz and a colleague developed a theory and performed experiments that showed how nonlinear control systems of the third-order operated: "The physical manifestation of this control process is a linear system with discrete variable proportional and derivative feedback around the linear member" (Flügge-Lotz and Taylor 1956, p 4). This work had similarities to the research performed soon after in a Federal Grant.

With Flügge-Lotz as the principal investigator, Stanford University received a $15.5 K Federal grant (NA-w-6655) in 1958 for an "investigation of third-order nonlinear control systems, including studies of the operation of quasi optimum control systems and the effects of imperfection in the systems on their operation (continuation NA w-6533)" (House of Representatives 1960, p 197). The experimental results demonstrated that a nonlinear control system performed more effectively than a linear one. Nonlinear systems depend less on past responses. Suggestions were made for the practical use of discontinuous coefficient variations introduced before nonlinear elements are added to a linear control system. (Flügge-Lotz et al. 1958). Four papers (House of Representatives 1960, p 197) shared specific findings from this grant's work, as shown in a Congressional sub-committee hearing for the 1961 NASA authorization.

Flügge-Lotz represented the United States in Russia in 1960 at the Moscow International Federation of Automation Control (IFAC) Congress. She presented "Synthesis of Third-order Contactor Control Systems." First, she described the problem as optimizing a third-order contactor control system, then showed the mathematics behind this system and how it might be optimized. She concluded:

> The most important result of these investigations on third order systems with velocity control is the fact that, after reaching an end-point, the motion of the controlled vehicle does not depend on the parameters of the uncontrolled system (Flügge-Lotz 1960, p 405).

Flügge-Lotz was the only female delegate from the United States at this first international controls conference. The technical world took notice of the level and quality of her work. She was a leader in discontinuous control systems and their use in aviation and space technologies. Thus, by 1960, her Stanford University position title (i.e., lecturer) came into serious question. "That she was the only female delegate from the United States surprised the other participants, that she did not have the rank of full professor left them in utter disbelief" (Stanford News 1969).

This disparity had reached too many: when she returned for the fall semester in 1960, it was as a full professor in Engineering Mechanics and Aeronautics. "Stanford had existed for 70 years, but she was its first woman professor in engineering" (Vogt 2005, p 204). Stanford University's web page (dated 2021) for Flügge-Lotz states she "joined the Stanford faculty in 1950 as the University's first female professor of engineering, " rewriting some history showing 1950 instead of 1960. They note:

> She made significant advancements in methods for predicting aerodynamic pressures on bodies, wings, and turbine blades, some of which were adopted as standard procedures worldwide. In automatic control theory, she developed the first theory of discontinuous, or on-off, control systems (Stanford University 2021).

Earth-pointing satellites needed effective control systems for long-term operation, particularly in an elliptical orbit. A 1967 paper by Busch and Flügge-Lotz (1967/2012) outlined a control system for this situation. Next year, with a colleague at Stanford, Theodore Reyhner, she explored laminar boundary layers and shock waves on a flat plate, looking at viscosity and heat conduction (Reyhner and Flügge-Lotz 1968). Increasingly focusing on control theory, Flügge-Lotz published a second book in 1968, *Discontinuous and Optimal Control.* She retired from Stanford that same year but continued her research. Ultimately, she published over 50 technical papers. (Johnsson et al. 2023).

24.5 Later Career

After retiring in 1968 as Stanford required (she was 65), Dr. Flügge-Lotz continued to research problems, notably "mission and satellite control problems and problems of heat transfer and drag of fast-moving vehicles [continuing to have] a life full of work…never boring" (SWE Records n.d.).

Flügge-Lotz was invited to present at the 1971 Theodore von Kármán lecture hosted by the American Institute of Aeronautics and Astronautics (AIAA): "Trends in the Field of Automatic Control in the Last Two Decades." She had met Theodore von Kármán, who worked with Prandtl at the ARI earlier, for the first time in 1934 at a conference in England. (Flügge-Lotz 2012).

In her talk, Flügge-Lotz related that since von Kármán had first suggested that engineers needed to consider nonlinearities in 1940, Flügge-Lotz became interested in exploring those. This led to a life's work exploring, defining, and sharing how nonlinear control systems would benefit various situations. While much of her work was used in aeronautics and astronautics, she saw control systems as any system.

> Built to relieve the human being from any dull and repetitive operation or may be to supervise a chemical process or may be to determine the path of a rocket, a missile, and finally to let satellites fulfill their task without manual interference (Flügge-Lotz 2012, p 721).

Irmgard Flügge-Lotz and her husband, Wilhelm lived in Los Altos, CA. She died in 1974, at the age of 70.

> While her teaching stopped with retirement, her research on problems of satellite control and of heat transfer, and drag of highspeed vehicles continued actively right to the end. This was the more remarkable because of the constant pain and discomfort with which she lived for many years as disabling arthritis attacked first one part and then another of her body. Through it all, she kept her good spirits and maintained frequent contacts with friends and colleagues, all of whom were sadly shocked by her sudden death (Spreiter, Van Dyke, and Vincenti 1975).

24.6 Awards, Recognition, and Professional Organizations

With over 50 papers published in her career and two books, Flügge-Lotz earned awards and recognition for her work and its impact. She received the SWE Achievement Award in 1970, *for her significant contributions to the field of fluid mechanics, in particular, wing theory and boundary layer theory: notably, computation of wing-lift distributions.* (Fig. 24.2.).

Fig. 24.2 Dr. Irmgard Flügge-Lotz. (Society of Women Engineers Records, Walter P. Reuther Library, Archives of Labor and Urban Affairs, Wayne State University)

Flügge-Lotz was awarded an honorary doctorate from the University of Maryland in 1973, that cited she had "a central role in the development of the aircraft industry...her contributions have spanned a lifetime during which she demonstrated, in a field dominated by men, the value and quality of a woman's intuitive approach in searching for and discovering solutions to complex engineering problems. Her work manifests unusual personal dedication and native intelligence" (Johnsson et al. 2023, p 4696).

Three professional engineering organizations recognized her: AIAA Fellow (first female Fellow) in 1970 (NAE n.d.) and first woman invited to give the annual Theodore von Kármán lecture in 1971 (NAE n.d.; Ogilvie and Harvey 2000); and Sigma Xi (honorary scientific research society); and IEEE Senior Member.

24.7 SWE Involvement

Irmgard became a member of SWE in 1957 (SWE Records n.d.). The Pacific Northwest Section nominated her for the 1970 SWE Achievement Award. In the nomination letter, Dr. Irene Peden (1973 SWE Achievement Award recipient), Electrical Engineering Professor, University of Washington, SWE Pacific Northwest Section shared: "In private life, the wife of Professor Wilhelm Flügge is a warm and compassionate woman who radiates interest in and concern for students. She has been very much respected and loved by them, in turn" (Peden 1969).

Flügge-Lotz was a member of the Santa Clara Valley section, one of the early SWE sections. In 1992, the section posthumously established an endowed scholarship in her honor, which is awarded each year. The scholarship citation reads:

> **Irmgard Flügge-Lotz Memorial Endowment Scholarship (established 1992)**
> The Irmgard Flügge-Lotz Scholarship was one of our first endowed scholarships, recognizing a true aeronautics pioneer. Born in Germany in 1903, she was one of the first women in the world to study aeronautics, earning her doctorate in 1929. She applied her mathematical skills to develop lifting line theory, a fundamental breakthrough to compute the lift distribution of an airplane wing. She left Europe in 1948 to teach at Stanford University, with a focus on fluid mechanics, numerical methods, and automatic control theory. In 1960, she became the first female full professor in Stanford's College of Engineering. In 1970, she was selected as the recipient of the SWE Achievement Award, the highest honor awarded nationally by SWE. (SWE Santa Clara Valley 2025).

24.8 Legacy

Flügge-Lotz consistently made her mark in the mathematics of aerodynamics, control systems, and fluid mechanics. She wrote two foundational books on control systems and was recognized by AIAA, IEEE, and SWE for her research accomplishments. She is known for developing the Lotz method, a process for calculating the lift of a three-dimensional wing, and for the theory of on–off control systems.

Her legacy, in her own words:

I wanted a life that would never be boring. That meant a life in which always new things would occur. I wanted a career in which I would always be happy, even if I were to remain unmarried. (Johnsson et al. 2023, p 4696).

References

Busch, R. E. and I. Flügge-Lotz. 1967/2012. Attitude control of a satellite in an elliptic orbit.1967 April/2012 May 23. Journal of Spacecraft and Rockets 4 (4). https://doi.org/10.2514/3.28883

Flügge-Lotz. I. 1931. Computation of the lift distribution of wings of arbitrary form. *Zeitschrift fur Flugtechnik und Motorluftschiffahrt* 22: 189 (in German)

Flügge-Lotz, I. 1953. *Discontinuous Automatic Control*. Princeton University Press.

Flügge-Lotz, Irmgard. 1960. Synthesis Of Third Order Contactor Control Systems at the First International Federation of Automatic Control (IFAC) Congress on Automatic and Remote Control, Moscow. *IFAC Proceedings* 1(1).

Flügge-Lotz, I. 1968. *Discontinous and Optimal Control*. McGraw Hill.

Flügge-Lotz, I. 1972. Talk at a meeting for the San Francisco BA Section of SWE. *Irmgard Flügge-Lotz*, Society of Women Engineers Records (LR001539). Box 188, Folder 9, Walter P. Reuther Library, Archives of Labor and Urban Affairs. Wayne State University.

Flügge-Lotz, I. 1972/2012. Trends in the Field of Automatic Control in the Last Two Decades. *American Institute of Aeronautics and Astronautics* 10 (6). 1972 June/2012 May 17. https://arc.aiaa.org/doi/abs/10.2514/3.50199?journalCode=aiaaj

Flügge-Lotz and C. Taylor. 1956 May. Synthesis of a nonlinear control system. *IRE Transactions on Automatic Control* 1(1): 3–9, doi: https://doi.org/10.1109/TAC.1956.1100823. May.

Flügge-Lotz, I., C.F. Taylor, and H. E. Lindberg. 1958. Investigation of a nonlinear control system. *National Advisory Committee for Aeronautic*s. https://ntrs.nasa.gov/citations/19930091023

House of Representatives, Committee of Science and Aeronautics, Subcommittee No. 2. 1960. Review of the Space Program: 1961 NASA Authorization, page 197. February 19. Accessed March 11, 2025. https://www.google.com/books/edition/Hearings/sLW-U8B-bUQC?hl=en&gbpv=1&bsq=flugge

Johnsson, Charlotta, Ulrika Jönsson-Belyazid, Eva Westin, Kristian Soltesz, Tore Hägglund, and Margret Bauer. 2023. Historical Female Influencers in Automatic Control. *IFAC Papers Online* 56 (2): 4693–4698. https://doi.org/10.1016/j.ifacol.2023.10.989

National Academy of Engineering (NAE). (n.d.) Irmgard Flügge-Lotz: Pioneer in Aviation Theory. *Engineer Girl*. https://www.engineergirl.org/125328/irmgard-flugge-lotz

Ogilvie, Margaret Bailey, and Joy Dorothy Harvey. 2000. *The Biographical Dictionary of Women in Science: Pioneering Lives from Ancient Times to the mid-20th Century*. Routledge.

Peden, Irene. 1969. Nomination Letter to Dr. Marguerite Rogers. February 18. *Irmgard Flügge-Lotz*, Society of Women Engineers Records (LR001539). Box 188, Folder 9, Walter P. Reuther Library, Archives of Labor and Urban Affairs. Wayne State University.

Proffitt, Pamela. 1999. *Notable Women Scientists*. Gale Group.

Reyhner, Theodore A., and I. Flügge-Lotz. 1968. The interaction of a shock wave with a laminar boundary layer. *International Journal of Non-Linear Mechanics* 3(2): 173–199.https://doi.org/10.1016/0020-7462(68)90014-0

Spreiter, John R., Milton D. Van Dyke, and Walter G. Vincenti. 1975. April. In Memoriam Irmgard Flugge-Lotz, 1903-1974. *IEEE Transactions on Automatic Control* CA-20 (2).

Stanford Evening News. 1969. A Life Full of Work - the Flügge. May 7. From Irmgard Flügge-Lotz, Society of Women Engineers Records (LR001539). Box 188, Folder 9, Walter P. Reuther Library, Archives of Labor and Urban Affairs. Wayne State University.

Stanford University. 2021. Irmgard Flügge-Lotz - aerodynamics and automatic control theory expert. Accessed March 14, 2025. https://engineering.stanford.edu/about/history/heroes/2014-heroes/irmgard-flugge-lotz

Society of Women Engineers (SWE) Records. n.d. *Irmgard Flügge-Lotz*. Society of Women Engineers Records (LR001539), Walter P. Reuther Library, Archives of Labor and Urban Affairs. Wayne State University.

Society of Women Engineers, Santa Clara Valley. 2025. Scholarships. https://scvswe.org/scholarships/.

Vogt, Annette. 2005. Women in Army Research: Ambivalent Careers in Nazi Germany (Chp. 7). In *Crossing Boundaries, Building Bridges,* edited by Annie Canel, Ruth Oldenziel, and Karin Zachmann. Routledge Studies in the History of Science, Technology and Medicine. Taylor and Francis. Kindle Edition.

Chapter 25
Alva T. Matthews, PhD (1933–2022) Achievement Award Recipient 1971

In recognition of her meritorious contributions to the field of engineering mechanics and applied mathematics in the area of shock analysis, elasticity and structural design

Deborah J. Fisher-Rebstock

25.1 Introduction

Dr. Alva T. Matthews (Fig. 25.1) made significant contributions in many areas of engineering mechanics and structures. She conducted fundamental research on the mechanical behavior of materials that have been subjected to dynamic impact loads. This work influenced the development of models for wave propagation in materials such as soil and rock and extended to nuclear weapons effects on structures. Other accomplishments included the development and adaptation of large computer codes, structural design of thin shell concrete, application of potential theory to underwater acoustics, and computer programming in the development of the Telstar tracking antenna and of helicopter blade design. Matthews earned her BS, MS, and PhD degrees from Columbia University. Matthews was an active member of the Society of Women Engineers (SWE) and received SWE's 1971 Achievement Award in recognition of her meritorious contributions to the field of engineering mechanics and applied mathematics in the area of shock analysis, elasticity and structural design.

D. J. Fisher-Rebstock (✉)
Albuquerque, NM, USA

C. D. Craig et al. (eds.), *Women Engineering Legends 1952-1976: Society of Women Engineers Achievement Award Recipients*,
https://doi.org/10.1007/978-3-032-00224-2_25

Fig. 25.1 Dr. Alva Matthews Achievement Award portrait from 1971. (Photograph courtesy of Society of Women Engineers Records, Walter P. Reuther Library, Wayne State University)

25.2 Early Life and Education

Alva Matthews was an only child who grew up in Great Neck, Long Island, New York (New Cannanite 2022, p 1). Her father took engineering courses at Cornell, got his professional license to practice, and then founded his own industrial construction firm of Brown and Matthews. Her mother was a stay-at-home mom with interest in the arts. Matthews decided to study engineering relatively early. In her words, "When I was fifteen, it never occurred to me that I couldn't be an engineer. My father was an industrial builder and tramping around construction sites with him seemed as natural as learning the domestic things my mother taught me (Merrill 1961, p 40)."

In the early 1950s, Massachusetts Institute of Technology (MIT) and Middlebury College in Vermont had a 3-2 program, wherein one spent 3 years at Middlebury and 2 years at MIT, graduating with both a Bachelor of Science degree with a specialization in engineering and a Bachelor of Arts. Alva Matthews entered Middlebury College as one of very few women, shortly after her father died. Luckily, her dean at the time, Wes Hennessey, was starting to encourage women to enter engineering, so it lessened Matthews' barriers with college entrance or with on-campus courses. However, she did have stumbling blocks. "The first week of school at Middlebury my physics teacher and advisor told me not to study engineering. He said it was too difficult for a girl, and nobody would ever hire me. I was hurt and surprised. After the hurt wore off I decided to ignore him" (Fitzroy 1973, p 62).

She transferred to Barnard College after completing her freshman year at Middlebury to be closer to her mother in the wake of her father's death. Barnard College had an association with the engineering school at Columbia University. She completed her sophomore year at Barnard, and then her junior and senior years were spent across the street at Columbia University, where she majored in civil engineering, like her father. In order to enter Columbia University in her junior year, she needed one course that was not offered at Barnard. She recalled, "I lacked one course that was available at the all-male Columbia College, but women were not permitted even to sit in their classes. I took the course during the summer, at my own extra expense. Neither Columbia [University] nor Barnard thought this unusual or even worthy of comment. This time I was not hurt, I was mad" (Fitzroy 1973, p 62).

She recounted another memory about completing the curriculum. One of the requirements for the civil engineering degree at Columbia University was a summer surveying camp in the hills of Connecticut. Her experience was the same as the male students, with one exception:

> ... but the boys stayed in dorms, and no girl had ever attended. They first said it was impossible, but then I had second thoughts, so I went a year later than normal. I stayed off-campus at a guest house. The setting was country, and the students lugged their transits over the hills and through the fields. One assignment included the sighting of the north star, Polaris, at a precise time near two o'clock in the morning. My squad partners were considerate enough to call for me and walk me the mile distance in the dark to the sighting area. It wasn't so bad for one night, but clouds covered that star for four nights running and for four nights I disappeared into the dark with two young men. The lady who owned the guest house registered her disapproval, never believing a word of my story about Polaris (Fitzroy 1973, p 62–63).

In retrospect, Matthews felt that her years at undergraduate school were lonely. She did not date other engineering students nor did homework with them because, "It was often perceived as a date situation. Almost all my work was done alone" (Kata 2003). She summed up her feelings about her undergrad experience this way:

> I was very aware of the uniqueness of my situation, afraid of upsetting the balance of what I thought were "privileges" being extended to me. As a result, I even avoided dating the male engineering students. This caused me to miss the give-and-take of information that students use as part of their learning process. While the attention I found flattering, I was scared of making mistakes and looking idiotic in the spotlight (Fitzroy 1973, p 63).

During summer breaks Matthews did secretarial work, assisted in the Graphics Department, and one summer worked in the field office of the contractor who was digging the Delaware Water Gap tunnels to increase the water supply to New York City. She gathered information from the field on quantities of construction materials used each day. She yearned to, but was prohibited from, going down into the tunnels. "Superstition about women causing the collapse of tunnels was very strong. It was, however, a rare opportunity to see how heavy construction on a remote job site operates" (Fitzroy 1973, p 63).

25.3 Further Education

Matthews completed her BS in Science with honors in 1955, majoring in civil engineering with a specialization in soil mechanics. After graduating from Columbia University, she joined the Civil Engineering Department as an instructor in the

Strength of Materials Laboratory and assisted the head of the Department, Professor Garrelts, in his structures courses. At the same time, she was going to school at night to work on her master's degree. Her MS took her 2 years, instead of the usual 1 year, due to her teaching load, but for her it was well worth obtaining the teaching experience.

She completed her MS in Science with honors in 1957, with a specialization in mechanical engineering in structures, while teaching at Columbia University. She then began her association with the Weidlinger firm in New York City designing unusual structures. Matthews worked on thin shell concrete design at first, but later became involved in the research that Mel Baron was directing within the group in shock and vibration studies of huge gas storage tanks. That research involved a great deal of applied math and knowledge of elasticity and vibrations of structures. At the same time, she continued taking courses at Columbia University because she felt the need for more theoretical background, and more mathematics, to enable her work (Merrill 1961, p 63).

Matthews married Abby R. Solomon in 1959. She had first seen her future husband at the surveying camp while she was an undergraduate. Matthews stated that, "None of my accomplishments could have occurred without an understanding husband, who I met viewing him in [i.e. through] a transit at surveying camp" (Merrill 1961, p 64).

When her husband had finished his master's degree in 1961, he was offered a job with the Xerox Corporation in Rochester, New York. That brought up a problem that faces many dual-career married couples.

> I felt, of course, that he should take the job. For one year my husband lived in Rochester, and commuted to New York [to visit me] on weekends. Then we moved to Rochester, and I started commuting to New York during the week, staying at my mother's home. The solution seemed to work. We felt unusual among professional people at that time, but subsequently we learned of more commuting solutions to maintaining a home and a marriage (Merrill 1961, p 64).

Matthews became an Associate Professor in the Mechanical and Aerospace Sciences at the University of Rochester, lecturing in applied mathematics and elasticity (descriptive geometry) on Mondays and Fridays. Two days a week (Wednesdays and Thursdays), she would commute to New York City and conduct laboratory courses in Strength of Materials at Columbia University (New Canaanite 2022, p 30). Finally, she taught an undergraduate class in Structural Mechanics at Swarthmore College on Tuesdays (Kata 2003, p 19). Although Matthews thought she was preserving her family life with her husband (Merrill 1961, p 64), she was actually commuting to three locations. "So I thought, 'You know this is crazy,' so I didn't do it again. That's youthful enthusiasm" (Kata 2003, p 20).

25.4 Career and Doctoral Studies

During this period, Matthews carried on original research and development studies and consultation services in a variety of fields. One such field of importance was that of nuclear weapons effects in relation to the vulnerability of structures (Merrill 1961, p 64). This was an interdepartmental research-oriented engineering endeavor conducted with the Weidlinger firm. It was only natural that she eventually became a full-time employee at Weidlinger in the early 1960s, due to the multiple roles teaching at three universities and conducting research at Weidlinger (Kata 2003, p 4).

Matthews eventually stopped teaching at the University of Rochester and Swarthmore, but continued to teach courses at Columbia in mechanics, dynamics, and indeterminate structures along with conducting laboratory courses in Strength of Materials. In addition, she began to pursue her doctorate from Columbia University.

Because she needed the knowledge in her daily work, taking courses in engineering mechanics at night was more manageable. As she described it:

> Sort of without my realizing it the course credits for my Doctor's degree were accumulated. I was then able to write a dissertation directly from work done at the office. I found this as the most convenient way of completing degree requirements. The whole process took seven years from the time I got my Master's. There were many times when continuing seemed too much. But I did finish, and found that having that degree opened many doors and opportunities (Merrill 1961, p 63).

Matthews completed PhD research on the mechanical behavior of materials which have been subjected to dynamic impact loads. This work influenced the development of models for wave propagation in soil and rock and extended to nuclear weapons effects on structures (New Canaanite 2022, p 30).

> My dissertation was on shock waves progressing through certain types of materials that could be used to model the actions of soils if a bomb went off. The company used that in underground silo design. And then with a stretch we went to shock waves in water and did a lot of submarine hull modeling, mathematical modeling (Kata 2003).

As her career progressed, Matthews worked in other areas including the development and adaptation of large computer codes, structural design of thin shell concrete, application of potential theory to simplify iterative calculation of wave propagation in underwater acoustics, and computer programming in the development of the Telstar tracking antenna and of helicopter blade design (New Canaanite 2022, p 30).

> It was always, to me, wonderful when something I worked on was published. And that was always a great satisfaction, enormous satisfaction (Fitzroy 1973, p 11).

One of her favorite work assignments at Weidlinger was analysis of missile silos, where she could apply her mathematical concepts to physical phenomena. When computers first arrived in the workplace, Dr. Matthews and other engineers had to teach themselves FORTRAN in order to program their mathematical models of the soils and materials that were subjected to the blast effects. She recounted, "Well, from the point of satisfaction, it was satisfying to see the mathematics take shape in something that was very important to the country" (Fitzroy 1973, p 26). Matthews' work helped national security efforts during the Cold War.

25.5 Retirement

Matthews retired and focused on family when her daughter Stephanie, who was born 2 days prior to Matthews' 44th birthday, was about 5 years old. Matthews picked up her mother's interest in art (faux finishes of furniture and gilding with gold leaf). She also did database programming for friends who were starting their own businesses, as she could do this from home without commuting into the city. She became interested in showing horses. She later mused, "I think we are living so much longer now that we can transition in our lives from one career to another. And I think a life can have three or four or five careers" (Fitzroy 1973, p 30).

Matthews became an avid equestrian and active member of the dressage community at Ox Ridge Hunt Club. Through her friends, she discovered decorative arts at the Isabel O'Neil Studios and taught several highly specialized faux finish classes at the studio in New York. In her later years, she became an enthusiastic member of the Garden Club, the Presbyterian Church Choir, and the New Canaan Artisans Group (New Canaan n.d.).

25.6 Professional Organizations

Matthews was a member of the American Association of Civil Engineers, the American Association of Mechanical Engineers, earned membership in the engineering honor society Tau Beta Pi and the scientific honor society Sigma Xi. Her other professional activities included serving on the Board of Directors of the Engineers Joint Council, and participating in both the American Society of Mechanical Engineers and the New York State Society of Professional Engineers.

Matthews became active in SWE in the early 1960s and helped plan the First International Conference of Women Engineers and Scientists (ICWES) held in New York City for women from all over the world. One of her responsibilities was to greet conference guests at John F. Kennedy International airport (Fig. 25.2) (Fitzroy 1973, p 21).

At that time, the female engineering mentor that everyone knew was Dr. Lillian Gilbreth, the so-called "first lady of engineering" and SWE's first Honorary Member. Matthews had the opportunity to meet her in 1964 at ICWES where Dr. Gilbreth gave the keynote speech. Matthews describes the experience as follows:

> I was the chairman of hospitality. It was my job to go pick her up and I mean, I was just so thrilled. Tall, very tall, slender, and stately, very with-it. And I don't know, she was either

Fig. 25.2 Dr. Alva Matthews (on the right) at John F. Kennedy International Airport in 1964 for the First ICWES. (Photograph courtesy of Society of Women Engineers Records, Walter P. Reuther Library, Wayne State University)

> eighty-nine or ninety. And it was just fun. And we talked. I drove her over to the hotel and we just talked in the car. And I asked her about her life (Fitzroy 1973, p 15).

Matthews saw the need for mentors and actively advocated for women to be educated as engineers and have engineering careers. In her oral history, she recognized Bea Hicks (first president of SWE and 1963 Achievement Award recipient) and Ruth Shafer, the ICWES chair. She said these women "really saw that SWE had an important role to play in educating women and giving them a place to share thoughts and maybe even use it for networking" (Fitzroy 1973, p 14).

Matthews had many fond memories of those meetings in SWE's early days. She recalled they talked about the value of SWE:

> …the Society of Women Engineers was the only organization that existed to put itself out of business. You know, we thought that was wonderful. Once you educate everybody to think about engineering as a profession for women, and you interest women in engineering and math and science, there would be no more need for the organization (Fitzroy 1973, p 34).

Dr. Matthews put her creative talent and whimsical side to work for SWE. While working at Paul Weidlinger, she recalls putting together a fashion show as a fund-raiser for SWE (Fig. 25.3). She co-opted other female engineer colleagues into modeling with her. "There was a designer called Baba. And we talked her into... designing clothes. And then the women engineers were the models. And it was fun" (Kata 2003).

25.7 Honors and Advocacy

In 1971, Matthews received SWE's Achievement Award in recognition of her meritorious contributions to the field of engineering mechanics and applied mathematics in the area of shock analysis, elasticity and structural design.

Fig. 25.3 Alva Matthews modeling attire to wear in the field (1959). (Photograph courtesy of Society of Women Engineers Records, Walter P. Reuther Library, Wayne State University)

During the 1970s, Matthews helped plan and attended national SWE conferences (Figs. 25.4 and 25.5) and advocated for SWE, visiting sections at various universities. She conducted research on early women engineers, including Emily Roebling, who was instrumental in the building of the Brooklyn Bridge (Fitzroy 1973, p 18).

SWE enabled women to get together and share information regarding discrimination in the industry. Matthews described it as follows:

> If there were any problems of discrimination, it was salary discrimination. Those problems were common to all of the women in SWE. We would get together and talk about these problems, as well as the desire to spread out and educate (Fitzroy 1973, p 25).

Fig. 25.4 Dr. Alva Matthews (second from left) working on planning a conference. Nancy Fitzory (1972 Achievement Award Recipient) at far right. (Photograph courtesy of Society of Women Engineers Records, Walter P. Reuther Library, Wayne State University)

Fig. 25.5 Dr. Alva Matthews (right) with President Arminta Harness at a 1976 SWE Conference. (Photograph courtesy of Society of Women Engineers Records, Walter P. Reuther Library, Wayne State University)

25.8 Legacy

The story of Alva Matthews is a different journey than other Achievement Award recipients. Matthews' research in shock wave propagation was critical during the Cold War years during the 1950s and 1960s for national security, and she produced many technical publications on that subject in relevant journals. Her story is marked by a frenetic career in which she earned advanced degrees while teaching and performing research, followed by an early retirement. She had a child late in life, and made the decision to step out of the workplace and pursue other interests and hobbies.

By all accounts, including her own, Matthews had no regrets about her career choices, including the early retirement. She followed her own path and defined success on her own terms, providing a role model for those women in engineering who have made similar choices.

References

Kata, Lauren. 2003. Interview with Alva Matthews Solomon, Profiles of SWE Pioneers Oral History Project. Society of Women Engineers (SWE) Records (LR001539), n.d. Alva Matthews. Box 190, Folder 59, Walter P. Reuther Library, Archives of Labor and Urban Affairs, Wayne State University. https://ethw.org/Oral-History:Alva_Matthews_Solomon. May 14.

Fitzroy, Nancy. 1973. "Engineering as an Ideal Woman's Career," CAREER GUIDANCE FOR WOMEN ENTERING ENGINEERING, Proceedings of an Engineering Foundation Conference, New England College, Henniker, New Hampshire.August 19-24.

Merrill, Sam. 1961. WOMEN IN ENGINEERING, Society of Women Engineers, Reccrds Box 190, Folder 59, Walter P. Reuther Library, Archives of Labor and Urban Affairs, Wayne State University. July 13.

New Canaanite 2022. Alva Matthews Solomon, 88. https://newcanaanite.com/alva-matthews-solomon-88-6638359. September 12.

New Canaan, n.d. Connecticut Reuther Library Oral History ID: LOH 001952.23 at the Walter P. Reuther Library, Archives of Labor and Urban Affairs, Wayne State University.

Society of Women Engineers. 1971. Newsletter, "Award Winner," Vol. 18, No. 1. August.

Part V
1972–1976

Chapter 26
Setting the Stage: 1972–1976 The Dawn of Personal Computers and Women's Rights

Holly J. Teig and Janet L. Williams

26.1 Introduction

The use of personal computers gained momentum during the mid to late 1970s. Integrated circuits and microprocessors characterized this fourth generation of computers. Miniaturization of components evolved rapidly, with many inventors generating groundbreaking technology and approaches. Components began to shrink, with chip and microprocessor designers striving to embed as much as possible into tiny packages. With increased computing power, peripheral innovations were needed. From data storage to data encryption to data networking, hardware and software inventions in this time period were numerous and long-lived.

No time period generated more personal gains for women, thanks to the burgeoning women's movement. From financial barriers falling to rising personal prospects for education, athletics, and political power, women's rights were growing. Technical career opportunities in the computer, defense, and space industries opened up the need for engineers and scientists, and Title IX allowed women to share in these scientific endeavors.

H. J. Teig (✉)
Peoria, IL, USA

J. L. Williams
Albuquerque, NM, USA

C. D. Craig et al. (eds.), *Women Engineering Legends 1952-1976: Society of Women Engineers Achievement Award Recipients*, https://doi.org/10.1007/978-3-032-00224-2_26

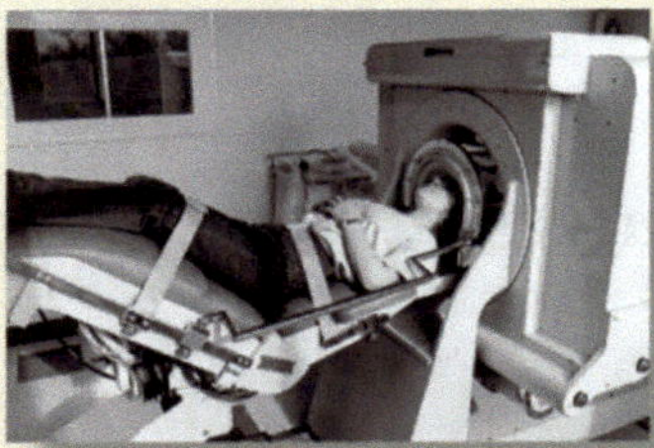

In 1972, biomedical engineer Godfrey Hounsfield invented the computed axial tomography scan. It was considered the biggest diagnostic breakthrough since the X-ray, later earning him a Nobel prize.

Nancy D. Fitzroy received the Achievement Award in recognition of significant contributions to the fields of heat transfer, fluid flow, properties of materials, and thermal engineering. Her inventions included a thermal protection system for use in the US early-warning defense system.

In 1972-1973, Motorola's Dr. Martin Cooper invented the first hand-held cell phone and made the first cell phone call to his archrival at AT&T. Called "the brick," it weighed 1.1 kg.

1972

By the early 1970s, women had made gradual progress in obtaining additional rights in the workplace. Women's participation in the workplace continued to grow and would exceed 50% by 1980.

Congress passed Title IX of Education Amendments in 1972, banning discrimination based on gender in any educational setting receiving federal funding.

In 1973, IBM's UPC barcode concept was adopted as the industry-wide standard. The next year, a supermarket in Troy, Ohio, sold the first officially scanned barcode item — a pack of gum.

0 36000 29145 2

America's first research space station, SkyLab, was launched in 1973. It hosted three crews over 151 days, collecting valuable data on the sun and conducting 270 experiments.

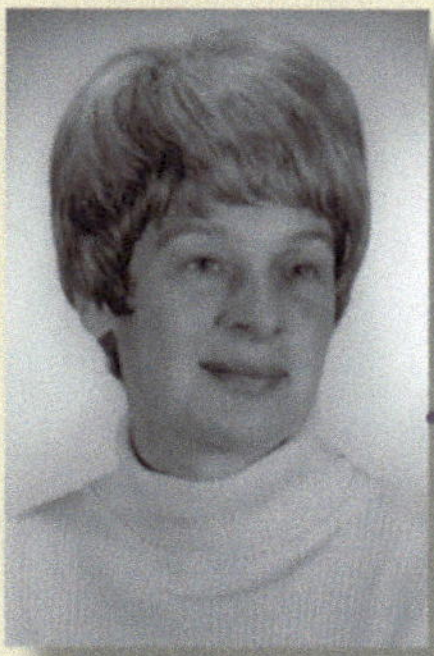

Irene Carswell Peden, PhD received the Achievement Award in recognition of significant contributions in the fields of radio wave propagation research and electrical engineering education, including research on radio propagation over long paths in polar regions.

Barbara Crawford Johnson received the Achievement Award in recognition of her significant engineering contributions in support of manned spaceflight programs.

1973 **1974**

- The US Supreme Court rendered its landmark decision in *Roe v. Wade*, confirming a woman's right to an abortion.
- The Equal Credit Opportunity Act was passed, prohibiting discrimination against applicants based on sex, race, and many other factors.

In 1975, Kodak engineer Steve Sasson invented the first digital camera. Very low resolution and weighing 10 pounds, it represented a remarkable turning point in photography.

Sheila E. Widnall, PhD received the Achievement Award in recognition of her significant contributions to the fluid mechanics of low-speed aircraft and hydrofoils. She designed MIT's advanced wind tunnel facility and holds multiple patents in airflow technology.

1975

- The United Nations General Assembly proclaimed 1975 as International Women's Year. Mexico City hosted the first UN World Conference on Women that summer.
- There were 19 women serving in the US House of Representatives, constituting 3.5% of Congress.

In 1976, college dropouts Steve Jobs and Steve Wozniak started Apple based on a new concept of how people and computers interface.

Ada I. Pressman received the Achievement Award in recognition of her significant contributions in the field of power control systems engineering. She developed emergency safety systems for fossil-fuel and nuclear power plants.

In 1976, Seymour Cray, considered the Father of Supercomputers, unveiled the next-generation supercomputer. Its unique circular design minimized wire lengths, making it 10 times faster than any other computer at the time.

1976

Between 1970 and 1980:

- The percent of women attorneys increased from 3% to 8%.
- The percent of women doctors increased from 9% to more than 13%.
- The number of women engineers inched up from less than 3% to nearly 5%.

26.2 State of Technology

26.2.1 Computer Hardware and Software

Programming Languages Dennis Ritchie, a Unix inventor, developed the programming language C at Bell Laboratories in 1972 (Britannica 2024a). Through multiple iterations, Ritchie created a flexible and fast programming language. Version "C" caught on quickly with early computer programmers because of its structured function flow using the now ubiquitous *if/then/else* and *do/while* statements.

Microprocessors Intel released the *8008* microprocessor in 1972 (Wood 2008). It was an 8-bit microprocessor envisioned for calculators, computer terminals, and industrial robots. Within a year, it was incorporated into the Sacramento State *Sac State 8008* microcomputer to handle thousands of patient medical records, complete with an IBM disk operating system, a color display, hard drive, keyboard, modem, audio/paper tape reader, and printer (Terdiman 2010). By 1974, Intel had upgraded to the *8080* microprocessor, which was so versatile for all types of applications that it kick-started the microcomputer industry. Motorola countered with the *MC6800*, an easier-to-implement control microprocessor targeted for industrial use (Electronics 1974).

Personal Computers The first commercially successful hobby computer was released in 1975, the Micro Instrumentation and Telemetry Systems (MITS) *Altair 8800*. Kits were purchased through *Popular Electronics* magazine and swarms of electronics geeks did just that, with over 10,000 ordered at $397 each (Garland 1977). Computer clubs sprang up throughout the country so people could share ideas on how to put together electronic components and use their creations. It was the equivalent of a do-it-yourself, in-person help desk.

Computer Pioneers Inspired by the *Altair 8800* kit, Bill Gates and Paul Allen collaborated and sold their BASIC interpreter software to MITS in 1975, the first product of the Microsoft company (Allan 2001). Steve Jobs and Steve Wozniak unveiled the *Apple 1* computer a year later, the first computer with a single-circuit board and read-only memory (ROM) (BBC 2006, Reuters 2006). It was an all-in-one unit designed for affordability, boasting a keyboard and output video that went to a TV screen. Wozniak demonstrated their invention at the Silicon Valley Homebrew Computer Club meeting in July 1976 (NPR 2006).

Software and productivity gadgets Designed to improve business efficiency, these inventions gained traction first in academia and then flowed into home use. Hewlett-Packard introduced the first scientific hand-held calculator, the HP-35 in 1972 at a price point of $395 (note: virtually the same price as an *Altair 8800* personal computer kit three years later) (IEEE 2009). While these calculators were unaffordable for most college students, the phaseout of the slide rule had officially begun. The release of *Electric Pencil* in late 1976 was the first word processor for home computers, originated by Michael Shrayer (Freiberger 1982).

Video games In 1972, entrepreneur Nolan Bushnell, engineer Allan Alcorn, and Atari released *Pong*, the world's first commercially successful video game (Britannica 2024d).

Ethernet In 1973, Robert Metcalfe, a researcher at the Xerox Palo Alto Research Center, and David Boggs, a computer scientist, invented a high-speed data transmission protocol using coaxial cables as the *ether* that was revolutionary in connecting computers and other hardware assets. The ability to join multiple computers and other hardware into a system's network was a huge business advantage, saving money on assets and allowing data sharing. Metcalfe and Boggs also created the first high-speed network interface card, including an Ethernet port (Goodrich 2023). Ethernet quickly became the standard wired local area network in businesses, academia, and homes worldwide. Harriett B. Rigas (Achievement Award 1982) subsequently created technology that made it feasible to update computer operating systems through the use of automatic patches (now called *upgrades*) (ETHW n.d.).

Data storage Portable data storage 5.25 inch floppy disks were introduced in 1976. When this product reached the personal computer market it caused an exponential increase in digital information storage (Britannica 2024c). In a nod to their ancestry, software programs and apps today continue to use pictorial icons that look like floppy disks to designate saving data.

Supercomputers The Cray-1 supercomputer was invented by Seymour Cray in 1976 (Britannica 2024e). Its unique circular physical shape and 80-MHz processing power filled a need for high computer power used by research institutions (NCAR 2015). The amount of heat given off by the components required a unique refrigeration system and large temperature-controlled rooms.

Data encryption In 1976, IBM developed the first National Security Agency-approved data encryption standard for protecting sensitive, unclassified electronic government data (Simmons 2025).

26.2.2 In Space

Space station NASA launched the first US laboratory in space *Skylab* in May 1973 (Uri 2023). The micrometeoroid shield was damaged at liftoff and affected the climate control of the laboratory, delaying the first astronaut crew from launching. The investigation afterward generated new processes and governance protocols for subsystems in future spacecraft. It also demonstrated the ingenuity of the engineering team to repair the laboratory so it could be occupied for 24 weeks. The laboratory enabled research regarding the astronauts' physiological responses to long-duration space flight, earth sciences, solar physics, materials, and astronomy (Uri 2023).

Skylab also gave students the opportunity to fly experiments in space, awakening in many young children the desire to become astronauts or choose a space-related career. Barbara Crawford Johnson (Achievement Award 1974) contributed her engineering expertise and leadership on manned space travel to the Skylab program (ETHW n.d.).

Concorde The Concorde supersonic passenger transport entered commercial service in 1976 with scheduled passenger service on British Airways flights from London to Bahrain and Air France flights from Paris to Rio de Janeiro. The Concorde had a maximum cruising speed of 1354 miles per hour, or Mach 2.04 (more than twice the speed of sound), allowing the aircraft to fly from London to New York in about 3 hours. A technological marvel, the Concorde's massive development costs outweighed its operational revenue, resulting in financial unprofitability (Britannica 2024b).

Space Shuttle Space Shuttle *Enterprise* rolled out in September 1976, marking a new era of space travel technology. *Enterprise* was the first space shuttle, although it never flew in space. Constructed without engines or a functional heat shield, it was used to test critical phases of landing and other aspects of the shuttle technology. The shuttle orbiter and external fuel tanks were specifically designed for reusability, with the orbiter gliding back to Earth using a winged design rather than landing with parachutes (Ostovar 2024). The space shuttle program inspired many young women to pursue engineering degrees with the dream of becoming astronauts.

> Therefore, there is a real requirement for an efficient earth-to-orbit transportation system, an economical space shuttle.... The shuttle ideally would be able to operate in a mode similar to that of large commercial air transports and be compatible with the environment of major airports.... The cockpit of the space shuttle would be similar to that of the large intercontinental jet aircraft, containing all instrumentation essential to complete on-board checkout. George Mueller, head of the Office of Manned Space Flight, National Aeronautics and Space Administration (Heppenheimer 1999).

26.2.3 *Communications*

Mobile phones The first handheld cellular mobile phone was demonstrated by John F. Mitchell and Martin Cooper of Motorola in 1973, using a handset nicknamed *the brick* weighing 4.4 pounds (Teixeira 2010). It had a talk time of 30 min and took 10 h to recharge.

26.2.4 In the Home

Video Cassette Recorder (VCR) Before 1972, if you were not at home when your favorite TV program came on, there was no way to see it except to hope for a rebroadcast at a later date. Philips' invention of the Video Cassette Recorder allowed consumers to record programs and movies to watch later, on their own schedule (Britannica 2023). It was an immediate success, forever changing the home TV viewing experience and the movie industry. The VCR cassette *format wars* were spawned, pitting the JVC VHS open standard cassettes versus the higher quality Sony Betamax cassettes.

Bar Codes In 1973, a committee chose IBM's rectangular design of lines as the universal standard barcode for price coding in retail and grocery stores (Laurer 2001). Later, bar codes were used for storing goods and tracking packages.

Digital cameras Texas Instruments patented the first filmless electronic camera in 1972 (Britannica 2025b). In 1975, Steven Sasson at Eastman Kodak built a self-contained electronic camera that used a charge coupled device (CCD) image sensor (Dobbin 2005). For hobbyists, the Cromemco Cyclops was an all-digital camera published as a home kit project in the February 1975 *Popular Electronics* magazine (Walker et al. 1975).

Post-it Notes Microsphere adhesive technology, invented by Spencer Silver of 3M, was applied to bookmarks Arthur Fry wanted to use (and reuse) in his church hymnal in 1974 (3M 2025).

26.2.5 Medical and Safety

Medical imaging In 1972, Raymond Damadian patented the use of nuclear magnetic resonance to create internal body imaging, called a magnetic resonance image (MRI). Damadian went on to design and build a full-body MRI machine in 1974, eventually testing it on his postgraduate assistant Larry Minkoff in 1977 (Garet 2020). The first computerized tomography (CT) scanner was also invented in 1972 and was recognized with a Nobel Prize in 1979 awarded to Godfrey N. Hounsfield and Allan M. Cormack (Britannica 2025a). These two inventions revolutionized non-invasive medical diagnostic imaging.

Airbags GM's Oldsmobile Toronado was the first domestic US vehicle to include an electromechanical automotive passenger airbag in 1973 (Bellis 2019).

26.3 State of Women

No time period generated more personal gains for women than from 1972 to 1976. The women's movement was in full swing. Women became vocal in pointing out inequalities, sometimes with consequences to their jobs, themselves, and their families. Title IX opened up opportunities in education and athletics, including scholarship programs to encourage college enrollment across all degrees. Change was in the air, and women of the 1970s pressed forward to break down long-standing barriers to gender equality.

Women gained a political voice—individually and as a group—on issues that affected themselves and their families. The National Organization for Women (NOW) focused on passage of the Equal Rights Amendment (ERA) along with advocating for women's rights. The National Women's Political Caucus (NWPC), founded in 1972 by Gloria Steinem, Shirley Chisholm, and others, focused on increasing women's presence in politics and the number of female officeholders. The National League of Women Voters, in existence since 1920, was heavily engaged in the passage of *Title IX of the Education Amendments* in 1972 and various anti-discrimination laws and programs. Betty Friedan organized the *Women's Strike for Equality* on August 26, 1970, the 50th anniversary of women's suffrage. SWE, as a women's organization, found its first voice in political advocacy coming out of this time period (Lewis 2019).

Financially, a woman could finally apply for credit in her own name, without having a male co-signatory. Marriage was no longer required in order to have a credit card or own a home. She could get a degree, have a job with her own income, and be an independent career woman. Jobs provided financial security, but not yet equal pay, to male counterparts (Britannica 2025c; Britannica 2025e).

Women also had new rights to control their bodies throughout their lifespan—from contraception and family planning to pregnancy rights and even abortion choices.

26.4 State of SWE[1]

SWE was shaped in the 1970s by the ERA movement, the legislation of Title IX, rapid membership growth, and SWE's advocacy for a diverse workforce.

In 1972, SWE reported that out of an estimated 8000 women engineering graduates in the workforce, along with tens of thousands of other women doing engineering work, only 15% were SWE members. A new grade of membership was introduced: r*etired member*. The Society had been around long enough that some of its older members were retiring, and the Society wanted to retain this revered

[1] This section includes personal recollections about the time period and SWE from a survey of SWE Fellows and Late Career/Retiree Affinity group members conducted in February 2025.

portion of the membership. Dues were set at 50% of regular dues. In January, Lillian Gilbreth, the *first lady of engineering* passed away at age 93. In May of that same year, Ruth Shafer, chair of the first International Conference for Women Engineers and Scientists, also died. The torch was indeed passed to the next generation (SWE Records 2000).

By 1975, the Society had swelled to 3000 members, nearly triple the number it had five years earlier. The organization was growing, especially in collegiate sections, where women wanted the support and community SWE sections provided. Thirty-seven student sections were chartered during the fall of 1972. The rate at which students applied for membership in SWE was 5 to 1 over the same period, the year before. The influx of students continued through the remainder of 1973. With greater numbers, the students wanted a more significant voice in the organization. They expressed the desire to be consulted on policy changes affecting them. They also wanted closer contact with their member sections and requested a student newsletter (SWE Records 2000).

An executive committee was formed to evaluate student representation and male membership. After all, discrimination worked both ways, and if SWE were to conform to the law it needed to accept males as members. Contrary to the wishes of the conservative factions of SWE, in 1976 the Council of Section Representatives (CSR) approved student representation in that body, and male membership. Men's Auxiliary (MASWE) members automatically became SWE members, with full rights and privileges of the membership grade in which they were eligible. Past Rodney D. Chipp Award recipients were also given honorary SWE memberships in recognition of their support of women engineers and/or SWE. The MASWE organization was officially disbanded. The issue of admittance of two-year college students was also considered, but did not pass (Kindya 1990; SWE Records 2000).

The student sections were elated with the vote. Each of the four student section regions could send an elected representative to the council to voice student issues and opinions. The first student conference was held just prior to the national convention in 1976, reflecting the rapid increase of students within the Society and nationwide of women pursuing engineering (SWE Records 2000).

SWE diligently worked with several corporations to obtain funds for student scholarships, which were to be administered by SWE and meant to encourage women to enroll in engineering curricula. In 1974, SWE increased the number of scholarships it offered from one to 13 and fully endowed the Gilbreth Scholarship (SWE Records 2000).

In 1972, the Engineering Foundation and SWE continued their conferences on women in engineering. *Henniker II*,[2] had a theme of "Women in Engineering and Management." This conference focused on achieving executive status for women engineers and managers, receiving the nickname *Keys to the Executive Powder Room*. It addressed the need for women to take the reins of their own destiny and

[2]The *Henniker Conferences* were named after the meeting location of Henniker, New Hampshire, where they were held.

prepare themselves for advancement in their profession (Kindya 1990; SWE Records 2000).

Henniker III followed in 1973 under the chairmanship of Nancy Fitzroy, entitled, "Career Guidance for Women Entering Engineering," it was the most successful of all the conferences and the one with the most long-lasting effects. It was a role model showcase demonstrating that there are women engineers who have made it in their careers. Proceedings from *Henniker III* became the principal career guidance tool of SWE during the 1970s, and well into the 1980s. There were additional conferences in following years. The last conference, *Tidewater,*[3] was held in October 1976. It dealt with upward mobility and continued content of *Henniker II and III* (Kindya 1990; SWE Records 2000).

SWE was also entering the computer age. The influx of student memberships to be processed at headquarters had become unmanageable. In 1975, the Society purchased computer time at a service bureau to relieve the organization's administrative tasks. Naomi McAfee, SWE President 1972–1974, admitted about the nearly all-volunteer organization, "SWE has been so busy with the simple problem of survival, it has been unable to achieve its objectives completely" (Kindya 1990; SWE Records 2000).

In 1973, SWE signed a memorandum of understanding to affiliate with the National Society of Professional Engineers (NSPE), cementing their place at the table as a legitimate technical society on par with the longest established technical societies (Kindya 1990).

In a controversial vote, the SWE Council of Section Representatives endorsed the Equal Rights Amendment in 1973. SWE through its history had avoided commenting on public policy, afraid of violating its tax-exempt status. While the resolution passed the Council by a margin of 15 to 1, support among SWE's wider membership was mixed. Some members felt SWE was making a political statement, while others saw it as necessary to achieve complete equality in the workforce. Concerns arose amongst SWE members that associating with the women's liberation movement might negatively impact their reputation in the engineering community and the reputation of SWE (English 2023; Kindya 1990; SWE Records 2000).

Sexism in the workplace was common for women engineers until they proved themselves. Women engineers were either ignored, left out of leadership assignments, or given excessive negative attention. To fit in and be accepted, women had to excel in their assignments, perform over and above expectations, and without any considerations of favoritism. Suzanne Moore, SWE Fellow and Life Member stated:

> I always stopped to support a female engineer being harassed by male engineers so [that] by the time I left the area, they wished they had never initiated anything inappropriate; showing the female how to take care of herself the next time. I'm very proud I played a significant role in helping develop the culture for younger engineers to all work in support of one another. What I witnessed was [that] those men that had a daughter interested in

[3] The last conference, called *Tidewater*, was held at the Tidewater Inn located in Easton, Maryland.

> STEM careers were much more supportive of women on their team. It seemed to take a lot more time and more proven results to convince the others [of] the value of women on their team. (personal correspondence 2025).

Women engineers desired both professional respect and successful careers, but gaining that respect took time, while determination and patience were necessities:

> I worked at least twice as hard as my male counterparts to stay a step behind... Some clients of the company refused to consider a female engineer on their staff. SWE was a breath of fresh air to have the opportunity to interface with so many women who were breaking barriers and being firsts. It usually took 6 months to a year for my counterparts to realize I did have a lot to offer the team. - T. Meredith Ross, PhD, PE, SWE Fellow and Life member (personal correspondence 2025).

Practicalities, such as having available restrooms, were being addressed by legislation, but barriers to their success on the job remained. It was typical for women engineers to be directly asked in job interviews if they planned to have children and take time off to care for them. Meredith Ross related how domestic stereotypes were not always accurate. "I was asked to make coffee a lot, until I did, which was not in my skill set as I was not a coffee drinker and the coffee ranged from undrinkable to bad" (personal correspondence 2025). Another obstacle women engineers faced when performing their engineering roles was finding woman-sized safety shoes and equipment to wear on job sites and in plants.

> I am petite and trying to get full body PPE [Personal Protection Equipment] to fit was still a problem 15 years after graduating from college. At one training session with senior fire department and Coast Guard staff, the Class B full body suit used by the participants was so oversized that some of my classmates carried part of the suit like it was a coronation train when I had to put it on and attempt to collect a sample. Also the air tanks used with these suits were 50% longer than my back which made wearing the unit a safety hazard. - Laura Gimpelson, PE, SWE Fellow and Life member (personal correspondence 2025).

26.5 Laws and Culture

Women's rights movement Women take to the streets and call upon politicians to guarantee complete and equal rights for women in the US Constitution.

Gender equity Title IX legislation forces educational institutions receiving federal funds to provide gender-equitable programs. From locker rooms to classrooms, women experienced immediate impacts in terms of expanded educational and athletic opportunities. Many took advantage of newly expanded intercollegiate sports scholarships to fund their higher education.

Women's bodies New rights for contraception and abortion allowed options for women to plan their families.

Summary Table 26.1 highlights examples of these changes.

Table 26.1 Culture and Legal Highlights for 1972–1976

Date	Event
1971–1982	**Equal Rights Amendment (ERA):** After being passed by the US House of Representatives in 1971, the full Senate approved the ERA without changes with a vote of 84 to 8. A majority of the 50 states required ratification by 1982 to amend the US Constitution, but it was not attained. Congress has yet to invalidate the imposed deadline to allow it to stand as the 38th Amendment to the Constitution. The primary provision of the amendment reads, Equality of rights under the law shall not be denied or abridged by the USA or any State on account of sex.
1972–1979	**Title IX:** Title IX of the Education Amendments of 1972 bans sex discrimination in educational programs or activities receiving federal funds, dramatically increasing the number of women participating in intercollegiate athletics and attending higher education schools. The clause states "no person in the United States shall, on the basis of sex, be excluded from participation in, be denied the benefits of, or be subjected to discrimination under any education program or activity receiving Federal financial assistance". Despite multiple challenges, in 1979 the Department of Health, Education, and Welfare (HEW) mandated that educational institutions provide equal opportunity to men and women in athletic programs. To adhere to Title IX, each college and university must have a ratio of male-to-female athletes that was equal to its ratio of male-to-female students (Britannica 2025e; SWE Records 2000).
1972	**Contraception:** *Eisenstadt v. Baird, 405 U.S. 438 (1972).* Unmarried couples have the right to use contraception and the right to privacy, based on the Equal Protection Clause of the Fourteenth Amendment. The Court struck down a state law that allowed the distribution of contraception to married adults, but not unmarried adults (Justia US Supreme Court 2025c).
1973	**Roe v. Wade:** *Roe v. Wade 410 U.S. 113.* The Court affirmed women's decision to terminate a pregnancy as a fundamental right and that unduly restrictive state regulations were unconstitutional and violated their right to privacy in the Fourteenth Amendment (Britannica 2025d). Widely hailed as a watershed moment in the women's rights movement, it allowed women full control over their bodies.
1974	**Equal Credit Opportunity Act (ECOA):** This Act prohibited lenders and creditors from denying women mortgages or credit cards in their own name because of their sex or marital status. Women were often refused loans or credit cards unless they had a male cosigner on their agreements—a husband for married women or a father or brother for single women (McDonough 2024).
1974	**Maternity Leave:** *Cleveland Board of Education v. LaFleur 414 U.S. 632.* The Court determined it was illegal for companies to force pregnant women to take maternity leave before or after childbirth under the assumption they are incapable of doing their job (Justia US Supreme Court 2025b).
1974	**Equal Pay Challenge:** *Corning Glass Works v. Brennan 417 U.S. 188.* For the first time, the Court considered an Equal Pay Act claim based upon an employer paying women less than men for the same work, determining that the wage difference between Corning's female inspectors and male inspectors violated the Equal Pay Act. The American Civil Liberties Union files an amicus brief, authored by Ruth Bader Ginsburg and others (Justia US Supreme Court 2025a).

References

3M. 2025. History Timeline: Post-it Notes. 3M. https://www.post-it.com/3M/en_US/post-it/contact-us/about-us/. Accessed January 26, 2025.

Allan, Roy A. 2001. *A History of the Personal Computer.* Allan Publishing. Accessed January 24, 2025.

BBC News Online. 2006. Bill Gates: A Timeline. July 15. Web Archive.org. Accessed January 26, 2025.

Bellis, Mary. 2019. The History of Airbags. ThoughtCo. August 9. https://www.thoughtco.com/history-of-airbags-1991232. Accessed January 27, 2025.

Britannica, T. Editors of Encyclopaedia. 2023. Videocassette Recorder. *Encyclopedia Britannica*, December 15. https://www.britannica.com/technology/videocassette-recorder. Accessed January 20, 2025.

Britannica, T. Editors of Encyclopaedia. 2024a. 'C'. *Encyclopedia Britannica.* November 18. https://www.britannica.com/technology/C-computer-programming-language. Accessed January 20, 2025.

Britannica, T. Editors of Encyclopaedia. 2024b. Concorde. *Encyclopedia Britannica.* December 24. Accessed January 10, 2025.

Britannica, T. Editors of Encyclopaedia. 2024c. Floppy Disk. *Encyclopedia Britannica.* December 10. https://www.britannica.com/technology/floppy-disk. Accessed January 26, 2025.

Britannica, T. Editors of Encyclopaedia. 2024d. Pong. *Encyclopedia Britannica.* December 30. https://www.britannica.com/topic/Pong. Accessed January 26, 2025.

Britannica, T. Editors of Encyclopaedia. 2024e. Seymour Cray. *Encyclopedia Britannica.* October 1. https://www.britannica.com/biography/Seymour-R-Cray. Accessed January 23, 2025.

Britannica, T. Editors of Encyclopaedia. 2025a. Computed Tomography. *Encyclopedia Britannica.* January 24. https://www.britannica.com/topic/computed-tomography. Accessed January 27, 2025.

Britannica, T. Editors of Encyclopaedia. 2025b. Digital Camera. *Encyclopedia Britannica.* March 7. https://www.britannica.com/topic/Equal-Rights-Amendment. Accessed April 25, 2025.

Britannica, T. Editors of Encyclopaedia. 2025c. Equal Rights Amendment. *Encyclopedia Britannica.* January 19. https://www.britannica.com/topic/Equal-Rights-Amendment. Accessed January 23, 2025.

Britannica, T. Editors of Encyclopaedia. 2025d. Roe v. Wade. *Encyclopedia Britannica.* January 19. https://www.britannica.com/event/Roe-v-Wade. Accessed January 23, 2025.

Britannica, T. Editors of Encyclopaedia. 2025e. Title IX. *Encyclopedia Britannica.* January 21. https://www.britannica.com/event/Title-IX. Accessed January 23, 2025.

Dobbin, Ben. September 8, 2005. Associated Press. Kodak Engineer had a Revolutionary Idea: the First Digital Camera. https://www.seattlepi.com/business/article/kodak-engineer-had-revolutionary-idea-the-first-1182624.php. Accessed January 23, 2025.

Electronics. March 7, 1974. Motorola joins Microprocessor Race with 8-bit Entry. *Electronics.* 47(5). New York: McGraw-Hill: 29–30. Accessed January 26, 2025.

English, Troy Eller. 2023, What's Past is Prologue. https://swe.org/magazine/past-is-prologue/. Accessed January 26, 2025.

Engineering and Technology History Wiki (ETHW). n.d. SWE Awards. https://ethw.org/SWE_Awards. Accessed March 22, 2025.

Freiberger, Paul. May 10, 1982. Electric Pencil, First Micro Word Processor. *InfoWorld.* p. 12, books.google.com. Accessed January 25, 2025.

Garet, Lynette. December 26, 2020. History of MRIs and the Evolution of This Life-Saving Technology. Ezra. https://ezra.com/blog/history-of-mri-scans#strongthe-development-of-magnetic-resonance-imaging-mri-scansstrong. Accessed January 25, 2025.

Garland, Harry. March 1977. Design Innovations in Personal Computers. *Computer*, 10(3). IEEE Computer Society: 24. Accessed January 26, 2025.

Goodrich, Joanna. November 16, 2023. Ethernet is still going strong after 50 years. IEEE. https://spectrum.ieee.org/ethernet-ieee-milestone. Accessed 10 January 2025.

Heppenheimer, T.A. 1999. The Space Shuttle Decision NASAs Search for a Reusable Space Vehicle, https://www.nasa.gov/wp-content/uploads/2023/04/sp-4221.pdf?emrc=dade10. Accessed January 24, 2025.

IEEE. April 2009. HP-35 Scientific Calculator Awarded IEEE Milestone. https://archive.ph/20120905152953/http://www.ieee.org/about/news/2009/14april_1.html. Accessed January 25, 2025.

Justia US Supreme Court. 2025a. *Corning Glass Works v. Brennan.* 417 U.S. 188. https://supreme.justia.com/cases/federal/us/417/188/. Accessed January 23, 2025.

Justia US Supreme Court. 2025b. *Cleveland Board of Education v. LaFleur.* 414 U.S. 632. https://supreme.justia.com/cases/federal/us/414/632/. Accessed January 23, 2025.

Justia US Supreme Court. 2025c. *Eisenstadt v. Baird,* 405 U.S. 438 https://supreme.justia.com/cases/federal/us/405/438/.Accessed January 23, 2025.

Kindya, Marta. 1990. Four Decades of the Society of Women Engineers. Society of Women Engineers Records (LR001539). Box 180, Folder 9, Walter P. Reuther Library, Wayne State University. Accessed January 10, 2025.

Laurer, George J. October 2001. Development of the UPC Symbol. http://bellsouthpwp.net/l/a/laurergj/UPC/upc_work.html. Accessed January 26, 2025.

Lewis, Jone Johnson. February 21, 2019. Feminist Organizations of the 1970s. https://www.thoughtco.com/top-feminist-organizations-of-the-1970s-3528928. Accessed January 28, 2025.

McDonough, M., December 5, 2024. Second Wave of feminism. *Encyclopedia Britannica.* https://www.britannica.com/topic/second-wave-feminism. Accessed January 23, 2025.

NCAR. June 7, 2015. SCD Supercomputer Gallery. https://web.archive.org/web/20150607021044/http://www.cisl.ucar.edu/compute/gallery/cray/cray1.jsp. Accessed January 26, 2025.

NPR. September 29, 2006. A Chat with Computing Pioneer Steve Wozniak. NPR.org. https://www.npr.org/templates/story/story.php?storyId=6167297. Accessed January 26, 2025.

Ostovar, Michele. September 5, 2024. Space Shuttle. National Aeronautics and Space Administration. https://www.nasa.gov/space-shuttle/. Accessed January 10, 2025.

Reuters. September 27, 2006. Co-founder tells his side of Apple story. http://news.oneindia.in/2006/09/27/co-founder-tells-his-side-of-apple-story-1159346150.html. Accessed January 26, 2025.

Simmons, G. J. January 24, 2025 Data Encryption Standard. *Encyclopedia Britannica.* https://www.britannica.com/topic/Data-Encryption-Standard. Accessed January 26, 2025.

Society of Women Engineers (SWE) Records (LR001539). 2000. SWE 50 Years 1950-2000. Box 180, Folder 9, Walter P. Reuther Library, Archives of Labor and Urban Affairs, Wayne State University. Accessed January 10, 2025.

Teixeira, Tania. April 23, 2010. Meet Marty Cooper - the Inventor of the Mobile Phone. http://news.bbc.co.uk/2/hi/programmes/click_online/8639590.stm. Accessed January 20, 2025.

Terdiman, Daniel. January 2010. Inside the world's long-lost first microcomputer. https://www.cnet.com/culture/inside-the-worlds-long-lost-first-microcomputer/. Accessed January 26, 2025.

Uri, Jon. 14 May 2023. 50 Years Ago: The Launch of Skylab, America's First Space Station. Johnson Space Center. https://www.nasa.gov/history/50-years-ago-the-launch-of-skylab-americas-first-space-station/. Accessed January 10 2025.

Walker, Terry, Harry Garland, and Roger Melen. February 1975. Build Cyclops. *Popular Electronics.* Vol. 7, no. 2. Ziff Davis. pp. 27–31.

Wood, Lamont. Aug 2008.The True Origins of the Personal Computer. Computerworld. Accessed January 26, 2025.

Chapter 27
Nancy D. Fitzroy (1927–2024) Achievement Award Recipient 1972

In recognition of her significant contributions to the fields of heat transfer, fluid flow, properties of materials and thermal engineering

Holly J. Teig

27.1 Introduction

Nancy DeLoye Fitzroy (Fig. 27.1) was a widely recognized expert in thermal engineering, heat transfer, and fluid flow, gained over a long career with General Electric (GE). Fitzroy developed advanced heat transfer concepts for a diverse range of equipment, including space vehicles, satellites, home appliances, air conditioning systems, jet engines, gas turbines, steam turbines, nuclear reactor cores, motors, and many other types of equipment. Two of her significant achievements were the invention of a thermal chip to measure temperature in integrated circuits and a thermal protection system for use in US early warning systems. Her work reflects the interdisciplinary nature of engineering, as she can be considered both a chemical and a mechanical engineer. Equal to her technical accomplishments was her dedication to expanding the applied literature of her discipline, as editor of the GE *Heat Transfer and Fluid Flow Data* books for 20 years.

Fitzroy is recognized as the first woman to graduate in chemical engineering from Rensselaer Polytechnic Institute (RPI) in 1949 and attained nationwide prominence as the first woman president of the American Society of Mechanical Engineers (ASME). She was elected to the National Academy of Engineering and received numerous honors for her engineering achievements. Fitzroy contributed to the image of women as engineers by speaking and writing about her career and encouraging women to pursue engineering. A Renaissance woman in her time, Fitzroy was an accomplished engineer, orator, pilot, sailor, and sportswoman (Bailey 1998; Baylor 1988; RPI n.d.; Tarplee and Williams 2000).

H. J. Teig (✉)
Peoria, IL, USA

C. D. Craig et al. (eds.), *Women Engineering Legends 1952-1976: Society of Women Engineers Achievement Award Recipients*,
https://doi.org/10.1007/978-3-032-00224-2_27

Fig. 27.1 Portrait of Society of Women Engineers Achievement Award recipient Nancy D. Fitzroy, 1972. (Photograph courtesy of Society of Women Engineers Records, Walter P. Reuther Library, Archives of Labor and Urban Affairs, Wayne State University)

27.2 Early Life and Education

Nancy DeLoye Fitzroy was born in 1927 in Pittsfield, Massachusetts, the youngest of six children of Jules E. DeLoye, Jr., and Mabel E. Bush. Fitzroy's fascination with what makes things work took hold at a young age. As a child, when she wanted a record player, her father handed her parts and a diagram, and she built it (Fitzroy and Goldstein1987; Bailey 1998; SWE Records n.d.).

She was equally engaged in activities with both parents. "I helped Mom in the kitchen, then I ran off to fly with Dad." Her father's ideas of fun were adventurous. He flew gliders as a hobby, inspiring Fitzroy to become a pilot while in high school. Years later, she owned and operated a twin-engine Cessna and became a licensed helicopter pilot (Fitzroy and Goldstein 1987; Fitzroy Obituary 2024).

Fitzroy graduated from Pittsfield High School in 1945 (Fitzroy Obituary 2024). When asked how she became interested in engineering, Fitzroy shared she was "not interested in spending four years studying English":

> My father and [the] vice-principal of my high school had a talk with me. The discussion went down the usual route of teaching, nursing, [or teaching] mathematics. Finally, my vice-principal said, 'Have you considered engineering?' At first I thought he was being facetious; then we realized he was serious…My oldest brother was an engineer, and my father a building contractor, so I had [people to talk with about it]. (Fitzroy 1973).

Fitzroy applied to the Massachusetts Institute of Technology (MIT) and Rensselaer Polytechnic Institute (RPI), was accepted to both, and chose to attend RPI in Troy, New York. She was the first female student in RPI's chemical engineering program

(RPI n.d.), receiving her bachelor's degree in Chemical Engineering in 1949 (Baylor 1988; Bailey 1998).

> I thought if I studied Chem Engineering and I couldn't get a job as an engineer, I could fall back and get a job as a chemist. There were plenty of women chemist bottle washers, so I could at least [find] work (Fitzroy 2018 "ASME Oral History").

Fitzroy was an RPI student when she met her husband-to-be, Roland V. Fitzroy, Jr., in 1947. "The more I saw of him, the more I realized how simpatico we were, like two peas in a pod," she recalls. The Fitzroys were married in 1951 and settled in Massachusetts (Accolades 2006).

27.3 Career

In 1950, Fitzroy was hired into the Heat Transfer Group of the Knolls Atomic Power Laboratory (KAPL) in Schenectady, NY, operated by the GE Company. As an assistant Development Engineer, she was on a team working to solve a very difficult heat transfer problem: how to keep the incredibly high temperatures produced by an atomic reaction *inside* the nuclear reactor of an atomic generator. The team designed shielding that was used in nuclear power plants in submarines (SWE Records n.d.; Merrill 1976). She worked at KAPL for 2 years. Fitzroy recalled the difficulties in landing her first job:

> You can prove yourself after you are there, but how to get there? Jobs were scarce when I graduated; scarce for men, particularly so for girls. I was not able to convince anyone that I was serious about staying in an engineering job for at least five years. I was told that the norm for girls was two years and that in a professional job, it takes at least three years before anyone is worth his salt. (Fitzroy 1973).

Fitzroy's next opportunity was in the GE Research & Development Center, where her first assignment in 1952 was thermodynamic analysis on the Project Hermes Guided Missile Program. Soon after starting, she applied for another position in thermodynamic analysis.

> I met the same impasse. If you can do engineering work, why aren't you being treated as an engineer now? My prospective employer gave me a written test! I was terrified…I must have passed the test, because I was offered employment…as an engineering assistant…In six months time…since I was doing the same caliber of work as the other engineers…[I became] a development engineer…[with] a raise in salary too. (Fitzroy 1973).

Over the next several years, Fitzroy worked on a wide variety of heat transfer problems in the Research Center, performing heat transfer analysis, special thermal equipment design, and identifying thermal properties of materials. She consulted on thermal problems in the design of proposed products and the operation of existing equipment, including predicting the thermal behavior of freezing in the copper dip-forming process for wire production, advanced cooling concepts for future motor designs, and analyzing temperature and heat transfer mechanisms at a heavy water plant (Baylor 1988; Bailey 1998). These experiences cemented her expertise in heat transfer applications. In working with her male engineering colleagues, Fitzroy later noted:

> Working with all-male engineers seems to pose no more problems than it did in school. Engineers in general are some of the most even-tempered, steadiest people whom I have met. They treat you according to the caliber of engineer you are. But may I add that for a girl it helps to be just a little bit better (Fitzroy 1973).

In 1965, Fitzroy assumed responsibility for evaluating and disseminating new and existing heat transfer data and fluid flow information within the GE engineering community, establishing herself as an expert in the field.

By this time, Fitzroy had been promoted to heat transfer consultant, working on gas turbines, space satellites, and other technologies. She applied thermal design expertise to an icing detection system for J-79 aircraft engine inlets (Baylor 1988) and provided solutions for keeping the delicate electronic equipment in space satellites at room temperature while the skin of the satellite was being alternately superheated and supercooled (Merrill 1976).

Fitzroy also worked on more mundane projects such as the design of toasters and microwave ovens. "Toasters can present more problems than missiles and satellites. Whereas outer space is basically uniform, no two pieces of bread are alike" (Merrill 1976; Bailey 1998). Another accomplishment was a thermal design controlling temperature distribution in a steam-turbine wheel for GE's Steam Turbine-Generator Products Division (Baylor 1988).

During the 1970s, Fitzroy transitioned to administrative and management roles. She was named Manager of Heat Transfer Consulting, Thermal Branch at the GE Research and Development Center in 1971. This role leveraged her specialized skills in heat transfer analysis and thermal equipment design (SWE Records n.d.).

In 1974, Fitzroy took the job of Strategy Planner, GE Corporate Research and Development Center planning long-range goals for corporate research and development programs. She orchestrated the plans for advanced concepts for work to be carried out by the technical staff of the division (SWE Newsletter 1979).

In 1975, after 25 years of experience in industry, she was issued a Professional Engineer license in New York State, based on Eminence. By this time in her career, the quality and acceptability of her solutions to heat transfer problems arising in the design and developments of products had earned her a high level of respect as an engineer (Baylor 1988).

Fitzroy moved through several roles from 1976–1981: Manager Program Development, Gas Turbine Engine Division and Advanced Planner and Proposal Manager. In 1982 she was named Manager of the Energy and Environmental programs with GE's Turbine Market and Projects Division. In this role, she worked as a consultant involved with gas turbines, nuclear energy, and space vehicles. After an illustrious 37-year career with General Electric, Fitzroy retired in 1987 to spend time on her many outside interests while still doing public speaking about her career and support for women in engineering (Fitzroy Obituary 2024).

27.4 Publications and Patents

Fitzroy's most impactful legacy was as the principal author and editor of the GE *Heat Transfer and Fluid Flow Data Books* for 20 years. Starting as a contributor in 1956 and soon thereafter as chief editor, these books were continuously updated reference sources for GE engineers. Much of the book's subject matter was not covered in conventional textbooks, resulting in these books becoming outstanding technical resources with frequent supplements of additional information. First available only within GE, after 1970, they were offered for sale on subscription to engineers outside GE (SWE records n.d.; Baylor 1988; Fitzroy Obituary 2024).

Starting in 1953, Fitzroy originated a heat transfer and fluid flow newsletter (entitled *q's and p's)* within GE, that disseminated information on heat transfer. Its purpose was to keep GE engineers abreast of new or forthcoming developments in the field (Merrill 1976; Baylor 1988). The upkeep of the newsletter and books required Fitzroy to keep abreast of world literature in heat transfer. The specific topics in the books were selected based on their application to design and development work at GE. Her principal contribution as editor was to identify new areas of heat transfer where design data were lacking, and when sufficient experimental data became available in the open literature, to generate broadly applicable general correlations or design curves. These books are a compendium of design data of vital interest and value to practicing engineers working in heat transfer. Fitzroy served as author of many of the pages in these books, as well as editor and publisher. When GE required that the books be self-supporting, Fitzroy set up a billing system and handled all financial details, as well as ensuring the high technical quality of the books (SWE records n.d.; Baylor 1988; Fitzroy Obituary 2024).

Fitzroy wrote more than 100 technical reports, papers, and articles on a wide range of heat transfer topics. These ranged from material-oriented to application-oriented. They also included high-tech topics and more mundane ones. Her technical writing was prolific and reflected her expertise in the industry (SWE Newsletter 1972; SWE Record n.d.).

Fitzroy has multiple patents related to heat transfer topics. The first was a design for a unique advanced thermal protection system for nuclear-radiation hardened, phased-array radar for GE Heavy Military Department used to protect long-range radar from nuclear attack (Hardened Antenna Element Cover US Patent No. 3,896,450). The second is a unique heat-rejection mechanism for a small electronic power supply. (Male Plug Assembly Incorporating Power Supply US Patent No. 3,880,491). Additionally, Fitzroy invented special thermal design and analysis techniques for integrated circuit chips used in GE's semiconductor business (patent docket on file).

27.5 Technical Society Contributions

Throughout her career, Fitzroy served in engineering society roles at the local, regional, and national level. She worked to increase the pool of technological talent by initiating new programs to increase the participation of women and minorities in engineering. She chaired, sponsored, and addressed numerous conferences on career guidance for women entering engineering and established an ASME committee on women and minorities (Baylor 1988).

Fitzroy (Fig. 27.2) was elected as the first woman president of the American Society of Mechanical Engineers (ASME) in 1986. She had risen through the ranks of ASME, holding positions at the section and regional levels, and provided years of service to the organization while advocating for equal opportunities for women engineers. In an interview for the *Today* Show about her term as president, Fitzroy remarked on her hopes for women in engineering:

> If I am somewhat unique in this position as the first woman president of a major national engineering society, that's fine and dandy, but I hope that it will inspire more women engineers to take an active part in leadership roles in this country as they have in countries abroad. I also hope that it will inspire more women to become engineers. We need all the bright minds that we can get to tackle the technological challenges of the future (Fitzroy 1986).

Fitzroy's rise to the presidency of ASME began when she was elected chairman of the Hudson-Mohawk Section of ASME in 1963, the first woman in the history of that society to serve in that position. In 1971, she served on the National ASME

Fig. 27.2 Portrait of 1972 Society of Women Engineers Achievement Award recipient and Fellow Nancy D. Fitzroy, circa 1985–1990. (Photograph courtesy of Society of Women Engineers Records, Walter P. Reuther Library, Archives of Labor and Urban Affairs, Wayne State University)

Goals Program Work Party for *Goal 14—Equal Opportunity in Engineering*. In 1977, Fitzroy had the distinction of being the only woman member of the Policy Board Education (PBE), one of five ASME policy boards. Fitzroy helped develop educational policies, establish new academic programs, and develop accreditation guidelines for ASME (Krohn 1977). Fitzroy was also the senior vice president and chair of its Council on Public Affairs and head of the antecedent Policy Board on Professional and Public Affairs. She was an ASME National Nominating Committee member and was active in the ASME Heat Transfer Division. Fitzroy was elected to the ASME Board of Governors in 1983 (SWE Records n.d.; U.S. Woman Engineer 1986).

Fitzroy was equally engaged with the Society of Women Engineers. She was elected to the Executive Committee in 1977 after chairing the Long-Range Planning Committee the previous year, and served as Nominating Committee chair, Member-at-Large representative, and Achievement Award Committee chair (SWE Newsletter 1979). She also was a life member of the society (SWE Records n.d.).

A 1973 role as conference chair gave Fitzroy visibility at the Henniker III "Career Guidance for Women Entering Engineering" conference, sponsored jointly by SWE, the Engineering Foundation, and other major technical societies (Fig. 27.3). The conference aimed to inform young women of engineering career opportunities,

Fig. 27.3 Planners, Henniker III Conference; (*l to r*) Dr. Della Roy at Pennsylvania State University, SWE member Dr. Alva Matthews of Paul Weidlinger Construction Engineering in New York City, James A. Mason Jr. project administrator for the ASME, and Nancy Fitzroy of General Electric Research and Development Center and conference chairperson, work together to plan the 1973 Henniker III Conference. (Photograph courtesy of Society of Women Engineers Records, Walter P. Reuther Library, Archives of Labor and Urban Affairs, Wayne State University)

featuring women speakers from engineering, education, guidance counseling, and government fields (SWE Records n.d.).

Approximately, 100 women engineers participated in the conference, about half students. Also attending were guidance counselors and deans of engineering. Sixteen role-model woman engineers (including Fitzroy) made presentations on their careers and sought input from the students on building role model literature for high-school career guidance. The presentations were as varied as the women they represented, proving that the women who study engineering have diverse personalities and that there is no single stereotype of a woman engineer (SWE Newsletter 1973). Fitzroy compiled, edited, and published the proceedings *Career Guidance for Women Entering Engineering* post-conference. This work was a turning point in content for SWE's career guidance materials, showcasing actual working women engineers in their environment (SWE Records n.d.).

Several of the role-model presentations were published separately as pamphlets, including Fitzroy's *My Career as a Heat Transfer Engineer*, part of the *Women in Engineering—as told by Women Engineers* series. This excerpt shares Fitzroy's views on what engineers do, building from her helicopter pilot experience:

> Although the initial concept of the helicopter was conceived by Leonardo da Vinci [about] 400 years elapsed before a practical [vehicle] was designed by Igor Sikorsky, simply because of the lack of design technology and of engineers to carry out da Vinci's ideas. All disciplines of engineering are represented… mechanical drives, electrical control system, engine design and combustion processes, and [later] computers that take primary navigational information and provide flight commands and courses (Fitzroy 1973).

Fitzroy was also a member of the American Institute of Chemical Engineers (AIChE) and an affiliate member of the National Society of Professional Engineers (NSPE). She served on the Board of Governors of the American Association of Engineering Societies (AAES). She served on the Council on Public Affairs of the AAES and was chair of the New York Capital District Joint Council of Engineering Founder Societies (SWE Records n.d.; Baylor 1988).

27.6 National Advisory Boards

Fitzroy's contributions to technology included service on national advisory boards and committees. In 1972, Fitzroy was named to the National Science Foundation Advisory Committee for Engineering and subsequently served on the NSF Advisory Committee for Research (SWE Newsletter 1972; Baylor 1988). She was also on the Board of Engineering Manpower in Educational Policy of the National Research Council, a member of the NSF Panel for Evaluation of Research Initiation Grants, and chair of an NSF task group on industry/university interaction (SWE Newsletter 1979).

In 1975–1976, she served her alma mater (RPI) as a member of the Presidential Search Committee. She was a charter member of the advisory Rensselaer Council of the RPI (SWE Newsletter 1979). She was a committee member of the GE

Foundation on equal opportunity grants from 1974 to 1986, and a trustee of the Gear Research Institute in 1987–88 (Baylor 1988).

27.7 Personal Life and Retirement

Fitzroy was interviewed in 1976 for the *Cosmopolitan* Magazine article *Women in Engineering.* She described herself as "fairly home oriented." The Fitzroy's enjoyed doing projects together, like building their own home without professional help (Merrill 1976).

> [We] had outside help with the housework...and my husband has had help with yard and grounds maintenance. We...had time to devote to things we like to do together, like building our house single (two) handed without the help of architects, contractors, carpenters, masons, etc. ... [made] most of my clothes...and...made sails for a small sailing pram and...our Starboat....We both like to travel [taking] trips in our own twin-engine airplane. I continue to fly helicopters on occasion, too (Merrill 1976).

Fitzroy had plenty to keep herself busy as an active sportswoman, a licensed airplane pilot, a commercially licensed helicopter pilot, and a sailor. Flying was a passion. She and Roland often plotted out adventures. She owned and operated a twin-engine Cessna 310 called "Nancy's Fancy." Whenever she and Roland could manage time away, they cranked up their Cessna and put many miles between themselves and their jobs. She admitted, "travel keeps me from working all the time" (Fitzroy and Goldstein 1987; Merrill 1976; Fitzroy Obituary 2024).

Throughout her life, Fitzroy (Fig. 27.4) was engaged with *The Whirly Girls*, an International Association of Women Helicopter Pilots and the *Ninety Nines*, the International Organization of Women Pilots (SWE Record n.d.).

Fitzroy was a member of the Mohawk Golf Club in Schenectady, NY, for more than 50 years and a long-time member of the Coral Ridge Yacht Club in Fort Lauderdale, FL. She was also a member of the Hudson Mohawk Chapter of the American Theatre Organ Society and the Union College Annual Business Campaign (Community Scholars) Executive Committee (Fitzroy Obituary 2024).

She and Roland were soulmates ("we were sailors, skiers and flyers together") and intellectual partners until he passed away in fall 2004. They had been married 53 years. Fitzroy passed away in 2024 at the age of 96 (Accolades 2006; Fitzroy Obituary 2024).

27.8 Awards and Recognition

In 1972, Nancy D. Fitzroy [Fig. 27.5] received the Achievement Award of the Society of Women Engineers (SWE) *in recognition of significant contributions to the fields of heat transfer, fluid flow, properties of materials, and thermal engineering.*

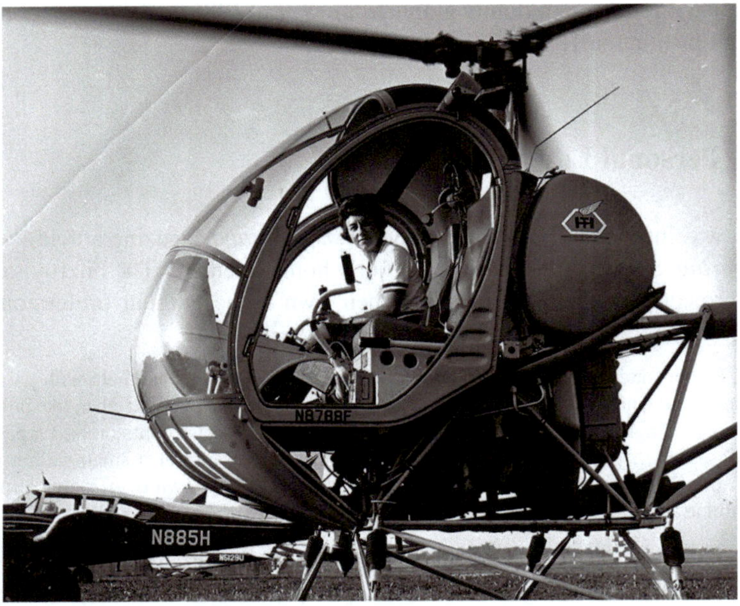

Fig. 27.4 Nancy Fitzroy, one of the earliest female helicopter pilots. (Photograph courtesy of miSci—Museum of Innovation and Science, Schenectady, NY)

Fig. 27.5 Portrait of 1972 Society of Women Engineers Achievement Award recipient Nancy D. Fitzroy, circa 1977–1978. (Courtesy of Society of Women Engineers Records, Walter P. Reuther Library, Archives of Labor and Urban Affairs, Wayne State University)

In 1975, Fitzroy's alma mater, RPI, honored her as the recipient of the Albert Fox Demers Medal from RPI. The citation on the award read:

> The Demers Medal is awarded to Nancy D. Fitzroy in recognition of dedication and loyalty to her alma mater by continued interest in numerous and varied campus projects, service on

> committees and to the Rensselaer Council; and in recognition of significant contributions and service to engineering, government, and professional societies that reflect credit on and add prestige to Rensselaer alumni and the Institute. A leader among the increasing number of women in the engineering profession and the first women to receive a Demers Medal, she is an inspiration to all Rensselaer alumni (SWE Records n.d.).

In 1979, Fitzroy was named an ASME Fellow, the third woman to receive this honor. Her predecessors were Lillian Gilbreth and Florence Buckland. In 1988, Fitzroy was awarded an Honorary Fellowship of the Institution of Mechanical Engineers (IMechE) of the United Kingdom, ASME's sister organization in the UK. Additionally, in 1990, Fitzroy was inducted as a Fellow of SWE (SWE Records n.d.).

One of the earliest women graduates of RPI, Fitzroy was awarded an honorary doctorate in engineering in the 1990 commencement ceremonies *for being in the vanguard as an engineer, breaking new ground in research and development in the field of heat transfer, and for having transcended the bounds which society had set for you* (U.S. Woman Engineer 1990). In 1996, she received the Distinguished Service Award, the highest honor given by the Rensselaer Alumni Association (RAA). Fitzroy was inducted into the Rensselaer Alumni Hall of Fame in 1999 (RPI n.d.). She had also received an honorary doctor of science degree from the New Jersey Institute of Technology in 1987 (SWE Records n.d.).

In 1995, Fitzroy was elected to the National Academy of Engineering for contributions to Mechanical Engineering. Elsa Reichmanis (Achievement Award 1993) was elected that same year, joining eleven other SWE Achievement Award recipients in this distinguished group of engineers at the time (SWE Magazine 1995; NAE 2025).

Fitzroy was the first woman to receive the prestigious Davies Medal for Engineering Achievement from RPI School of Engineering in 2014. This award recognizes a Rensselaer alumna with a distinguished career of engineering achievement, public service, and technical and managerial accomplishments (SWE Magazine 2014).

Other honors included the Centennial Medal from ASME (1980), the Achievement Award of the Federation of Professional Women (1984), the GE Power Systems Sector Engineering Award (1979), the RPI Hall of Fame (1999), and the Honorary Membership Award from ASME, recognizing a lifetime of service to engineering or related fields (2008) (Baylor 1988).

27.9 Legacy

Fitzroy's life was marked by many pioneering accomplishments. She was the first woman to graduate from RPI with a bachelor's degree in chemical engineering in 1949. In 1986, she was the first woman to be national president of ASME. She was the first American woman to be elected an Honorary Fellow of Britain's IMechE and the first woman to receive the RPI Davies Medal for Engineering Achievement in 2014 (SWE Records n.d.; Baylor 1988; RPI n.d.).

Fitzroy's most impactful technical legacy was her breadth of work in heat transfer at GE, as well as her role as principal author and editor of the GE *Heat Transfer and Fluid Flow Data* books for 20 years. These books became outstanding technical resources for future engineers at GE and in the industry (SWE records n.d.; Baylor 1988; Fitzroy Obituary 2024).

Fitzroy said of her 1995 election into the National Academy of Engineering, "that was a real feather in my cap" (EdisonTechCenter 2018).

ASME honored the Fitzroys in 2011, creating the Nancy DeLoye Fitzroy and Roland V. Fitzroy Medal. This honor recognizes their pioneering contributions to the frontiers of engineering, which have led to breakthroughs in existing technology, new applications, or new areas of engineering endeavor (Fitzroy Obituary 2024; ASME 2025).

In 2006, Union College held the dedication of the DeLoye-Fitzroy House in tribute to Nancy and her husband, Roland (Union College, Class of 1943). The Fitzroys also established the Roland V. and Nancy Fitzroy Scholarship for talented electrical engineering students. RPI renamed their Admissions building in 2019 to the Nancy DeLoye Fitzroy, '49 and Roland V. Fitzroy, Jr. Admissions Building (Fitzroy Obituary 2024; Accolades 2006).

References

Accolades. 2006. *DeLoye & Fitzroy: A perfect pair*. Union College Magazine (.pdf). Spring. miSci - Museum of Innovation and Science. https://nyheritage.org/collections/nancy-fitzroy-papers. Accessed March 17, 2025.

ASME. 2025. Nancy DeLoye Fitzroy and Roland V. Fitzroy Medal. https://www.asme.org/about-asme/honors-awards/achievement-awards/nancy-deloye-fitzroy-and-roland-v-fitzroy-medal. Accessed March 17, 2025.

Bailey, Martha J. 1998. *American Women in Science: 1950 to the Present—A Biographical Dictionary*. Santa Barbara, California: ABC-CLIO, Inc. Accessed March 17, 2025.

Baylor, Jill. 1988. National Medal of Technology nomination packet. Society of Women Engineers (SWE) Records (LR001539), Box 188 Folder 17, Walter P. Reuther Library, Archives of Labor and Urban Affairs, Wayne State University. Accessed March 17, 2025.

EdisonTechCenter. 2018. Full Interview with Mechanical Engineer Nancy Fitzroy. November 28. https://www.youtube.com/watch?v=qVfVW1sJpuQ. Accessed March 17, 2025.

Fitzroy, Nancy. 1973. *My Career as a Heat Transfer Engineer*, of the series *Women in Engineering—as told by Women Engineers*. John B. Parrish, Editor. Society of Women Engineers (SWE) Records (LR001539), Box 188 Folder 17, Walter P. Reuther Library, Archives of Labor and Urban Affairs, Wayne State University. Accessed March 10, 2025.

Fitzroy, Nancy. 2018. 'Big Old Male Syndrome' No Impediment to Nancy Fitzroy. ASME Oral History. July 12. http://videos.asme.org/detail/video/3028627865001/big-old-male-syndrome-no-impediment-to-nancy-fitzroy. Accessed March 10, 2025.

Fitzroy, Nancy. 1986. Today Show interview notes. Unpublished. miSci - Museum of Innovation and Science. September 18. https://nyheritage.org/collections/nancy-fitzroy-papers. Accessed March 17, 2025.

Fitzroy, Nancy; Stuart Goldstein. 1987. Challenges and Fun Attract Fitzroy to Mechanical Engineering. Unpublished. miSci - Museum of Innovation and Science. September 29. https://nyheritage.org/collections/nancy-fitzroy-papers. Accessed March 17, 2025.

Fitzroy Obituary. 2024. Dr. Nancy Fitzroy Obituary. Daly Funeral Home, Inc. January 15. https://dalyfuneralhome.com/tribute/details/2055/Dr-Nancy-Fitzroy/obituary.html. Accessed March 17, 2025.

Krohn, Barbara. 1977. Women Engineers: Extending their Influence through the Technical Societies. SWE Newsletter. January/February. Society of Women Engineers (SWE) Records (LR001539), Box 188 Folder 17, Walter P. Reuther Library, Archives of Labor and Urban Affairs, Wayne State University. Accessed March 10, 2025.

Merrill, Sam. 1976. *Women in Engineering*. Society of Women Engineers (SWE) Records (LR001539), Box 188 Folder 17, Walter P. Reuther Library, Archives of Labor and Urban Affairs, Wayne State University. Accessed March 10, 2025.

National Academy of Engineering (NAE). 2025. Dr. Nancy D. Fitzroy 1995. https://www.nae.edu/30092/Dr-Nancy-D-Fitzroy. Accessed March 17, 2025.

Rensselaer Polytechnic Institute (RPI) School of Engineering. n.d. Nancy DeLoye Fitzroy—1949. https://eng.rpi.edu/person/nancy-deloye-fitzroy-1949. Accessed March 17, 2025.

Society of Women Engineers (SWE) Records (LR001539). n.d. Fitzroy, Nancy DeLoye. Box 188 Folder 17, Walter P. Reuther Library, Archives of Labor and Urban Affairs, Wayne State University. Accessed March 10, 2025.

SWE Magazine. 2014. First Woman to Receive Rensselaer Davies Medal for Engineering Achievement. Fall. Society of Women Engineers (SWE) Records (LR001539), Box 188 Folder 17, Walter P. Reuther Library, Archives of Labor and Urban Affairs, Wayne State University. Accessed March 17, 2025.

SWE Magazine. 1995. Fitzroy, Reichmanis named to National Academy of Engineering. July/Aug. Society of Women Engineers (SWE) Records (LR001539), Box 188 Folder 17, Walter P. Reuther Library, Archives of Labor and Urban Affairs, Wayne State University. Accessed March 17, 2025.

SWE Newsletter. 1972. SWE Achievement Award Winner. September. Society of Women Engineers (SWE) Records (LR001539), Box 188 Folder 17, Walter P. Reuther Library, Archives of Labor and Urban Affairs, Wayne State University. Accessed March 10, 2025.

SWE Newsletter. 1973. Henniker III. October. Society of Women Engineers (SWE) Records (LR001539), Box 188 Folder 17, Walter P. Reuther Library, Archives of Labor and Urban Affairs, Wayne State University. Accessed March 10, 2025.

SWE Newsletter. 1979. Nancy D. Fitzroy named ASME fellow. January/February. Society of Women Engineers (SWE) Records (LR001539), Box 188 Folder 17, Walter P. Reuther Library, Archives of Labor and Urban Affairs, Wayne State University. Accessed March 10, 2025.

Tarplee, Jennifer; Jan Williams P.E. 2000. Extraordinary Women. *SWE Magazine*. June. Society of Women Engineers (SWE) Records (LR001539), Box 188 Folder 17, Walter P. Reuther Library, Archives of Labor and Urban Affairs, Wayne State University. Accessed March 17, 2025.

U.S. Woman Engineer. 1986. Nancy Deloye Fitzroy Named ASME President. March/April. Society of Women Engineers (SWE) Records (LR001539), Box 188 Folder 17, Walter P. Reuther Library, Archives of Labor and Urban Affairs, Wayne State University. Accessed March 10, 2025

U.S. Woman Engineer. 1990. Fitzroy Awarded RPI Honorary Doctorate. September/October. Society of Women Engineers (SWE) Records (LR001539), Box 188 Folder 17, Walter P. Reuther Library, Archives of Labor and Urban Affairs, Wayne State University. Accessed March 10, 2025

Chapter 28
Irene Carswell Peden, PhD (1925–2025) Achievement Award Recipient 1973

In recognition of her significant contributions in the fields of radio wave propagation research and electrical engineering education

Jill S. Tietjen

28.1 Introduction

The first American woman engineer or scientist to live and work in Antarctica, Irene C. Peden opened new frontiers for women. A woman of many firsts, Peden was the first woman to receive a PhD in electrical engineering (she said the first in any discipline) from Stanford University, the first woman faculty member in engineering at the University of Washington, and the first woman on the board of directors of the Institute of Electrical and Electronics Engineers. An expert in radio science and electromagnetic wave propagation scattering, Peden was a strong advocate for women in science, technology, engineering, and mathematics (STEM) and fought for equal treatment of women for years. The 1973 recipient of the Society of Women Engineers' (SWE) Achievement Award, Peden received many other awards in recognition of her technical expertise and contributions to professional organizations and was elected to the National Academy of Engineering in 1993.

28.2 Early Life and Undergraduate Education

The oldest of three girls, Irene Carswell Peden (Fig. 28.1) was born in Topeka, Kansas in 1925 and attended high school in Kansas City, Missouri. Her father was a businessman who had not finished college. Her mother was talented in mathematics and wanted to attend college, but her father (Irene's maternal grandfather) would not hear of it. So, Irene's mother and aunt went to a small town in Western Kansas

J. S. Tietjen (✉)
Technically Speaking, Inc., Greenwood Village, CO, USA

C. D. Craig et al. (eds.), *Women Engineering Legends 1952-1976: Society of Women Engineers Achievement Award Recipients*,
https://doi.org/10.1007/978-3-032-00224-2_28

Fig. 28.1 Irene Peden Portrait (Photograph courtesy of Society of Women Engineers Records, Walter P. Reuther Library, Archives of Labor and Urban Affairs, Wayne State University)

and alternated going to school and working, paying to send the other one to college. Although Irene's aunt did complete her college education in this manner, Irene's mother did not. However, because of her mother's experience, Peden grew up "knowing that if you wanted to do something, no matter what kind of roadblocks they throw in your way, you just do it" (Kata 2002).

Her mother was determined that Peden was going to become a musician. However, despite the discouragement of high school teachers and counselors, Peden discovered chemistry as a high school senior. One science was needed to graduate from her high school. She had been told that physics was too hard and girls didn't do physics, so she took chemistry. She says, "I really, really connected. I liked it a lot." She started college at Kansas City Junior College, the only college her family could afford coming out of the Depression, as a chemistry major, and took physics. She fell in love with physics (Kata 2002).

> I just loved it. And I wanted to do that. I'm talking at a time when they hadn't split the atom yet. There wasn't anything you could do with it in those days except teach it, and you had to get a PhD to do that, teach in a college. And I was a teenager. I didn't want to do those things. I wanted to get a four-year degree, get out of school and have a life. That was my ambition. (Kata 2002).
>
> So I decided that there must be some aspect of physics in which there was a practical aspect that a four-year degree would do. And I, in retrospect, realized, that I had a physics professor for that course who was very ahead of his time, because he was willing to sit down with me and talk over things like that. And together we reasoned that I enjoyed electricity and magnetism best. . . .And it was he who suggested that there was a four-year degree – he didn't know of any women that ever did it, but there was one, and it was electrical engineering. And gradually, I came to the realization that I wanted to try it. (Kata 2002).

Because Peden wanted to leave Kansas, she chose to pursue her engineering education at an out-of-state university. She wanted to go to the University of Illinois, but her allergist told her that her allergies would be worse in Urbana than they had been in Kansas City. One of her father's friends had earned his engineering degree at the University of Colorado. Peden thought the distance from home (by train) was manageable, and off she went to Boulder, Colorado (Kata 2002).

The University of Colorado at Boulder was very welcoming to the young women who were entering engineering in 1944. Peden relates that there were a dozen women in her entering class. Together, she and the other women formed a club, including some women architecture students, somewhat like SWE, but earlier. She also participated as a student member of the American Institute of Electrical Engineers, the predecessor organization to the Institute of Electrical and Electronics Engineers (IEEE). Although the electrical engineering department faculty were supportive, faculty in other departments were sometimes hostile. Peden says:

> I really believe that because I knew my mother had done what she darn well wanted to do and did it anyway, I never felt that I had to back down. Your mother can be a powerful role model. And even though she was not an engineer, I think she was probably mine. (Kata 2002).

Peden's other role model was Marie Curie: (Kata 2002).

> The only thing I could do was have inspirational thoughts because somebody like Marie Curie had done what she had done. I never had any illusions about winning a Nobel Prize. But just knowing that in the setting in which she lived and worked she was able to do something so spectacular made me feel that maybe I can do something that I wanted to do also.

28.3 Career and Graduate Education

Peden got married after she graduated from the University of Colorado with a BS in electrical engineering in 1947, and they moved to Wilmington, Delaware that year. She "pounded the pavements" and received many rejections: "We've never hired a woman before and we're not going to talk to you," until she got a job at Delaware Light and Power Company. She worked there from 1947 to 1949 until her then husband got admitted to Stanford for graduate school. As had happened in Delaware, no one in California wanted to hire a woman engineer either. Luckily for Peden, the Stanford Research Institute (SRI) had started an Antenna Lab and they needed scientific personnel. Peden was hired and worked there on antenna problems (Bailey 1998; Kata 2002; SWE Records).

As she worked at the Antenna Lab, Peden realized that she wanted more education. Her mentors at SRI helped her gain admission to Stanford's graduate school, and she began working on her master's degree. She worked part-time and attended school part-time until she decided to take a leave of absence to complete her master's degree (Kata 2002).

> And by the time I finished I didn't want to quit. There were more things I wanted to know. And so my relationship with SRI was just terminated and I became a full-time graduate student. I got a research assistantship. I have the first PhD that's in engineering that Stanford ever gave to a woman. And that was a long, lonely path. (Kata 2002).

Peden completed her MS in 1958 and her PhD from Stanford's Microwave Lab in 1962. Her research topic was measurement techniques for microwave periodic circuits. Her parents were somewhat mystified by her choice of major and her mother actively tried to discourage her from becoming a teacher. "She [Peden's mother] was very alarmed and negative when I finished my graduate work and said I wanted to go to a university and teach." In the end, however, they came around and were supportive of Peden's choices (Bailey 1998; Kata 2002; SWE Records).

Peden was divorced by the time she moved to Seattle, Washington in 1961. She accepted a position at the University of Washington's electrical engineering department as an assistant professor. She was the first woman faculty member in engineering (Bailey 1998; Kata 2002).

When she arrived at the University of Washington, she realized that she would need to shift her research focus from microwave work to something else. She discovered that a former colleague from SRI had a grant from the National Science Foundation (NSF) to do upper atmospheric research—polar research—in the Antarctic as it related to radio waves. She didn't intend to become an Antarctic researcher, but she was interested in what they were doing and had ideas to contribute. The problem of discerning the impact of ice on antenna signals fascinated her. Eventually, she became a co-project investigator on the NSF grant, which was through the Office of Polar Programs (ETHW 2025; Kata 2002; Perusek 2002).

Peden relates that the NSF policy in the Polar Programs Office required principal investigators of Antarctic research projects to spend adequate time on-site so that they could understand the region and the associated weather with its extreme difficulties for conducting research. The conditions included extremely cold temperatures, austere accommodations, isolation, long work hours, and a boring diet that consisted primarily of canned, frozen, and dried food (ETHW 2025; Kata 2002).

> If you're going to work in the Antarctic, you have to understand the environment to some degree, because it is so difficult to work in the polar environment. Your experiments have to be very rugged. They have to be relatively simple in a certain way, and yet very sophisticated in other ways, or else what you want to find out is already known. If it's research, it better be research. And that is not an easy criteria, necessarily, to meet. And so they wanted all their principal investigators to go to the Antarctic and experience the environment. (Kata 2002).

The problem for Peden, however, was that the US Navy didn't want a woman in the Antarctic, although women from other countries had been working on the Continent. She worked with NSF personnel and her colleagues at the University of Washington to put together a proposal for research that needed to be done in the Antarctic and could be completed within a limited time window, which she then had peer-reviewed and approved (Bailey 1998; Kata 2002).

> We needed to know more about the properties of the deep ice than was known by anybody at that time in order to make some progress on what we wanted to do. We were studying the

> lower boundary of the ionosphere. But the wavelength was so long that the ice at the point where the antenna was, was probably close to a quarter of the wavelength, just the ice. And electrically, that makes a lot of difference. But we had too many unknowns in the problem. We had to get some unknowns out . . .(Kata 2002).

Peden was targeting going to Antarctica with the season that opened in October 1970, and she envisioned a total of 4 weeks for her work there. After her proposal was written, the usual arguments surfaced—"no bathrooms for women," and many more. Then the admiral in charge decided that another female would need to accompany Peden on the same trip in case she was injured and required attention from a Navy doctor. The New Zealand woman geophysicist who was selected didn't pass the required physical, but a librarian from Christchurch, New Zealand (the hopping off point to the Antarctic), who was quite athletic and wanted to go, became her companion (Kata 2002).

Unfortunately, a piece of equipment that was needed for her experiment was taken from the team by the airlines (who decided it wasn't carry-on luggage) and lost. Despite the equipment issue, the NSF rep said the equivalent to Peden of "You can't fail. If you fail and are not able to complete your project and publish your results, there won't be another woman on the Antarctic Continent for generations." This put pressure on Peden as did the resentment and grumbling from some of the men at McMurdo, the American research station in the Antarctic, creating a somewhat hostile environment. Her experiment was saved when a graduate student from Stanford who was also in the Antarctic knew of a different piece of equipment already on the Continent that could be modified to do what needed to be done. Her experiment was a success and Peden opened the doors for generations of women in science and engineering to perform work in Antarctica. "I am pleased to know that women have taken their place in the Antarctic. It's what we were working for years ago" (ETHW 2025; Kata 2002; Perusek 2002).

Through her research, Peden developed new methods of characterizing the very deep ice at the Antarctic by examining the effects that it has on the radio waves that are transmitted through it. She did this through her study of the ionosphere which involved the development of mathematical models to analyze the data that her team collected. Peden and her students were the first to measure many of the electrical properties of the ice in Antarctica and to specify the characteristics of the very low frequency propagation over long paths of ice in the polar region. Later in her career, when her focus changed to subsurface exploration techniques using very high frequency radio waves, she again found the need to develop mathematical models and methodology to interpret the data collected (Bailey 1998).

Peden served as Associate Dean of Engineering at the University of Washington (1973–1977), associate chair of the electrical engineering department (1983–1986), and later completed a 2-year assignment as Director, Division of Electrical and Communication Systems at the National Science Foundation (Bailey 1998; Peden 1991). She retired from the University of Washington as Professor Emerita in 1994 (ETHW 2025).

28.4 Organizations

Peden's participation in organizations was extensive. She was quite active and rose to leadership positions in most. In some ways, that giving back can be traced to her early days at the University of Washington. The law librarian at the University of Washington took Peden under her wing. This older woman contacted Peden when she arrived on campus, and they went to lunch. "She impressed on me the importance of giving back as part of the responsibility that we have. And I believe that" (Kata 2002).

Through the Society of Women Engineers, Peden (Fig. 28.2) used to visit high schools and talk about career possibilities in STEM (Kata 2002). She attended the first International Conference of Women Engineers and Scientists in 1964 and met a lot of women, including women from the Pacific Northwest (Kata 2002). Peden helped form the collegiate section of SWE at the University of Washington, served as its Faculty Advisor, and served as President of the SWE Pacific Northwest Section (Peden 1991; SWE Records).

Peden's membership in IEEE served her well. A man whom she had met at IEEE conferences, Mac Van Valkenburg, served as her sponsor and mentor, introducing her to others and opening doors for her. Peden's initial connection with IEEE was technical through its societies—Antennas and Propagation, Geoscience and Remote

Fig. 28.2 President Lydia Pickup awards Best Student Section to Linda Ann Saunders, University of Washington Student Section Chairman under the watchful eye of Irene Peden University of Washington SWE Counselor, 1969 (Photograph courtesy of Society of Women Engineers Records, Walter P. Reuther Library, Archives of Labor and Urban Affairs, Wayne State University)

Sensing, and then Education. She was Vice Chairman of the Seattle Section of the IEEE and chairman of the Professional-Technical Group on Microwave Theory and Technique (SWE Records). Through the sponsorship of Mac Van Valkenburg, she ran for and was elected a Vice President of IEEE in 1975. She was the first woman member of the board of directors of IEEE. She was re-elected as vice president for educational activities after completing her first one-year term. At one point, the General Manager of the IEEE called her "Number One Woman in the IEEE." She also served on the Editorial Board of the *Proceedings of the IEEE* (Bailey 1998; Kata 2002; Krohn 1977; Newsletter 1973; SWE Records).

Peden served on the Advisory Committee to the Stanford School of Engineering and the National Drug Advisory Committee of the Federal Drug Administration. She was active in the University YWCA and served on the Education Committee of the Governor's Commission on the Status of Women (Newsletter 1973). Peden became an IEEE accreditation evaluator in 1973 and was elected in 1975 to ABET's Engineering Accreditation Commission. She chaired that Commission from 1981 to 1982 and served on ABET's Board of Directors from 1982 to 1988 (IEEE 1990; Krohn 1977; Newsletter 1973).

Peden was the first woman selected to the Advisory Board of the United States Merchant Marine Academy (1978). She also served on the United States Army Science Board (which she chaired 1986–1987), whose responsibilities include advising the assistant secretary of the Army. She was a member of the Committee on Space Physics of the Space Sciences Board, National Academy of Sciences. Peden was a consultant to UNESCO on the Access of Women to Technical Careers in the West Coast Region of the United States (Firsts and Onlys 1978; SWE Records).

28.5 Awards and Honors

Peden (Fig. 28.3) received SWE's 1973 Achievement Award "in recognition of her significant contributions in the fields of radio wave propagation research and electrical engineering education." This was Peden's first major honor, and she said, "It was wonderful. And I felt very privileged. I still do" (Kata 2002). She was elected a SWE Fellow in 1991. She was also a Fellow in several other organizations, including the American Association for the Advancement of Science, American Society for Engineering Education, ABET, and the Explorer's Club (Bailey 1998; People 2018).

Very active in the IEEE, she received many honors from that organization, including Fellow (1974), "for contributions to radio science in the polar regions and for leadership of women in engineering." Her other IEEE honors include: 1988 Haraden Pratt IEEE Award "for leadership and outstanding contributions to the IEEE in education, technical activities, and Institute Boards" (SWE Records), IEEE Antenna Society's "Man of the Year" Award (People 2018), the Centennial Medal (1984) (Bailey 1998), and Meritorious Achievement Award in Accreditation

Fig. 28.3 Irene Peden speaks at the 1983 SWE Convention (Photograph courtesy of Society of Women Engineers Records, Walter P. Reuther Library, Archives of Labor and Urban Affairs, Wayne State University)

Activities, "for sustained leadership in support of the accreditation process in IEEE and ABET" (IEEE 1990).

Peden received many additional honors and awards and was elected to the National Academy of Engineering in 1993 (Bailey 1998; People 2018). The myriad honors include: U.S. Army's Outstanding Civilian Service Medal (1987) (Bailey 1998); Centennial Medal, University of Colorado (1988) (Bailey 1998); University of Colorado Distinguished Engineering Alumnus (1974) (SWE Records); Hall of Fame, American Society for Engineering Education (Bailey 1998); Diamond Award for Distinguished Service from the University of Washington, 2018 (People 2018); Honorary doctorates from Michigan State University and Southern Methodist University (People 2018); Sigma Xi, Tau Beta Pi, and Theta Sigma Phi "Woman of Achievement" (Newsletter 1973).

28.6 Advocacy

For decades, Peden advocated for women in STEM and fought for equal treatment of women. Along with two other tenured women faculty at the University of Washington, Peden collected data and performed the first salary study comparing male and female faculty salaries. Disparities were found, and deans and department chairs were informed. Peden states that the effort, which required the intervention of the Lieutenant Governor, led to changes but also had long-term negative ramifications for her and the other two women faculty members involved (Kata 2002).

A role model, mentor, advisor, champion, and confidante, Peden (Fig. 28.4) worked to expand pre-college engineering education and advocated for women engineers so they could succeed on the campus at the University of Washington. She fought for equal opportunity for her work and research. She was so dedicated to students and faculty that she invited new faculty members to stay at her house until they found their own housing (People 2018; SWE Records).

Of the women engineers and women faculty she has known, Peden said, “The faculty women that I have known match the women engineers that I have known in the quality of sheer dogged persistence” (English 2020).

28.7 Personal

When Peden moved to Seattle, there was a significant amount of publicity about her arrival as the first female faculty member in engineering at the University of Washington. Due to this publicity, she met an attorney, Leo Peden, who was in private practice as a trial lawyer. He was divorced with two teenage daughters when they married in 1962. The girls lived with their mother except for some holidays and the summer until the ex-wife and her second husband moved out of state and the older girl decided to move in (Bailey 1998; Kata 2002).

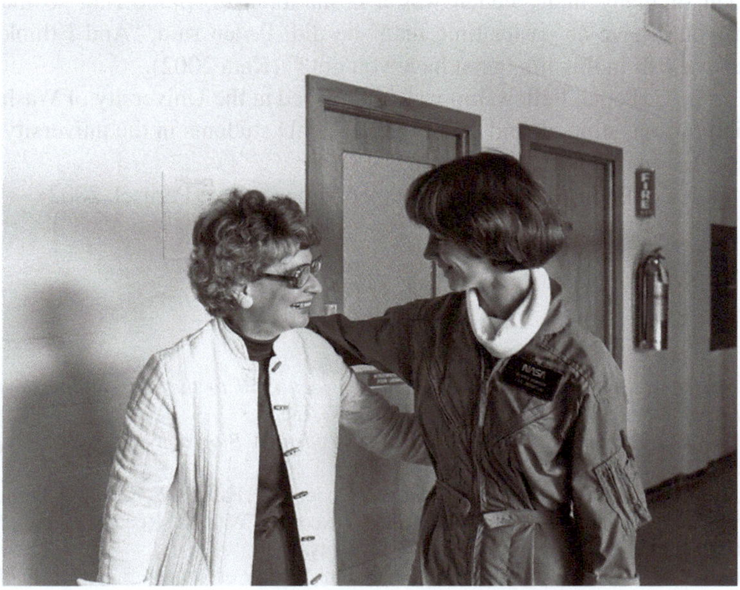

Fig. 28.4 Irene Peden and Bonnie Dunbar (1993 SWE Resnik Challenger Medal recipient) in conversation in 1980. Bonnie Dunbar graduated from the University of Washington (Photograph courtesy of Society of Women Engineers Records, Walter P. Reuther Library, Archives of Labor and Urban Affairs, Wayne State University)

> She was going to be a senior in high school, which is an important time in a kid's life. You've built your friendships, you've got your activities. . . . And she decided all by herself that she wanted to go to the University of Washington anyway, and she had to change high schools anyway, that she wanted to come and live with us. . . . I really wasn't sure myself how you balance everything, but I figured one day at a time we'll figure this thing out. Because she and I enjoyed each other. She now refers to that period as 'Breaking in a new mother.' And I'm sure it was stressful for her too (Kata 2002).

Peden had a very supportive husband who helped around the house and provided her with significant encouragement.

> I did not do any housework, we hired that done. Leo hired the help, and if somebody else had to be replaced, he took care of that. If we had a period when there was nobody, he would run the vacuum cleaner, whatever he had to do to relieve me, so I didn't do that. All I did was cook the meals, which is something I thoroughly enjoyed. He did all the dishes, and just did everything to take the load off, because he knew I was under a lot of pressure. And with that kind of support, you know, it's possible to manage (Kata 2002).

She and her husband enjoyed swimming, water skiing, and boating on Lake Washington. Her hobbies included music, clay modeling, and cooking (Newsletter 1973).

28.8 Legacy

A woman of many firsts, Peden was a groundbreaker, pioneering so that other women would have an easier time than she did. Peden said, "And I think if you won't take a risk in this life, what have you got?" (Kata 2002).

The Irene C. Peden Fellowship was established at the University of Washington, providing tuition, stipend, and fees to female PhD students in the university's electrical engineering department (People 2018).

The Peden cliffs in Antarctica are named in her honor (Perusek 2002).

References

Bailey, Martha J. 1998. *American Women in Science: 1950 to the Present—A Biographical Dictionary*. Santa Barbara, California: ABC-CLIO, Inc.

Engineering and Technology History Wiki. *Irene Peden: A Biography.* https://ethw.org/Irene_Peden. Accessed March 18, 2025.

English, Troy Eller. 2020. Deciphering the Faculty Code. SWE Fall, *SWE Magazine.* From Society of Women Engineers Publications (LR002487) [excerpts]. Walter P. Reuther Library, Archives of Labor and Urban Affairs, Wayne State University.

"Firsts and Onlys," 1978. *SWE Newsletter.* May. From Society of Women Engineers Publications (LR002487) [excerpts]. Walter P. Reuther Library, Archives of Labor and Urban Affairs, Wayne State University.

IEEE. EAB honors seven members with awards for 1989. 1990. *IEEE Institute.* March. From Society of Women Engineers Publications (LR002487) [excerpts]. Walter P. Reuther Library, Archives of Labor and Urban Affairs, Wayne State University.

Kata, Lauren. 2002. Irene Peden Interview. *Profiles of SWE Pioneers—Oral History Project*. Reuther Library Oral History ID: LOH001952.28. March 2.

Krohn, Barbara. 1977. Women engineers: extending their influence through the technical societies. *SWE Newsletter*, January/February. From Society of Women Engineers Publications (LR002487) [excerpts]. Walter P. Reuther Library, Archives of Labor and Urban Affairs, Wayne State University.

Newsletter. 1973 SWE Achievement Award Winner. *Society of Women Engineers*. September. Volume 20, Number 1. From Society of Women Engineers Publications (LR002487) [excerpts]. Walter P. Reuther Library, Archives of Labor and Urban Affairs, Wayne State University.

Peden, Irene C. 1991. "Dr. Irene C. Peden," U.S. Woman Engineer, September/October. From Society of Women Engineers Publications (LR002487) [excerpts]. Walter P. Reuther Library, Archives of Labor and Urban Affairs, Wayne State University.

People. 2018. SWE Conference, *SWE Magazine*. From Society of Women Engineers Publications (LR002487) [excerpts]. Walter P. Reuther Library, Archives of Labor and Urban Affairs, Wayne State University.

Perusek, Anne M. 2002. Women in Antarctica. *SWE Magazine*. December/January. From Society of Women Engineers Publications (LR002487) [excerpts], Walter P. Reuther Library, Archives of Labor and Urban Affairs, Wayne State University.

Society of Women Engineers Publications (LR002487) [excerpts]. Walter P. Reuther Library, Archives of Labor and Urban Affairs, Wayne State University (n.d.).

Chapter 29
Barbara Crawford Johnson (1925–2005) Achievement Award Recipient 1974

In recognition of her significant engineering contributions in support of manned spaceflight programs

Cecilia D. Craig

29.1 Introduction

Born in 1925 in Sandoval, Illinois, Barbara "Bobbie" Crawford was the youngest of five children, with two brothers and two sisters. Her parents were college-educated; her father, Clarence, was a school superintendent, and her mother, Nellie, was a teacher. Johnson and her siblings were raised knowing they would attend college. Her eldest brother became an electrical engineer at the Naval Academy, Johnson became an engineer, and their other siblings became a librarian or a teacher (Society of Women Engineers (SWE) 2003; Wolters 2018).

Johnson was a paper girl for the St. Louis Dispatch, packing candy and nuts for a local supplier, waitress, and in college, "counted bugs in the biology department, checked hats in the student union. I worked in a bar for about three days" (Society of Women Engineers (SWE) 2003).

> I was probably a tomboy, I guess you'd call it, and from the time I was a little kid, I had jobs. I picked berries and sold them to the neighbors, and I delivered papers. I was the first papergirl the *St. Louis Post Dispatch* ever had. (Society of Women Engineers (SWE) 2003).

Amelia Earhart was Johnson's role model. Johnson cut her hair like her hero and wore an Amelia cap (Wolters 2018). Role models often significantly influence career choice or personality traits (Craig 2024), and Amelia had that influence on her. Johnson loved to watch planes on a "dirt landing strip near her hometown of Sandoval IL" (Wolters 2018). She begged for rides and would ask pilots about flying. Learning more and more about planes, "her love for flight grew" (Wolters 2018).

C. D. Craig (✉)
San Jose, CA, USA
e-mail: Ceal@Research.Druai.com

C. D. Craig et al. (eds.), *Women Engineering Legends 1952-1976: Society of Women Engineers Achievement Award Recipients*,
https://doi.org/10.1007/978-3-032-00224-2_29

Another role model arrived in the form of an engineer boarding with her family. He was overseeing the construction of a high school that her father headed (Wolters 2018). Between this most likely civil engineer and her brother, Johnson explored and learned engineering techniques (e.g., surveying) and tools (e.g., a slide rule). Her parents also fully supported her interests, another key factor in that career interest forming so strongly (Society of Women Engineers 2003; Wolters 2018).

Scholarships were important, with five siblings in the family all planning for college, so Johnson sought them. She decided that the University of Illinois (U of I) made economic sense since it was in-state. She started at the U of I in 1943 (Wolters 2018), carrying a heavy course load to finish as fast as possible to save money.

Johnson was homesick at the university, living in a dormitory, and wondered if she should change her major to business. Some people suggested that a job might be hard for a woman engineer to come by. Both her parents recommended that she stick with her engineering choice. Faculty and peers supported her, though she was the only woman in engineering (Wolters 2018). She was one of only a few civilians; most students were servicemen during WWII (Society of Women Engineers (SWE) 2003).

Learning more about flying remained a passion during college. Johnson would attend mixers at a local Army Air Corps base during World War II. It was bustling. Meeting pilots, she gained their trust and they were eventually able to take her flying, which ultimately helped her learn to fly (Wolters 2018).

After graduating in 1946 with a bachelor's degree in general engineering, Johnson began her career at North American Aviation (NAA) in California, which would later become Rockwell International's space division (Wolters 2018). She had four job offers: one at the University of Illinois to earn a master's degree and teach, and three others from "large firms, one in Chicago, one in Pittsburgh, and one in Los Angeles" (Johnson 1977).

29.2 Early Career

Johnson began her career at North American Aviation in Los Angeles, where she performed numerous calculations. She was skilled at these, although she was a bit bored. She approached her supervisor and asked for engineering tasks: "I want a real engineering assignment" (Society of Women Engineers (SWE) 2003). The next week, she began working on a supersonic inlet design and did not look back! (Society of Women Engineers (SWE) 2003). "Within 5 years [1946–1953] of employment at Rockwell, Johnson moved up from the position of mathematician to senior engineer, aerodynamics" (SavvyExacta 2012).

Around 1958, Johnson was part of the NAA team pursuing the Hound Dog contract from the Air Force, an "air-to-surface missile, designed for nuclear warhead delivery" (Air Force Historical Foundation 2017). Johnson was involved in the Hound Dog project from the first design concept through the final design selection

(Society of Women Engineers (SWE) Records n.d.). She was the Rockwell Hound Dog Project Leader (Society of Women Engineers (SWE) Records n.d.).

The NAA Hound Dog team periodically went to Wright-Patterson Air Force Base in Dayton, OH, to meet with the Air Force team. In the 1950s, women's travel on project teams was restricted. So, practicing a tiny bit of deception, Johnson made her reservations using her initials. She had an impact at the first meeting she attended, showing her technical understanding and composure. When she arrived back in California, a new procedure had been developed to allow women to travel for NAA with a President's or Vice President's approval. She traveled frequently after that (Wolters 2018).

Barbara married Robert H. Johnson in 1956. Her husband was also in the aerospace industry at Rockwell (Society of Women Engineers (SWE) Records n.d.). They had a son, and she worked 3 days a week for the first few years, spending time with her son. Then, when the mission to land on the moon started, she went back to work full time (Wolters 2018).

29.3 Mid and Late Career 1963–1982

29.3.1 Apollo Program and Other Earlier Space Programs

Johnson's assignments included the "flight dynamics studies for Dyna-Soar, the Recovery of Hypersonic Gliders, Lunar Reentry Vehicle Research, Orbital Rendezvous and the performance analyses for the XSM-64 phase of the Navaho missile" (Society of Women Engineers (SWE) Records n.d.). Then, from 1961 to 1968, for NAA, she was the Entry Trajectories supervisor for the return of Apollo missions. From 1968 to 1973, she was the Systems Engineering Manager for "mission and flight operations activities...[for] Command and Service Modules" (Society of Women Engineers (SWE) Records n.d.).

In April 1963, Johnson shared her work on the Apollo Project with a reporter. "After all, it's never been done before. No two days are ever the same, and you can't find the answers in a handbook. There's a lot we don't know yet." Johnson described why Apollo vehicles must stay within a corridor when making an entry back to Earth, known as *capture*. If there is not enough atmosphere to slow the vehicle down, and if the vehicle cannot stay in the corridor, it would need many orbits to land (Sammis 1963). In 1968, Johnson (Fig. 29.1) was the Mission Requirements and Integration manager for Rockwell's Apollo Landing NASA contract, focused on atmospheric entry performance and design trajectories (Wolters 2018).

> A key accomplishment [during this period] was the definition of the entry corridor through which the Apollo must be guided during entry to avoid excessive deceleration or an uncontrolled skip out of the atmosphere (Society of Women Engineers (SWE) Records n.d.).

George E. Mueller wrote to Johnson at Rockwell Space Division, extending special congratulations

Fig. 29.1 Johnson by an Apollo spacecraft simulator near her office at North American Aviation in 1963, where she managed the department responsible for plotting the return course of the Apollo moon rocket (Courtesy of University of Illinois Archives)

> ...for her special efforts and your significant contribution to this epoch of space voyaging will never be forgotten by those of us who had the pleasure of joining with you in this, the greatest of all quests yet undertaken by man. (Mueller 1973).

This was echoed by G. W. Jeffs, Executive Vice President of Space Division, mentioning an Apollo 11 medallion given to Johnson and others at Space Division by Chris Kraft, NASA Director. Chris Kraft, a well-known NASA leader, sent a heartfelt thanks and a medal to Rockwell Space Division:

> 60 stainless steel medallions, [were] made in commemoration of the first lunar landing mission. A special die was flown on Apollo 11 and this die was utilized in fabricating a limited number of medallions. I would like you to have one in recognition of the major role you played in achieving this national goal (Kraft 1973).

The team developed a display for the Apollo crews to monitor progress in entry back to Earth, helping the crew be aware of any issues and allowing the crew to take manual control if required. The astronauts were not immediately receptive to this graphical display. It was included in the flight simulator to gain their acceptance, and only then was the new idea accepted by the astronauts (Wolters 2018). During the Apollo 13 challenges,[1] Johnson led a team focused on conserving the vehicle's consumables and helped "extend life support and system capabilities" (Society of Women Engineers (SWE) Records n.d.).

29.3.2 Space Shuttle

From June 1972 to October 1981, Johnson was Mission Analysis and Integration Manager for the Space Shuttle Program (Society of Women Engineers (SWE) Records n.d.). In this role, Johnson was responsible for the Space Transportation System (Space Shuttle) Orbiter vehicle, directing more than 100 engineers. They were responsible for the development and integration of Shuttle system performance analyses. These included the "ascent, on-orbit, and descent phases" (Rub and Romm 1988). Her team assessed payload and Shuttle mission interactions in communication with NASA (Rub and Romm 1988).

During her career, Johnson (Fig. 29.2) had more than 100 engineers working for her on multiple space projects: Lunar Landing, including Apollo 11, Skylab, and Apollo-Soyuz programs (SavvyExacta 2012). She worked for NAA and then Rockwell for 36 years, 25 of which were spent as a manager (Wolters 2018). Eventually, she had to take a medical disability retirement and retired in the early 1980s (Harness 1988).

29.3.3 SWE Achievement Award

In 1974, Johnson (see Fig. 29.3) received SWE's Achievement Award with a citation that read: *In recognition of her significant engineering contributions in support of manned spaceflight programs.*

One recommender, in a letter to the Associate Dean, Joseph McNamara, then President of Rockwell Space Division, asserted he "consider[ed] her to be among our top technical personnel in this Division...[describing] her complete dedication

[1] Apollo 13: Intended to be the third moon landing, launching on April 11, 1970, an oxygen tank exploded putting the crew at risk. After a tense and incredibly harrowing experience with water and oxygen shortages, fitting lithium hydroxide canisters from the command module to the lunar module, and more technical challenges requiring the on-ground team and the crew to work collaboratively together, the crew safely landed back on the Earth on April 17 (National Air and Space Administration 2020).

Fig. 29.2 Barbara "Bobbie" Johnson portrait (Courtesy of University of Illinois Archives)

Fig. 29.3 Barbara Johnson at work, circa 1970–1974 (Photograph courtesy of Society of Women Engineers Records, Walter P. Reuther Library, Archive of Labor and Urban Affairs, Wayne State University)

and her superb effort in helping make ...programs successful." G. B. Merrick, Vice President and Program Manager for Rockwell Space division, shared Johnson's recent contributions:

> ...in the area of propellant utilization during the proposed Apollo-Soyuz joint mission. She established that it was practical and safe to use the command module reaction control system to complete a backup deorbit and thus permit the use of more service module reaction control propellant for normal mission operations...[and] significantly enlarged the mission performance potential of the USA- USSR joint mission (Merrick 1974).

Also, G. W. Jeffs, Executive Vice President and Space Shuttle Program Manager for Rockwell Space shared that Johnson:

> ...was the project leader for a study of recovery characteristics of hypersonic gliders, and she was on the proposal teams that captured Paraglider recovery system contracts for NAA, Inc. She was then assigned to the Apollo program where she was responsible for the conceptual design and operational support of the entry monitor system used successfully on the Apollo flights (Jeffs 1974).

29.4 Honors and Professional Organizations

Barbara Johnson was recognized throughout her career with awards and medals. While she was apparently assertive in her work, she expressed surprise when she received recognition:

> I think people...are highly motivated by and grateful for the recognition of the work we perform, the responsibilities we undertake, and the products we produce, but when an unexpected recognition occurs [like this award], it is overwhelming (Johnson 1974).

Significant honors she received included:

- 1971: *Medallion of Honor Award* in recognition of activities in her professional career as an engineer, her participation in the space program and the safe return of the astronauts, and her ability to combine the dual role of professional woman and homemaker. Presented by the Mothers Association of the University of Illinois (Society of Women Engineers (SWE) Records n.d.)
- 1973: *NASA Medallion* commemorating first Lunar Landing mission, Apollo 11 (Society of Women Engineers (SWE) Records n.d.)
- 1974: Gamma Epsilon Distinguished Alumni Award, University of Illinois (Society of Women Engineers (SWE) Records n.d.)
- 1974: SWE Achievement Award *in recognition of her significant engineering contributions in support of manned spaceflight programs* (Society of Women Engineers (SWE) Records n.d.)
- 1975: MEDAL from the *University of Illinois College of Engineering*, Distinguished Alumni Award for Distinguished Service in Engineering (Drucker 1974).
- 1976: *Institute for the Advancement of Engineering*—IAE Outstanding Engineer Merit Award for Contributions to Aeronautical Engineering (Society of Women Engineers (SWE) Records n.d.).

- 1978: *The American Astronautical Society* granted her the Dick Brouwer Award in 1978 (1977?) (SWE Newsletter December 1977). to honor her *significant technical contributions to space flight mechanics and astrodynamics* (Wolters 2018).
- 1989: Fellow, Society of Women Engineers.

Firsts

- First woman graduate from the University of Illinois General Engineering program.
- First woman awarded University of Illinois Distinguished Alumni.

Professional Organizations

- *Institute for the Advancement of Engineering* (IAE) Fellow in 1978 (Society of Women Engineers (SWE) Records n.d.)
- *American Institute of Aeronautics and Astronautics*. Associate Fellow.

29.5 SWE Involvement

After receiving the SWE Achievement Award in 1974, Johnson became active in SWE at the national level in her later career years. She served as Achievement Awards Chair, the first Chair of the new SWE Fellow Selection Committee (the first SWE Fellows were awarded in 1980), and a member of the SWE Nominating Committee. Although elected to the SWE Executive Committee, she resigned early in her term due to health issues. Later, she continued to be active in her local community, speaking at various functions and sharing her experiences.

29.6 Legacy

In her 36 years at Rockwell Space Division, Bobbie Johnson was the Project Leader for several special projects, including the Lunar Reentry Vehicle Research Study (guidance and corridor needs for vehicles hitting the atmosphere at escape speed) and the Orbital Rendezvous Study (rendezvous between a non-maneuverable orbital vehicle and earth launch vehicle) (Society of Women Engineers (SWE) Records n.d.). She was known for her "leadership and guidance in operational philosophies and equipment designs for US-manned space programs (Apollo, Skylab, Apollo-Soyuz, Shuttle)" (Williams and Tarplee 2000). In her work on the Apollo Entry Monitoring System and as the Mission and Integration Manager for Space Shuttle, she brought strong technical acumen and leadership to all the teams she led.

In a eulogy for Johnson by Mary McCarty, Mary shared that Johnson had been featured in *Time Magazine* in 1968 as a *woman on the move* (McCarty 2005). In Johnson's own words, her final thoughts:

> The only reason you'd take up engineering is because you think you'll like it. ...Sometimes it takes a little longer for a woman to be recognized...It's all a question of your interest and ability, not your sex (Sammis 1963).
>
> I think of space as a broadening and a growth for man (Richardson 1974).
>
> I think women have a distinct advantage over men in engineering. They are observed more closely, their work is appraised more carefully. Most people know who you are…Engineering is a highly challenging field…not to be overlooked (Johnson 1977).

References

Air Force Historical Foundation. 2017. On this Day in History: October 16, 1958. October 16. Accessed on March 19, 2025. https://afhistory.org/october-16-1958/

Craig, Cecilia D. 2024. Supportive Relationships: A Spectrum of Heroes for Women in STEM. In Cecilia D. Craig, ed., *Mentoring in STEM Through a Female Identity Lens: Heroes Make a Difference for Women,* Vernon Press.

Drucker, D. C. 1974. Letter to Barbara Johnson. May 15. Society of Women Engineers Records (LR001539). *Barbara Crawford Johnson.* Box 189, Folder 49, Walter P. Reuther Library, Archives of Labor and Urban Affairs. Wayne State University.

Harness, Arminta J. 1988. Letter to SWE in support of Barbara C. Johnson Fellow Nomination. November 25. Society of Women Engineers Records (LR001539). *Barbara Crawford Johnson.* Box 189, Folder 49, Walter P. Reuther Library, Archives of Labor and Urban Affairs. Wayne State University.

Jeffs, G. W. 1974. Letter to H.L. Wakeland. January 24. Society of Women Engineers Records (LR001539). *Barbara Crawford Johnson.* Box 189, Folder 49, Walter P. Reuther Library, Archives of Labor and Urban Affairs. Wayne State University.

Johnson, Barbara C. 1974. Remarks in Dallas TX. Society of Women Engineers Records (LR001539). *Barbara Crawford Johnson.* Box 189, Folder 49, Walter P. Reuther Library, Archives of Labor and Urban Affairs. Wayne State University.

Johnson, Barbara C. 1977. The Human Side of Engineering. Talk at Long Beach State University to students. March 3. Society of Women Engineers Records (LR001539). *Barbara Crawford Johnson.* Box 189, Folder 49, Walter P. Reuther Library, Archives of Labor and Urban Affairs. Wayne State University.

Kraft, Chris. 1973. Letter to Joseph McNamara, President of Rockwell Space Division. August 3. Society of Women Engineers Records (LR001539). *Barbara Crawford Johnson.* Box 189, Folder 49, Walter P. Reuther Library, Archives of Labor and Urban Affairs. Wayne State University.

McCarty, Shirley. 2005. In Memory: Barbara Crawford Johnson. *SWE Summer 2005.* From Barbara Crawford Johnson, Society of Women Engineers Records (LR001539). Box 189, Folder 49, Walter P. Reuther Library, Archives of Labor and Urban Affairs. Wayne State University.

Merrick, G. B. 1974. Letter to H. L. Wakeland. January 25. Society of Women Engineers Records (LR001539). *Barbara Crawford Johnson.* Box 189, Folder 49, Walter P. Reuther Library, Archives of Labor and Urban Affairs. Wayne State University.

Mueller, George E. 1973. Letter to Barbara Johnson). October 17. Society of Women Engineers Records (LR001539). *Barbara Crawford Johnson.* Box 189, Folder 49, Walter P. Reuther Library, Archives of Labor and Urban Affairs. Wayne State University.

National Air and Space Administration. 2020. Apollo 13: The Successful Failure. April 6. https://www.nasa.gov/missions/apollo/apollo-13-the-successful-failure/

Richardson, Barbara. 1974. Engineer touts exploration. Dallas Times Herald. June 28. Society of Women Engineers Records (LR001539). *Barbara Crawford Johnson.* Box 189, Folder 49, Walter P. Reuther Library, Archives of Labor and Urban Affairs. Wayne State University.
Rub, Sharon J. and Laurel Romm. 1988. Nomination for Fellow Grade. November 29. Society of Women Engineers Records (LR001539). *Barbara Crawford Johnson.* Box 189, Folder 49, Walter P. Reuther Library, Archives of Labor and Urban Affairs. Wayne State University.
Sammis, Constance Sharp. 1963. Apollo is her project, but Barbara Johnson Loves Home. *Christian Science Monitor*. April 30. Society of Women Engineers Records (LR001539). *Barbara Crawford Johnson.* Box 189, Folder 49, Walter P. Reuther Library, Archives of Labor and Urban Affairs. Wayne State University.
Society of Women Engineers (SWE) Records (LR001539). n.d. *Barbara Crawford Johnson.* Box 189, Folder 49, Walter P. Reuther Library, Archives of Labor and Urban Affairs. Wayne State University.
Society of Women Engineers. 2003. Barbara 'Bobbie' Johnson, interviewed by Lauren Kata. *Profiles of SWE Pioneers Oral History Project,* LOH001952.20. May 3. Society of Women Engineers Records (LR001539), Walter P. Reuther Library and Archives of Labor and Urban Affairs, Wayne State University.
SavvyExacta. 2012. Barbara Crawford Johnson: Women Pioneer in the U.S. Space Program. *Women of the Week Blog*. November 15. Accessed March 19, 2025. https://cr4.globalspec.com/blogentry/21434/Barbara-Crawford-Johnson-Woman-Pioneer-in-the-U-S-Space-Program
Wolters, Angela S. 2018. Touching the Sky. In *Women and Ideas in Engineering: Twelve Stories from Illinois*, edited by Laura D. Hahn and Angela W. Wolters. University of Illinois Press.
Williams, Jan and Jennifer Tarplee. 2000. Extraordinary Women. June. Society of Women Engineers Records (LR001539). *Barbara Crawford Johnson.* Box 189, Folder 49, Walter P. Reuther Library, Archives of Labor and Urban Affairs. Wayne State University.

Chapter 30
Sheila E. Widnall, PhD (1938–) Achievement Award Recipient 1975

In recognition of her significant contributions to the fluid mechanics of low-speed aircraft and hydrofoils

Vicki S. Johnson

30.1 Introduction

Dr. Sheila E. Widnall blazed a trail for women from the time she became a freshman at the Massachusetts Institute of Technology (MIT) in the fall of 1956 until she retired as Institute Professor Emerita in September 2020. Her most visible accomplishment was being appointed as Secretary of the Air Force by President Bill Clinton in 1993 (the first woman to lead a branch of the service). At MIT, Widnall had a lasting impact on the percentage of women students at MIT when she used a colleague's data to show that math Scholastic Aptitude Test (SAT) scores underpredicted the academic performance of women students at MIT. MIT was able to increase the number of women applicants they admitted with no impact on academic success of those women. She had a similar effect on the percentage of women in the graduate school program when she pointed out that there were no women in the admitted group; they were only in the waitlist group or the declined group. Pointing that out to her colleagues caused them to be embarrassed and rethink their choices. Widnall and Dr. Mildred Dresselhaus (1977 SWE Achievement Award recipient) co-taught a "What is Engineering" class which targeted women and minority students understanding career paths in engineering. Hoping for 15 students, they had 100 students each semester. Widnall had a successful research program that initially focused on the life of wingtip vortices. As a result of the understanding her research provided, there was improved safety and increased

V. S. Johnson (✉)
Fountain Hills, AZ, USA

C. D. Craig et al. (eds.), *Women Engineering Legends 1952-1976: Society of Women Engineers Achievement Award Recipients*,
https://doi.org/10.1007/978-3-032-00224-2_30

airport capacity for aircraft operating near each other while landing and taking off. She discovered the Widnall instability, one of the factors that helped determine how the wingtip vortices broke down. Later research included helicopter noise, unsteady flow, and turbulent flow.

30.2 Early Life and Education

Sheila Marie Evans Widnall (born in 1938) grew up under the final approach path into McCord Air Force Base in Tacoma, Washington. Her father Rolland Evans started as a rancher in Colorado before being sent away with a horse and a saddle after the ranch failed. He wound up in Tacoma and met Widnall's mother, Genevieve. Rolland earned a bachelor's degree during World War II by attending night school while working full-time at the shipyards. While Widnall was in grade school, her father sold insurance and taught mathematics in a vocational school in the evenings. While Widnall was in high school, Rolland was a planner at Boeing. Later, he earned a master's degree in business and taught at a junior college (Negroni 2020; Widnall 1977).

Widnall's mother Genevieve was a juvenile probation officer her entire career. She had a full-time job when only one-third of US women were working. She demonstrated to Widnall that women could be successful as wives and mothers while holding full-time employment. When she was in first grade, Widnall had to send her mother back to work because "she was driving me crazy!" Widnall said, "I was never convinced that raising children had anything to do with staying home. In fact, I was convinced that raising children was better done if you didn't stay at home." The family had live-in help for Widnall's early life. Starting at age 12, Widnall cooked dinner for the family while her mother worked. Because of her mother, Widnall always knew that she would work (Negroni 2020; Widnall 1977).

Widnall remarked that, typical of families with only daughters, the oldest girl becomes the father's right-hand boy. Widnall did a lot of tinkering with her dad. They worked together building things and fixing things around the house. One day, they painted the front porch; it was the same day her mother was having a dinner party. The guests had to use the back door (Widnall 2010). Her dad taught her self-reliance. Widnall said, "I was 20-years old before I realized you could hire people to do work on your house" (Negroni 2020).

Widnall's neighborhood in Tacoma had boys only, so she played with the boys and was a tomboy until the eighth grade. She was a Brownie and Girl Scout and a voracious reader. Her childhood hobbies included horseback riding, piano, harpsichord, and special painting classes. She still plays the harpsichord (Widnall 1977).

Very early education was pushed hard by Widnall's parents. By the time she was in high school, Widnall was entirely self-motivated. She attended a mostly girls parochial grade school and an all-girls parochial high school (Aquinas Academy) (Widnall 1977).

By high school, Widnall thought she would teach mathematics and science. When she was a junior in high school, an uncle gave Widnall a small chunk of uranium. She used the uranium along with models of atoms to explain radioactive decay for a local science fair project which won first prize. She then participated in the national science fair in Cleveland, Ohio where Arthur Anderson, a Tacoma businessman with civil engineering degrees from MIT, was a judge. He encouraged Widnall to go to MIT (Negroni 2020; Widnall 1977).

Having decided that she would go to MIT, Widnall discovered that a class in physics was an entrance requirement. Her high school did not offer physics. As a senior in high school, she rode the bus to the public high school in the afternoon and took physics. From there, she went to her after-school job at a drugstore. Her high school also did not offer calculus or trigonometry. Those were not entrance requirements, though their absence in her academic background put her at a disadvantage compared to other students. She taught herself trigonometry and caught up with the other students in calculus (Widnall 1977).

The Tacoma MIT alumni chapter gave Widnall a scholarship so that she could afford to attend MIT. She arrived at MIT as an incoming freshman in the fall of 1956 along with 22 other first-year women students. Of the 6000 total MIT students, 120 were women (Widnall 1977).

Although Widnall arrived at MIT thinking she would major in physics, she realized that she did not have the academic preparation to do that (Widnall 2010). Knowing that she loved airplanes, Widnall decided to major in Aeronautics and Astronautics. Widnall said, "When you come from Tacoma, it seems that aeronautical engineering is an obvious thing to do" (Widnall 1977).

Widnall worked at Boeing in Washington the summer after her freshman year in structural test—reducing flight-test data on vibration of wings during flight and designing a test to measure loads on clamps that held fuel lines together. She continued getting summer jobs throughout her undergraduate career. Her plan was to get her bachelor's degree in aeronautical engineering and then go to work for Boeing. It didn't quite work out that way (Widnall 1977).

The difference between being in the aeronautical engineering department and the general student population was huge. As a freshman female coed, Widnall experienced harassment by upperclassmen. Her freshman advisor asked her, "Why are you here?" An economics professor told dirty jokes to see if she would blush (Widnall 2010). Arriving as a sophomore in the aeronautical engineering department, life changed for Widnall. She rated the treatment she received from faculty and students inside the department as excellent (Negroni 2020; Widnall 1977).

Widnall had a very practical philosophy when deciding on the extracurricular activities in which she would participate. She chose activities very carefully—those that didn't require a lot of work but gave a big return for the effort (Widnall 1977). One activity Widnall was devoted to during her entire academic and working career was mentoring and advising new freshmen women. She wanted to do everything she could to help them be successful (Widnall 2010).

Due to her outstanding academic performance, Professor Holt Ashley planted a seed that she should think about getting a PhD (Widnall 2010). The decision to pursue a PhD resulted in her admittance to a special 5-year program leading to a master's degree. Also in her junior year, Professor Martin Landahl, a fluid dynamics expert, joined the faculty. Widnall knew that she wanted to work with him (Widnall 1977).

Widnall received her bachelor's degree at the end of the spring semester in 1960 and married William Widnall, an electrical engineering student who had finished his first year of graduate school (Widnall 2010). Widnall spent the summer after her senior year working at the Aeronautical Research Institute in Sweden while her husband sailed in the Transatlantic Race which finished in Sweden (Widnall 1977).

Widnall earned her master's degree in 1961, and her doctorate in 1964. Widnall's thesis was on unsteady aerodynamics (Widnall 1977). She was the only woman working on a doctorate in Aeronautical Engineering at MIT. As planned, their first child Billy was born in January 1964 while she was in the final stages of writing her thesis. Widnall said it was easy to combine having a child and graduate school in her younger life (Widnall 1977). Later she reflected that it was the opposite—very challenging to have a child while finishing her doctorate, though her husband was very supportive (Widnall 2010). Four years later, their family was complete with the birth of Ann Marie (Widnall 1977).

30.3 Career

Sheila Widnall joined the faculty of the Aeronautics and Astronautics Department at MIT after she received her PhD (Widnall 2010). Because her husband was still working on his doctorate at MIT and working at Draper Labs, the easiest path was to stay at MIT. Sheila Widnall was the first woman faculty member hired in the School of Engineering at MIT (Negroni 2020; Widnall 1977).

Widnall was the first (and only) woman to be awarded a Ford postdoctoral fellowship at the MIT School of Engineering. That gave her a 2-year position as an assistant professor. She was promoted to Associate Professor in 1970 and to Professor in 1974 (Widnall 1993).

Widnall's initial research was on fluid mechanics, especially the analysis of vortices trailing from the wing tips of aircraft, which helped determine the hazards of wake turbulence. Smaller aircraft could fly into the invisible wake turbulence of a larger aircraft and experience an upset and a crash. Widnall's work helped determine when the wake turbulence would be diminished enough to make it safe for another aircraft to fly through that identical space, increasing safety and improving airport capacity. Two things help diminish the wake turbulence: Crow and Widnall instabilities. Widnall (Fig. 30.1) was honored to have an instability named after her. Widnall's other research included helicopter and aircraft noise, unsteady aerodynamics, wing theory, and turbulent flow. She had four patents over the course of her career (AeroAstroMIT 2020; Widnall 1977).

Widnall studied aerodynamic noise for the Army Research Office at a time when the usefulness of helicopters was limited because of their distinctive noise. Widnall

Fig. 30.1 Young professor Dr. Sheila Widnall (Courtesy of MIT Museum)

designed and managed the development of a new MIT wind tunnel facility so that she could use it to study the noise of vertical and short takeoff and landing (V/STOL) aircraft in forward flight (Hoffman 1975).

Widnall maintained an interest in trying to help women. She said:

> I think there are two roles that I can play in the women's movement. One of them is by being very successful at what I'm doing, which means I have to be competent, and I have to go out of my way to make sure that I stay visible so that other women will be encouraged and that men will be used to working with women. The other thing I can do is a more local thing. I can do things at MIT which make life at MIT better for women faculty and students. But my real clout and power is likely to be at MIT, in terms of really making change. So that is where I tend to put my active energy (Widnall 1977).

In 1974, MIT physicist and electrical engineer Mildred "Millie" Dresselhaus (SWE Achievement Award recipient 1977) engaged Widnall to start the course "What is Engineering" for freshmen engineering students, aimed at particularly women and minorities. The goal was to help students with no background in engineering understand what a career path as an engineer looked like. Their goal was to attract 15 students in the class every semester. They had more than 100 students every semester. Their impact was that several women and minorities taking the course decided to major in engineering. (Negroni 2020).

As part of her service to the department and the university, Widnall served as chair of the MIT Admissions Committee. The group was interested in recruiting more women students and examined data showing that the Scholastic Aptitude Test (SAT) score in math underpredicted the actual academic performance of women undergraduate students. Widnall recommended a change and a year later, the percentage of women admitted had increased from 26% to 38% (Negroni 2020).

Widnall had an opportunity to impact the number of women admitted to MIT as graduate students. In the early 1990s, when faced with applications sorted as either yes, no, or waitlist, Widnall pointed out that all the women had been waitlisted while they accepted many of the men:

> Every time there was a question, 'Is this candidate capable?' the men were given the benefit of the doubt and the women were not. The women went to the waitlist pile, says Hastings, head of the Aeronautics and Astronautics department. We felt collectively ashamed and we went back to correct that (Negroni 2020).

Widnall had this same influence on pay discrepancies between male and female faculty. She was involved with the early years of the Women's Forum at MIT, focused on providing female faculty with the information necessary to get tenure. Widnall was elected Chair of the Faculty at MIT and served from 1979 to 1981. She was the first woman elected to that position. Under her guidance, the faculty undertook a major reform of the MIT undergraduate curriculum (Widnall 2010). She supported starting a student section of SWE at MIT; by 1980 that section had been organized (Society of Women Engineers (SWE) Massachusetts Institute of Technology (MIT) n.d.).

Widnall became a full professor in 1974 and became the Abby Rockefeller Mauzé Professor of Aeronautics and Astronautics in 1986. She served as the Associate Provost of MIT from 1992 to 1993. That service was interrupted by her appointment to be Secretary of the Air Force in 1993 (Tietjen 1999). Upon her return to MIT in 1997, she immediately became an Institute Professor (the highest honor given to a faculty member by the administration and other faculty; there are only about a dozen Institute Professors at any given time at MIT) (AeroAstroMIT 2020).

By 2010, Widnall had transitioned her work to focus on teaching and service, not research (Widnall 2010). In 2018, Widnall served as co-chair of the National Academy of Science/National Academy of Engineering/National Academy of Medicine (NAS/NAE/NAM) report on the Sexual Harassment of Women in Science and Technology. The report examined the costs and consequences of sexual harassment in these fields (Widnall n.d.; Negroni 2020).

Over a career as a faculty member of 56 years, Widnall served on a plethora of university committees. Widnall saw MIT as a base with her professional colleagues around the world. She was happy to spend her entire career at MIT (Widnall 1977). Widnall (Fig. 30.2) officially became professor emerita in 2020 after 64 years at MIT (AeroAstroMIT 2020).

Fig. 30.2 Dr. Sheila Widnall in 2018 (Courtesy of Nina Gallant Photography)

Fig. 30.3 Secretary of the Air Force Sheila Widnall (Courtesy US Air Force)

30.4 Secretary of the Air Force

In 1993, President Bill Clinton nominated Widnall to be Secretary of the Air Force, the first woman to lead a military branch (Fig. 30.3). Widnall served two terms (appointed by President Carter) on the Board of Visitors of the US Air Force

Academy, including 2 years as the Chair. Widnall served on Air Force advisory boards. Additionally, Widnall served as a trustee of the Carnegie Corporation, including Vice Chair of the Board and Chair of the Board Nominating Committee. She was on the Carnegie Commission on Science, Technology, and Government (Tietjen 1999). The Chair of the Board called Widnall one day to tell her about an idea that was being kicked around—that she should become the Secretary of the Air Force. Widnall's response was a quick "I think that is a great idea." Widnall recounted that she had many flight experiences while the Secretary of the Air Force, including flying supersonic and pulling nine g's (Negroni 2020; Widnall 2010).

Widnall made two major contributions to the Air Force during her 4-year term. She started the program to develop the expendable launch vehicle for the Atlas 5 and Delta 4 rockets. In 2020 Widnall noted that "these vehicles still provide the majority of the launch capability for national security launches, and there has never been a launch failure." She also defined the character of the Air Force. When she arrived, the Air Force did not have core values. She promoted the core values of the Air Force Academy: "Integrity first. Service before self. Excellence in all we do." Once she saw those core values engraved on a granite wall, she knew they would stick (Negroni 2020).

30.5 National and Professional Organizations

Widnell was requested by MIT engineering professor Robert Canon to be the first director of the Office of University Research for the US Department of Transportation in 1974–1975. Widnall oversaw the distribution of $6.5 million for nationwide academic research projects. In arranging to go to Washington to do this job, the people in Washington told then Associate Professor Widnall:

> "It sure would be a lot easier to qualify you for this particular GS level if you were a full professor. Do you think you could see about doing that?" [Widnall] said, "Yes. I think I could probably do that." So she called up the Dean and said, "It would be a lot easier if you would just make me a full professor." He effectively said, "Yes. I think we can do that." So they just put through something special, and it sailed through" (Widnall 1977).

Widnall was very involved in service to the National Research Council (NRC) and its discipline-specific academies: National Academy of Engineering (NAE), National Academy of Science (NAS), and National Academy of Medicine (NAM). Widnall was a member of the Space and Aeronautics Board at the NRC. She served on the NRC/NAS Committee on NASA Aeronautics Program Reductions, and on the NRC/NAE Panel on the Education and Utilization of Engineers. Widnall served on the NAS Committee on Scientific Responsibility and on the NAE Board on Engineering Education. Widnall served on the NRC Governing Board. Widnall served on the NAS Committee on Law and Science and on the NAE Peer Committee Aerospace Section (Widnall n.d.).

Widnall was also very involved in service to professional societies. She served as associate editor of the American Institute of Aeronautics and Astronautics (AIAA) *Journal of Aircraft*, the *Physics of Fluids*, and the American Society of Mechanical Engineers (ASME) *Journal of Applied Mechanics*. She served on the Editorial Board of *Science*. She was also a member of the ASME Committee on Fluid Dynamics (Widnall n.d.).

Widnall is a member of the AIAA; she served on the AIAA Board of Directors and as the president from 1999 to 2000 (she was the first female president of AIAA). Widnall was a member of the Executive Committee of the American Physical Society (APS) Division of Fluid Dynamics. Widnall served on the Board of Directors of the American Association for the Advancement of Science (AAAS) she was President from 1988 to 1989 (Widnall n.d.).

30.6 Honors and Awards

The National Engineering Honor Society did not accept women as members when Widnall was an undergraduate. She was nominated for a "Woman's Badge." The same with Tau Beta Pi which was for all engineering majors, where she received a woman's award. Another honor society that excluded Widnall was Beaver Key. Again, they figured out how to give her a woman's award. Widnall hypothesized that if she had made straight A's and been off in the corner being quiet, she would have been ignored. But she was extremely active and a class officer. She could not be ignored (Widnall 1977).

Widnell received the Lawrence Sperry Achievement Award from AIAA in 1972 for "contributions to the understanding of vortex flows." In 1977, Widnall was chair of the Lawrence Sperry Award Committee. The Society of Women Engineers honored Widnall with the Achievement Award in 1975 for "significant contributions to the fluid mechanics of low-speed aircraft and hydrofoils." In 1977, Widnall was the chair of the SWE Awards Committee (Tietjen 1999).

Widnall was elected to the NAE in 1985. She served as NAE vice president, as a member of the NAE Council and COSEPUP (Committee on Science, Engineering, and Public Policy), as a member and later chair of the Draper Prize Committee, as a member of the Membership Committee, and as a member of the Nominating Committee (National Academy of Engineering n.d.).

In 1987, Widnall received the Washburn Award from the Boston Museum of Science. In 1993 she received the National Academy of Engineering (NAE) Distinguished Service Award. Widnall won the ASME Applied Mechanics Award in 1996. Also, in 1996, Widnall was inducted into the Women in Aviation Pioneer Hall of Fame (Tietjen 1999). In 1997, Widnall received the Defense Distinguished Service Medal. The Goddard Award from the National Space Club was awarded to Widnall in 1998, and the Reed Aeronautics Award was presented to Widnall in 2000. She was inducted into the National Women's Hall of Fame in 2003. Widnall received the Spirit of St. Louis Medal from ASME in 2001. In 2005, Widnall received the NASA Public Service Medal (Widnall n.d.).

In 2009, Widnall received the Arthur M. Bueche Award from NAE. The award citation read "for a remarkable academic career in fluid dynamics combined with the highest level of public service, and for championing the role of women in engineering" (National Academy of Engineering 2009).

Widnall was the first woman to receive the Daniel Guggenheim Medal in 2019 (90 years after the medal was established). The medal was established in 1929 by four professional societies "as an international award for the purpose of honoring an individual who makes notable contributions to aeronautical research and education, the development of commercial aircraft and equipment, and the application of aircraft to the economic and social activities of the nation" (Guggenheim Medal Recipients n.d.).

Widnall was elected Fellow in APS, American Academy of Arts and Sciences, and AIAA. She was a corresponding member of the Puerto Rican Academy of Sciences, and a member of the International Academy of Astronautics and the NAE (Tietjen 1999). Widnall has 18 honorary degrees (Widnall n.d.).

30.7 Legacy

Widnall was the first undergraduate alumna appointed to the MIT faculty (1964), the first woman appointed to the MIT School of Engineering faculty (1964), the first and only woman to receive the Ford postdoctoral fellowship when appointed to the MIT faculty (1964), and the first woman elected Chair of the Faculty at MIT (1979) (AeroAstroMIT 2020). Widnall's service at MIT resulted in significant and measurable increases in the number of women undergraduate and graduate students (Negroni 2020).

Widnall worked in many areas of fluid mechanics. Her work on vortex flow resulted in the Widnall instability (a low frequency instability) being named after her; she was honored by the naming of her discovery (Widnall 1977).

Widnall was the first woman to serve as the head of a US military branch when she was appointed Secretary of the Air Force (1993–1997) (AeroAstroMIT 2020). She was the first woman to serve as President of the AIAA (2000) (AIAA Presidents n.d.). She was the first woman to receive the Guggenheim Medal in its 90-year history (2019) (Guggenheim Medal Recipients n.d.). It is because of "these firsts and a lifetime of work that ensured she won't be the last" (AeroAstroMIT 2020).

Widnall saw herself as having two roles in the women's movement: (1) being very successful and visible at what she was doing so that women would see it is possible to be successful and men would get used to working with women; and (2) making life better for women at MIT (Widnall 1977).

References

AeroAstroMIT. 2020. Celebrating the Career of Sheila Widnall. 30 Sept. https://www.youtube.com/watch?v=1IFJAjrWvAE. Accessed 30 Mar 2025

AIAA Presidents. n.d. American Institute of Aeronautics and Astronautics (AIAA). https://aiaa.org/wp-content/uploads/2024/12/aiaa-presidents-history.pdf. Accessed 31 Mar 2025.

Guggenheim Medal Recipients. n.d. American Institute of Aeronautics and Astronautics (AIAA). https://aiaa.org/guggenheim/medal-recipients/. Accessed 30 Mar 2025.

Hoffman, Ellen. 1975. Letter dated 1 Apr to Dr. Irene Peden. From Society of Women Engineers Records (LR001539). *Sheila Widnall.* Box 193, Folder 38, Walter P. Reuther Library, Archives of Labor and Urban Affairs, Wayne State University.

National Academy of Engineering. 2009. Arthur M. Bueche Award. https://www.nae.edu/55353/Dr-Sheila-E-Widnall. Accessed on 4 Apr 2025.

National Academy of Engineering. n.d. Member Directory. https://www.nae.edu/20412/MemberDirectory?qln=Widnall. Accessed on 4 Apr 2025.

Negroni, Christine. 2020. Sheila Widnall: A lifetime exploring the unknown. MIT News 9 Oct. https://news.mit.edu/2020/sheila-widnall-lifetime-exploring-unknown-1009. Accessed 10 Feb 2025.

Society of Women Engineers (SWE) Massachusetts Institute of Technology (MIT). n.d. https://swe.mit.edu. Accessed on 4 Apr 2025.

Tietjen, Jill. 1999. Letter dated 15 Mar to National Women's Hall of Fame for Nomination for Dr. Sheila Widnall. From Society of Women Engineers Records (LR001539). *Sheila Widnall.* Box 193, Folder 38, Walter P. Reuther Library, Archives of Labor and Urban Affairs, Wayne State University.

Widnall, Sheila. n.d. Sheila Widnall Online biography on AeroAstroMIT. https://aeroastro.mit.edu/people/sheila-widnall/. Accessed on 4 Apr 2025.

Widnall, Sheila. 1977. Sheila Widnall interview transcript. March. https://hdl.handle.net/1721.3/190552. Massachusetts Institute of Technology. Libraries. Department of Distinctive Collections. Accessed 10 Feb 2025.

Widnall, Sheila. 1993. Resume received by SWE Headquarters on 12 Feb. From Society of Women Engineers Records (LR001539). n.d. *Sheila Widnall.* Box 193, Folder 38, Walter P. Reuther Library, Archives of Labor and Urban Affairs, Wayne State University.

Widnall, Sheila. 2010. Sheila Widnall interview. 17 Nov. MIT Infinite History Project https://infinite.mit.edu. https://youtu.be/9BDTuGNFCX8. Accessed on 30 Mar 2025.

Chapter 31
Ada Pressman (1927–2003) Achievement Award Recipient 1976

In recognition of her significant contributions in the field of power control systems engineering

Jill S. Tietjen

31.1 Introduction

Recognized as one of the nation's outstanding experts in power plant controls and process instrumentation throughout her career, Ada Pressman didn't originally intend to go to college, much less get a degree in mechanical engineering. But life holds surprises for all of us. Capitalizing on her bachelor's degree, Pressman began her career as a design and application engineer for a metering company where she had worked summers during college. Joining Bechtel after four years in that line of work, she solidified her expertise in control systems for every type of power plant, from coal to natural gas to nuclear, and served as a significant trailblazer. At Bechtel, she was involved in early design toward more automatic controls of equipment and systems, new packaging techniques, and breakthroughs in improved precision and reliability of sensors and controls. With these developments came the emergence of the systems engineering approach—which later became its own discipline. She designed controls to deal with unique environmental and efficiency issues for the fossil-fuel and nuclear plants with which she was involved. At Bechtel, she actively encouraged women to join the Society of Women Engineers (SWE) and ensured Bechtel corporate support for the organization. Active throughout her career (and after retirement) in SWE, Pressman became the first "three-pin" President—being selected as a Fellow, serving as Society President, and being honored as an Achievement Award recipient. Pressman, who served on the Society's Board of Trustees for many years, was also an avid golfer; her golf clubs were always close at hand.

J. S. Tietjen (✉)
Technically Speaking, Inc., Greenwood Village, CO, USA

C. D. Craig et al. (eds.), *Women Engineering Legends 1952-1976: Society of Women Engineers Achievement Award Recipients*,
https://doi.org/10.1007/978-3-032-00224-2_31

31.2 Early Life and Education

Ada Irene Richardson (later Pressman) was born in Sidney, Ohio. She grew up in that small town, attending primary and secondary schools. Pressman planned to become a secretary after she graduated from high school. Her father thought college was a better option, but there was only enough money for 1 year. Pressman selected an option that would be useful and interesting in a 1-year dose—engineering! (Seltz-Petrash 1980).

> Becoming an engineer never entered my mind until after I had graduated from high school. My high school curriculum included three years of typing, two years of shorthand, three years of bookkeeping all in preparation for a secretarial career. I did have four years of math —algebra and trigonometry —only because they were fun courses for me. In June when I graduated from high school I started working in the local bank. In September my parents suggested that I try college for a year and in looking through the catalog, engineering looked like the easiest courses since I only planned to attend one year. I should note that it was easier to get into the University at that time and within a month, I was a freshman in Engineering at The Ohio State University. Because I had not planned to attend college, I had not taken Chemistry or Physics in high school and did have to take one class without credit to make up the high school deficiency. I did continue in engineering and received a Bachelor of Mechanical Engineering in 1950 (Pressman n.d.-a).

She stayed more than 1 year, graduating with a bachelor's degree in mechanical engineering in 1950. However, her education was not over. Pressman earned certificates in accounting and business from the University of California, Los Angeles (1969 and 1972, respectively) before earning her Master's in Business Administration from Golden Gate University in 1974 (Pressman n.d.-b).

31.3 Career

Pressman graduated from Ohio State and went to work as a project engineer for Bailey Meter Company in Cleveland, Ohio.

> My first job after graduation came as a result of summer employment and my attending an American Society of Mechanical Engineers (ASME) conference. While attending a seminar I asked a question about opportunities for women in mechanical engineering. Mr. Bailey of Bailey Meter was the speaker and responded that there were plenty of opportunities at his company for women —so he practically had to hire me for the summers at Bailey. On graduation day, after working summers at Bailey, I was offered a full-time position — my only job offer (Seltz-Petrash 1980).

During her 4 years at Bailey as a design and application engineer, she performed hydraulic flow calculations for meters and flow elements and piping arrangements for control systems. She served as the liaison between the office and clients on field installations, analyzing the application of equipment to service functions and panel layout. The equipment was used for fossil-fired power plants and US Navy ships (Pressman n.d.-c).

Fig. 31.1 Ada Pressman (Photograph courtesy of Society of Women Engineers Records, Walter P. Reuther Library, Archives of Labor and Urban Affairs, Wayne State University)

Once she realized no promotional opportunities would be available to her for many years at Bailey, she began to look around for another job. One of her clients at Bailey was Bechtel, and they had an opening for a process instrumentation engineer. She applied, was selected for the job, and moved to California. Pressman (Fig. 31.1) stayed at Bechtel for the rest of her career during which she rose from the position of instrument engineer through a variety of steps to chief control systems engineer and, lastly, engineering manager. She was the first woman to be named chief engineer at Bechtel (1974) and the first to become an engineering manager (1979) (Pressman n.d.-a, n.d.-b; Seltz-Petrash 1980).

> The work at Bechtel has been bad and good. Bad only in the sense that all of us at certain times must do some work that is uninteresting and boring and, of course, I have had my share. It has been good in that I have had an opportunity to learn. Even though all my work has been on power plants, each one has been different —different sizes, different fuels including coal, gas, oil, and nuclear and best of all each project has some new, interesting and different people to work with (Pressman n.d.-a).

From 1955 to 1964, she performed instrument and control engineering on power plants, including those given in Table 31.1. Pressman was involved in the early design of more automatic controls for equipment and systems, new packaging techniques, and breakthroughs in improving the precision and reliability of sensors and controls. With these developments came the emergence of the systems engineering approach—which later became its own discipline. She participated in all this growth and its expansion in the power generation field (Pressman n.d.-c).

From 1964 to 1968, Pressman served first as Instrument Engineer and later as Instrument Group Leader for four Southern California Edison generating

Table 31.1 Power plants on which Ada Pressman worked 1955–1964

Unit	Client	Fuel	Capacity (Megawatts—MW)
Agua Fria 1, 2	Salt River Project	Gas	100
Combined Cycle	Dow Chemical	Gas	50
Mandalay 1, 2	Southern California Edison	Oil/Gas	200
Agua Fria 3	Salt River Project	Gas	165
Huntington Beach 3, 4	Southern California Edison	Oil/Gas	210
Etiwanda 3, 4	Southern California Edison	Oil/Gas	310
El Segundo 3, 4	Southern California Edison	Oil/Gas	325

units—Alamitos 5 and 6 and Redondo Beach 7 and 8. Each was rated at 450 MW and were supercritical (operating above the critical pressure and temperature for water such that boiling is not required) once-through drumless units that could burn oil or natural gas. Power plant efficiency and precision of instruments and controls were becoming critical elements of design and operation due to increased scrutiny and regulation of the utility industry in California and nationwide. Pressman performed many of the normal responsibilities associated with instrument and control design for power plants as well as combustion control and burner management. Each of the units in which she was involved was reported among the country's most efficient oil- and gas-fired power plants (Pressman n.d.-c).

From 1968 to 1971, Pressman served as the supervisor of the control systems engineering group for the Sacramento Municipal Utility District's Rancho Seco 1 nuclear unit. This 900 MW unit was the first nuclear power plant built in an area remote from a major body of water. Unique and creative design solutions were required to address the many unusual design and environmental problems and constraints mandated by the location. Elaborate control systems were required for each of the evaporative cooling system, the radiological waste removal and disposal system, and the water discharge system (Pressman n.d.-c).

All design facets posed major engineering challenges, not only for the usual power plant primary and secondary systems. This nuclear unit incorporated the first electro-hydraulic turbine control system, seven special closed-cycle cooling systems, a safety feature actuation system, special sampling systems, and cut-off systems for the wastewater retention basin (Pressman n.d.-c).

Pressman was also responsible for the review, verification, and control system integration for vital systems designed by others, including reactor protector systems, makeup and purification of reactor coolant, core flooding systems, high-pressure injection systems for emergency purposes, and gaseous and liquid effluents systems (Pressman n.d.-c).

All these designs and the construction itself were subject to intense regulatory oversight by the Atomic Energy Commission (since 1974, the US Nuclear Regulatory Commission). Additional regulatory oversight came from the Environmental Protection Agency, the Occupational Safety and Health Act, and California oversight for seismic design following the February 1971 San Fernando earthquake (Pressman n.d.-c).

When she was Chief, Control Systems Engineer, Pressman (Fig. 31.2) described her job responsibilities as follows:

> I have technical and administrative responsibility for more than 100 engineers assigned to ten nuclear and coal-fired projects. . . . control systems engineering is the official title within our company for people who are involved with instruments and control. As a simple example, within your home, you probably have at least one thermostat that starts and stops the furnace and maintains the temperature within the comfort range.
>
> Within a power plant, we are responsible for the simple systems such as a level controller on a tank that will open and close a valve from a tank to maintain level to complicated systems such as boiler control systems that will regulate the fuel, air, and water to a boiler to provide steam to a turbine while maintaining safety and environmental requirements such as minimum smoke from the stacks. We also are responsible for design of control panels, annunciators, and process computers.
>
> The technical responsibility involves keeping the engineers up-to-date on the various industry codes and Nuclear Regulatory Commission guides in addition to the Bechtel Corporate standards. All documents, specifications and drawings issued by the engineers on the project require the approval of the Chief Engineer. I do sign all the documents; however, most of the documents are reviewed by my staff of seven senior staff specialists. Each one of my staff has at least 25 years of specialized experience in the area of instruments and controls.
>
> Administratively, I am responsible for hiring and unfortunately layoffs, promotions, salary increases, assignment of the technical personnel to the various projects, performance evaluations and counseling. . . Each day is unplanned, not because I am not a planner but because there is always a new personnel problem, a new technical problem on a project or some other new problem that requires that I be responsive (Pressman n.d.-a).

Pressman (Fig. 31.3) said:

> My experience included the engineering management of millions of individual hours of power generation plants and various economic studies and proposals for potential projects. My responsibilities required continual monitoring of the costs associated with each project in addition to the technical engineering details developed as the design progressed (Pressman n.d.-d).

Fig. 31.2 Ada Pressman and field engineer Georgine Kabler outside at a Bechtel job site (Photograph courtesy of Society of Women Engineers Records, Walter P. Reuther Library, Archives of Labor and Urban Affairs, Wayne State University)

Fig. 31.3 Ada Pressman (Photograph courtesy of Society of Women Engineers Records, Walter P. Reuther Library, Archives of Labor and Urban Affairs, Wayne State University)

Fig. 31.4 Ada Pressman and field engineer Georgina Kabler in a power plant control room (Photograph courtesy of Society of Women Engineers Records, Walter P. Reuther Library, Archives of Labor and Urban Affairs, Wayne State University)

Additionally, she contributed to the *Instrument Engineers' Handbook* and the *Environmental Engineers' Handbook* and wrote or presented technical papers relating to instrumentation and controls (Pressman n.d.-c).

A pioneer in combustion control and burner management for supercritical power plants, Pressman (Fig. 31.4) managed 18 design teams for more than 20 generating plants worldwide. She served as the principal engineer for the San Onofre nuclear power plant, located south of Los Angeles (Society of Women Engineers 2024).

31.4 Society of Women Engineers' Involvement

Pressman joined SWE in 1954 as a charter member of the Cleveland Section, becoming a senior member in 1961. She served in many roles within the Los Angeles Section of SWE, including membership chairman, delegate to the Council of Section Representatives, secretary, and vice chairman. Nationally, she served on the Executive Committee and as Second Vice President and First Vice President before serving as Society President in 1979–1980 (Pressman n.d.-b, n.d.-d; Society of Women Engineers 2024, 2025).

During her term as Society President, the organization experienced some challenges and changes. One contentious issue was the Equal Rights Amendment (ERA) and whether the Society should hold its annual Convention in states that had not ratified the ERA. Changes included changing the name of the magazine, which was revised during her tenure, and selecting the first recipients of the Distinguished New Engineer Award, Pressman communicated directly with the Board of Trustees, opening up new avenues of communication and a new relationship (Pressman n.d.-d).

Pressman was SWE's representative to the American Association of Engineering Societies in 1980 and later served within SWE as Achievement Award Chair, as a member and Chair of the Nominating Committee, and as a member of the Advisory Committee and the Advisory Board. Pressman remained active in SWE as long as she was able to do so. She was awarded Fellow grade in 1983 (Pressman n.d.-b, n.d.-d; Society of Women Engineers 2024).

Pressman served on the SWE Board of Trustees from 1983 to 1998, including eight years as Treasurer, starting in 1991, during which she took the Trustees from handwritten ledgers into the computer age. On the Board of Trustees, Pressman brought not only her engineering expertise and problem-solving skills, she also brought knowledge of economic analysis and financial decision making. She kept the Trustees clear on their fiscal responsibilities, handled the interface with the brokerage firm, kept informed on tax items, and dealt with the SWE auditors. Her probing questions, including "Do we really want to do that?" and "Why did SWE do that?", made others think, change, and improve. As Treasurer, she did an immense amount of work for a volunteer. The amount of work was so detailed and time-consuming that someone had to be hired to replace her. Her dedication to SWE was legendary. She was forthright and fair, and her love for SWE was crystal clear. She was always a great source of reasonable advice as well as a dedicated worker. Pressman received SWE's Distinguished Service Award in 2000, the first year it was awarded (Pressman n.d.-b, n.d.-d; Society of Women Engineers 2024).

Pressman was the first "three-pin" SWE President—a SWE Fellow's pin, a SWE Past Presidents Pin, and a SWE Achievement Award recipient pin (Fig. 31.5).

Fig. 31.5 Ada Pressman wearing all three pins, ICWES VII, Washington, DC, 1984 (Photograph courtesy of Society of Women Engineers Records, Walter P. Reuther Library, Archives of Labor and Urban Affairs, Wayne State University)

She recognized the need for SWE as she progressed upward in her career and, throughout her career, encouraged women to become SWE members and get involved in the Society. In addition, Pressman was a champion of corporate support from Bechtel for the Los Angeles Section of SWE and nationally for SWE. SWE benefited from Bechtel's support during her years of employment—a legacy sustained today (Pressman n.d.-d.).

31.5 Associations, Honors, and Awards

Pressman was a member of the Instrument Society of America (ISA), served as a national Vice President, and was elected a Fellow (1979). She was the first woman to be elected President of any section of the ISA (1969). She received the Los Angeles Section of the ISA's W.A. Kates Award in 1971 and 1972 for outstanding participation. She was also a member of the American Nuclear Society and the Southern California Meter Association. As a student and early in her career, she was a member of the American Society of Mechanical Engineers (Pressman n.d.-b, n.d.-c; Society of Women Engineers 2024).

Pressman was elected a Fellow of the Los Angeles Institute for the Advancement of Engineering (1971) and also received that organization's Outstanding Engineer Merit Award (1975). The Engineer of Southern California selected her as Engineer of the Month in 1968. The Long Beach Engineering Council named her 1979 Engineer of the Year. Pressman was also recognized with the Outstanding

Engineering Merit Award from Bechtel Power Corporation. She was a registered professional engineer in Arizona and California. Pressman appeared before the State of California Board of Registration of Civil and Professional Engineers and presented a statement supporting control systems engineering as a separate discipline. It was established as such in 1975; she became the first person to be registered under it (Pressman n.d.-b, n.d.-c, n.d.-d; SWE Newsletter 1976).

In addition to SWE's 1976 Achievement Award (Figs. 31.6 and 31.7), Pressman received the 1974 Distinguished Alumni Award from The Ohio State University and the Twin Award from the YWCA in 1976 (Pressman n.d.-b; Society of Women Engineers 2024).

Pressman was an avid golfer—always ready to play. She was a member of the Rio Hondo Women's Golf Club and the Griffith Park Women's Golf Club. Additionally, she was a member of the Order of the Eastern Star and the Delta Zeta Sorority. She married Marvin Pressman on October 4, 1969 and was widowed in 1970. Her other outside interests included traveling, bowling, bicycling, and photography (Pressman n.d.-b, n.d.-c; Society of Women Engineers 2024).

Fig. 31.6 Ada Pressman (r) receiving her Achievement Award plaque from President Carolyn Phillips (Photograph courtesy of Society of Women Engineers Records, Walter P. Reuther Library, Archives of Labor and Urban Affairs, Wayne State University)

Fig. 31.7 Ada Pressman (l) receives a Steuben Achievement Award prism in 1991 in honor of receiving the 1976 Achievement Award (Photograph courtesy of Society of Women Engineers Records, Walter P. Reuther Library, Archives of Labor and Urban Affairs, Wayne State University)

31.6 Legacy

In 2014, The Ohio State University Department of Mechanical Engineering established the Ada Irene Pressman Award. It is "presented to alumni who have made noteworthy contributions to their chosen professions while overcoming significant obstacles or barriers to the completion of their education and/or obstacles in their careers" (The Ohio State University, 2025).

When Ada Pressman died, she left a substantial bequest for the SWE scholarship endowment. The SWE Board of Trustees established the Ada I. Pressman Memorial Scholarship utilizing the funds generated annually from the bequest endowment. Multiple renewable scholarships are available each year to engineering undergraduate and graduate students (Society of Women Engineers 2024).

Because of her significant service to SWE and the Board of Trustees, *The Ada Pressman Society* was established for those individuals who tell SWE that they have a planned gift to SWE after their death through a bequest (Society of Women Engineers 2025).

References

The Ohio State University, Department of Mechanical and Aerospace Engineering. n.d. Department Awards: Alumni. https://mae.osu.edu/department-awards-alumni. Accessed January 1, 2025.

Pressman, Ada. n.d.-a Untitled and Undated Speech of Ada Pressman, From Society of Women Engineers Records (LR001539). Box 191, Folder 54, Walter P. Reuther Library, Archives of Labor and Urban Affairs, Wayne State University.

Pressman, Ada. n.d.-b Ada Pressman Biographical Information, Undated. From Society of Women Engineers Records (LR001539), Box 191, Folder 54, Walter P. Reuther Library, Archives of Labor and Urban Affairs, Wayne State University.

Pressman, Ada. n.d.-c Society of Women Engineers, Achievement Award Nomination—Ada Pressman, from Society of Women Engineers Records (LR001539), Box 191, Folder 54, Walter P. Reuther Library, Archives of Labor and Urban Affairs, Wayne State University.

Pressman, Ada. n.d.-d Society of Women Engineers, Distinguished Service Award Nomination—Ada Pressman, from Society of Women Engineers Records (LR001539), Box 191, Folder 54, Walter P. Reuther Library, Archives of Labor and Urban Affairs, Wayne State University.

Seltz-Petrash, Ann. 1980. Straight Talk. Graduating Engineer. Spring 1980, pp. 23-26. From Society of Women Engineers Records (LR001539). Box 191, Folder 54, Walter P. Reuther Library, Archives of Labor and Urban Affairs, Wayne State University.

Society of Women Engineers. 2024. Ada I. Pressman Memorial Scholarship (Est. 2004). https://swe.org/scholarships/ada-i-pressman-memorial-scholarship-est-2004/. Accessed December 31, 2024.

Society of Women Engineers. 2025. "Planned Giving." https://swe.org/support-swe/planned-giving/. Accessed January 1, 2025.

SWE Newsletter. 1976. Ada Pressman Selected for SWE's Top Achievement Award. September, p. 3. From Society of Women Engineers Records (LR001539), Box 191, Folder 54, Walter P. Reuther Library, Archives of Labor and Urban Affairs, Wayne State University.

Afterwords from the Editors

The six of us dreamed up this adventure at WE24 in Chicago: after the awards banquet, in a Late Career and Retiree Affinity Group dinner, in emails and chats among roommates at the conference. We quickly formed our team of six and met for the first time on November 4, 2024. Meeting weekly thereafter, our project formed shape and substance. It has been a marvelous adventure.

We did not all know each other as well as we do now. We live in three time zones, and all are retired (sort of) and SWE Fellows. In one of the early virtual meeting calls, while discussing the *setting the stage* sections of the book, it was clear our own lived experiences varied across a range of happenings, but over a relatively short time. Ceal, the earliest graduate, 1974, used a slide rule in college that cost over $50 (and her apartment rent was $95). Jill used a slide rule at first in college, then moved to calculators (~$150) before she graduated. Vicki used a slide rule she bought for $1 in high school. Debra used a handheld TI-59 programmable calculator; Holly used a handheld HP 15C engineering calculator she still owns today. Ceal studied engineering from the beginning. Jill, by her sophomore year, moved from mathematics to engineering. Vicki, who developed a love of flying in high school, began studying aerospace engineering from the start. Jan studied engineering after graduating with a liberal arts degree. Debra enrolled at a community college to study accounting before switching to an Engineering Science degree. Holly studied mechanical engineering during the rise of the personal computer era. She bookended the team, graduating over 15 years after Ceal.

The six of us are a microcosm of women engineers growing up in the 1950s, 1960s, 1970s, and early 1980s. We came to engineering through different paths with unique experiences and perspectives. Researching and writing about these 25 legendary women recognized by SWE has been a learning experience. We hope that when reading about these incredible 25 women and our journeys shared in these afterwords, you will find your own story and path into the field we all love and have dedicated our lives to. Enjoy!

C. D. Craig et al. (eds.), *Women Engineering Legends 1952-1976: Society of Women Engineers Achievement Award Recipients*, https://doi.org/10.1007/978-3-032-00224-2

Afterword—Cecilia (Ceal) D. Craig

In 1969, the USA landed humans on the moon. As a 16-year-old female, I watched this on television in the basement dorm common room of Taylor Tower at The Ohio State University (OSU), the summer before I started my high school senior year. A young teen woman from Minnesota, I was at OSU for a National Science Foundation 8-week summer program exploring engineering, taking calculus, statics, circuits, and Fortran IV (as I remember!). That year, the program had about seven young women and 40-plus young men, mostly from Ohio, with three undergraduate engineering student counselors chaperoning us. Using punch cards for computer programs, taking those introductory classes, and staff inspiring us, that one event propelled me into studying engineering when I graduated from high school in 1970 at the age of 17.

In my high school senior year, I applied to the Massachusetts Institute of Technology (MIT), Carnegie-Mellon, and OSU, along with one or two others; I was accepted at all, but OSU offered me a small scholarship, tuition was cheaper, and I was the oldest child with three brothers. I became a Buckeye. OSU was formative for me in those undergraduate years. I was a counselor for that same NSF program the summer after my freshman year. I met my husband; he invited me out at the beginning of the summer, and we married in December of my sophomore year when I was 19.

Junior year: Wow! I spent 11 weeks in the University hospital in traction after a motorcycle accident. Professors from my classes and professors' wives visited me most days. I took three finals in traction and then three more after I left the hospital on crutches. The Society of Women Engineers student section had about 30 members, and I was President that eventful junior year. However, the Vice President (and my maid of honor, Susan) did all the work as my accident took me out for months. Navigating crutches in snow and ice is an experience I hope never to repeat.

Throughout college, I used punch cards and a K&E slide rule (a wedding present from my husband). My friend Susan taught me to do card weaving with half of a punch card; I still have some of those belts I made! We could not afford a calculator; the price was about three months' rent! I borrowed one for the Engineering-in-Training Professional Engineering exam. Each year in the mechanical engineering (ME) degree program sparked more interest in engineering. I thrived in the ME classes after a tough freshman year in calculus. The ME department had wonderful instructors and a department chair, who later became the OSU Dean. As a student, I met Admiral Grace Murray Hopper at a Boston SWE conference; I still remember her talking about the origin of the term *software bug* (and, decades later, seeing the journal with the moth affixed to it, now in the Smithsonian).

My husband finished his bachelor's degree in aeronautical engineering and then earned a master's in mechanical engineering while I finished my bachelor's (in less than 4 years). We took positions in Southern California, driving across the country to work for Rockwell International (where we bought our first calculator). I quickly started night school and earned a master's in engineering from California State University at Fullerton.

Over the next years, I experienced few issues with discrimination. Title IX had occurred, and companies were seeking female engineers and women in management. I moved up in management quickly. At night, we taught Fortran IV to high school students in an after-school Rockwell program. I became involved in the SWE Los Angeles/Orange County Section, sometimes Section Representative, and then as the first Scholarship Co-Chair, establishing its processes. My husband bought an Apple II, serial number 28 (I still have it), and I moved into the world of small computing.

As I remember, I met Ada Pressman on a panel speaking about engineering to college students. At OSU, they had told me I was the first female mechanical engineer to graduate there since Bertha Lamme (in the late 1800s), and I used that in my introduction on the panel. Ada said, "Ceal, I hate to burst your bubble, but I graduated from OSU in ME in 1953!" I quipped, "The year after I was born!" We became friends and worked with OSU to identify a few more between Lamme and Pressman, though still only a couple. SWE awarded its first Achievement Award to Mária Telkes in 1952 (my birth year) and Ada Pressman in 1976: appropriate bookends for this current book (Admiral Hopper received the Achievement Award in 1964)!

We left Southern California after 5 years, moving to Dallas for a new opportunity for me with Xerox. In Dallas, we observed racism and sexism that I had not seen much of until then. We decided not to have children; I worked long hours as a second-level manager, and my husband was a consultant. I was active in the Dallas SWE section and recruited for Xerox at SWE conferences. We enjoyed our time in Dallas, but the heat and culture were not our thing and when Xerox offered to move us to Northern California, we jumped.

Northern California was at the center of high tech, and we loved it. I began years of international travel, traveling over two million miles with American Airlines and visiting Europe, Japan, Hong Kong, and Singapore. Computers and software became my industries: fault-tolerant computing, network switches, disk drives, and circuit board assemblies. I built a career in manufacturing in larger companies like Tandem, Maxtor, and Siemens. I then shepherded new products into fruition, finally running a program office for a spin-off. Unfortunately, I was typically the most senior technical woman in the organizations where I worked.

At the SWE level, in the 1980s, I was Achievement Award Chair for 2 years. Back decades ago, I was a Region Deputy Director for SWE. Locally, I was active in SWE Santa Clara Valley section as an officer and committee chair until the mid-1990s (and again from 2010 to now). I began to think that women were being accepted, and the numbers were growing. I had moved up to the Director level at work, traveling and focused more on my career, with less time to volunteer. However, I wanted to try something different after being part of seven manufacturing layoffs as a manager or director.

Looking at women's graduation numbers in engineering and computer science was disheartening; they had barely broken 20% in 2000, and the needle had not moved since then. Why? How could young women be more directly inspired to pursue engineering careers? After retiring (the first time) from engineering in 2000, I became a high school math teacher for a year, then an adjunct online college math instructor, consultant, and not-for-profit board of directors (BOD) volunteer. Eventually, as a Senior Director, I worked in high tech for 7 years to earn money for doctoral program tuition. At age 55, I began earning a PhD in Education, graduating at age 61 (and fully retiring from paid work at 58). We are never too old to learn something new. Career theory and mentoring research are my specialties. Along the way, I became involved with For Inspiration and Recognition of Science and Technology (FIRST), a high school program FIRST Robotics Competition (FRC) mentor initially, then FIRST LEGO League (FLL), and another not-for-profit BOD. FRC was the program I explored in my dissertation: its influence on young women's career decisions.

I'm now a widow; my husband died in 2023 after our 51 years of marriage. I continue to work almost full-time as a volunteer: as a nonprofit Board Officer, Judge Advisor, Judge, and peer reviewer, and teaching bird watching, I manage to find time to photograph birds, garden, travel a bit, and take care of two cats. In 2024, a book I edited and wrote for was published: *Mentoring in STEM Through a Female Identity Lens: Heroes Make a Difference for Women.*

It was a journey during difficult times, my husband's final year battling cancer. SWE has become a bigger part of my life, where I've been helping to reshape its awards and recognition program for 3 years and helped the Late Career and Retiree Affinity Group get off the ground in its first years. I'm proud of being a female engineer, leading the Silicon Valley Engineering Council and the San Francisco Bay Wildlife Society, and nurturing the next generation to take over. Who knows what the next years will bring? Women need heroes in Science, Technology, Engineering, and Math (STEM). Hands-on experiences inspire girls and young women. Those ideas have and will form the foundation for the remainder of my life.

Afterword—Jill S. Tietjen

I am a product of Betty Friedan's book *The Feminine Mystique* and the resulting women's rights movement of the 1960s. I am the oldest of four children born to a PhD engineer father and an elementary school teacher mother. Mom was determined that her daughters would have professional choices she didn't have; she felt her choices were limited to secretary, nurse, or teacher. She had chosen to be a teacher. I became an engineer—my sister was a PhD food biotechnologist. Both of my brothers were trained as engineers.

I grew up in Hampton, Virginia, told by my parents that I would attend an in-state school as my parents intended to put four of us through college. I wanted to attend the University of Virginia. I am thankful that a young woman sued the University after they refused to admit her because she was female. That lawsuit led to women being admitted as undergraduates in the Fall of 1970. I was admitted as part of the third class of women in the Fall of 1972—the first class that didn't have a cap limiting the number of women who could attend.

In spite of my engineering aptitude, no one encouraged me to major in engineering—I started college as a mathematics major. I had started working jigsaw puzzles at age 2, played the violin starting at age 8, and began doing needlework at age 7. When I was 16, my father brought home from work a two-volume FORTRAN self-study guide. I taught myself FORTRAN while I was babysitting and working at the library. Fortunately, I found engineering, transferred, and graduated from the School of Engineering and Applied Science at the University of Virginia. I was one of the first ten women to graduate in engineering from UVA.

During my time at UVA, we used IBM punch cards—80-column paper cards—each of which had one line of computer code. The programming cards were fronted by JCL cards (Job Control Language)—also 80 columns and also with one line of computer code—to tell the mainframe computer how to start. The program was interspersed with comment cards that didn't have programming on them and were used to explain processes and keep me straight.

We submitted decks of computer cards (numbered by hand in case someone dropped the deck—you only had to learn that once!) to the "in" desk at the computer center. Hours later—or overnight—we got the results back. If the program hadn't worked properly, changes needed to be made and the "job" resubmitted. The time cycle until the next set of results were obtained could be 12–24 h.

When I started at UVA, we were using slide rules to do math calculations. The HP-35 (1972) and the HP-45 (1973), the earliest calculators that I was aware of, had just become available—but they cost a lot of money—$395 (almost $3000 in today's dollars). With the introduction of the TI SR-50 in 1974, calculators were deemed to be affordable (about $150) and we were allowed to purchase them and use them in our college classes.

When I started work in the electric utility industry, we were still using punch cards. Our job decks ran overnight—but payroll always had first priority! That process evolved to "dumb" terminals and eventually to what today operates at lightning speed compared to my college and early work experiences.

There were seven women in my engineering graduating class of 220 at UVA and no collegiate section of the Society of Women Engineers. I found SWE in 1979 in Raleigh, North Carolina at a card table in a gymnasium at a career fair at North Carolina State University where I was doing on-campus recruiting for my first employer, Duke Power Company. I joined immediately, participated as I could in activities based in the Research Triangle Park area of North Carolina and helped found the Charlotte-Metrolina Section. I became very active in the Denver Section (now Rocky Mountain Section) after I moved to Denver, Colorado in 1981.

The SWE Magazine at the time contained names and contact information for section presidents. For the Denver Section, it said Alexis Swoboda. I called Alexis about a week after I arrived in Denver and said something like "I'm here. Put me to work." She did. We became colleagues, friends, co-authors, and even served on the National Board of Directors together. In the 1980s, after SWE established the Fellow grade, I worked with a member of the Denver Section, Helen Huckenpahler, to nominate the early Denver Section members for SWE Fellow. Helen taught me a lot about the nomination process and how to write nomination packages.

In 1987, Alexis came back to Denver from the Kansas City SWE National Convention and said that she had found an outreach program at the Convention that the New York Section was doing and she thought we should implement it in the Denver Section. I didn't know that the program, an essay contest on Great Women in Engineering and Science, would change my life. There were so few books that had information on women in science and engineering but we did the needed research and discovered many more than the only woman that I knew in 1987—Marie Curie (by the way, she is still often the only woman that audiences that I speak to know from engineering and science). Our work on putting the essay contest together led to articles, conference presentations, nominations, and books.

In 1988, I was elected to the SWE National Board of Directors. At the first meeting, then-President Suzanne Jenniches held up the nomination forms for the National Medal of Science and the National Medal of Technology—the U.S. equivalent of the Nobel Prize, awarded by the President of the USA—and asked for a volunteer to submit nominations. Seventeen women in that room and no one raised her hand. So I did. The first nominations that I prepared were not for the right candidates. But Alexis and I were continuing our research for the essay contest and I realized that if I did a good enough job on her nomination that Admiral Grace Murray Hopper was a viable candidate and that she would receive the National Medal of Technology. I worked on that nomination for 2 years. I contacted the pioneers in the computer industry. I contacted senators and Secretaries of the Navy and anyone I could find who could help me get the information that I needed. And I was successful.

Admiral Hopper's selection for the National Medal of Technology bookended my year as President of SWE. I learned just before I was installed that she would receive the Medal—but I couldn't tell very many people because I couldn't upstage the President of the United States. Admiral Hopper asked me to receive the Medal in her stead—she was too frail to go to the ceremony. In September 1991, on the hottest day of my whole life, I was at the White House Rose Garden, where I received the National Medal of Technology for Admiral Hopper from President George H.W. Bush. She was the first individual woman to receive that Medal. As I looked around me at the recipients of the National Medal of Technology and National Medal of Science, I said to myself—where are the women? Why aren't they being recognized? Here was the genesis of what has become my life's work. Admiral Hopper died on January 1, 1992. The Board of Directors decided that we were going to do the fundraising effort necessary to endow a SWE scholarship in her name. By the end of my term, not only had she received the National Medal of Technology but in addition The Admiral Grace Murray Hopper Scholarship had been endowed. And I had a mission.

Then in 1992 or 1993, Alexis found the National Women's Hall of Fame—in Seneca Falls, New York—the birthplace of women's rights. We noticed that not very many science, technology, engineering and mathematics (STEM) women had been inducted. Our mission was clear—get the STEM women into the Hall to serve as role models. And we did. My first successful nomination was Admiral Hopper.

She was inducted posthumously in 1994, and the family asked me to accept the honor. That was my first trip to Seneca Falls and my first induction. I attended every induction from then through 2017, have nominated about 40 women successfully to the Hall, and served as CEO in 2015.

One thing led to another, led to another. In 1995, a woman named Betty Reynolds sent a letter to SWE headquarters telling them she was writing a book about women's entry into non-traditional professions and that she was looking for information about women in engineering. SWE headquarters sent me the letter and I started corresponding with Betty. In the end, she and I wrote three books, including one on the history of women in engineering, and I learned about women through history in all fields of endeavor.

In 2003, I was introduced to Charlotte Waisman. At lunch, she said that we were going to write a book together. In 2008, the hardback version of *Her Story: A Timeline of the Women Who Changed America* was published. The paperback version came out in 2013. I discovered that it wasn't only STEM women who weren't being acknowledged and written about—it was all women. Thus, I became an advocate for ALL women. And I continued to write technical articles, non-technical articles, make conference presentations, and write books.

In February 2014, I received an email from Tiffany, an editor at Springer. She had seen an encyclopedia article I had written on using energy storage technologies for integrating renewable energy technologies into the electric grid. She wanted to know if I would write an ebook on distributed energy for Springer to publish and if I was willing to have a conversation with her. I said yes I would have a conversation. In that conversation, I indicated that I wasn't interested in writing a book on distributed energy, but I had this book on the history of women in engineering (the one written with Betty Reynolds and published in 2001) that I would like to update and have Springer publish. Was that something of interest to her?

In October 2014 I heard back from Tiffany. She thought I wouldn't remember her. I absolutely remembered her. She said that Springer wanted me to update my book on the history of women in engineering and that they would publish it as the foundational volume in the Springer Women in Engineering and Science series. Springer wanted me to serve as series editor. That my job would be to recruit volume editors and I could write books in the series and serve as a volume editor, co-volume editor, or co-author. As I write this afterword, 27 volumes have been published in the series including that foundational volume on the history of women in engineering. I am delighted that my SWE colleagues and I have an outlet for our book on SWE Legends—to tell the stories of women who, as I have learned, are often invisible, marginalized, their ideas stolen, and ignored.

SWE is a family unit to me. A safe place. A place of significant skill development—where I started nominating women for awards. SWE provided me with an important outlet for honing my writing and speaking skills and served as the springboard for my participation in nonprofit, university, and corporate boards.

The changes in technology since I joined the professional workforce in 1976 are almost mind boggling. However, despite many years of work by many people—including me—the composition of the engineering workforce still does not reflect gender parity. Although this discourages me, it does not stop me. The outreach efforts are so important—we must continue to write women into history and ensure girls—and boys—and men and women—can see and understand the value that women bring. I believe that when women are valued—all of society will be better.

Our world faces so many challenges for which engineering minds are needed. When I think about this, I am always reminded of the IBM print advertisement from the 1980s that showed a pair of pink baby booties and a pair of blue baby booties. Its lament was if we aren't encouraging the occupants of the pink baby booties to pursue STEM careers, we might not discover the cure for cancer.

We must ensure that all available minds are applied to solving humanity's challenges. And, like the founders of SWE, I want to see a time SOON when a person says "engineer" the immediate reaction is not that the person referred to is male.

Afterword—Debra A. Kimberling

Women were just being admitted into military training academies, the Three Mile Island Nuclear Accident occurred, and *Bionic Woman* was on television when I graduated from high school in 1976. Although neither of my parents had the opportunity to attend college, they made sure my siblings and I did. My brother has a degree in business and my sister a degree in psychology.

I explored a potential mathematics-related field, accounting, at a local community college. A mathematics professor I was grading papers for saw my potential and suggested that I consider a career in engineering. He was certain that although the number of women in the field was small, the government incentives encouraging women to enter STEM fields would change that. For the first time, I felt like I had a career direction. As an engineer I could be creative, help solve some of the world's problems, and be independent. My parents were surprised, but they always encouraged me to follow my dreams.

I had no engineering role models while I was a student. I was either the only woman in my engineering classes, or there might be one other. I chose a general "Engineering Science" B.S. degree from Arizona State University. I attended my first Society of Women Engineers (SWE) event on campus during my senior year. Two people from the Naval Ocean Systems Center (now NIWC Pacific) delivered a career talk. I started working at the same lab four months later in submersible research and development.

I made a conscious decision to sign my reports using only initials for my first name. When a director called to compliment my report one day, he was surprised that a woman answered the phone. I enjoyed the analytical analysis of structures, and went back to school to obtain my MS in Aerospace Engineering three years later at the University of Washington.

My division head invited me to obtain a Navy civilian scuba diving certification to more fully participate in the submersible testing. This 1-month long training included water and endurance exercises and a submarine hatch escape drill. While I passed the swim portions, I did need to retake the physical. The dive master was extremely supportive; I was only the second woman to have completed this course. He coached me through my first ever one-mile run and the dreaded eight pull-ups. Even with this experience, I was prevented from participating in the half-day submarine excursion offered to all of the other new graduates. It wasn't until 2010 that the Navy would allow women to board and serve on submarines.

The workplace, fabrication and manufacturing workshops had a distinct casual feel to them. It was not uncommon to run into a framed sexually suggestive French curve set or an explicit pin-up calendar. When I went to work in my first aerospace company in the mid-1980s, I sat in a drafting bullpen consisting of 30 drafting tables. Most designers sat at a drafting table actively working on vellum. The younger engineers had both a drafting board and an early model ComputerVision CADD on their desk. While there were few personal items to decorate the room, ours included a fully-functional life-sized inflatable doll hanging from the ceiling rafter.

The 1991 nationally televised congressional testimony by law professor Anita Hill provided the nation with a backdrop for addressing sexual harassment in the workplace. When I switched jobs to work at Solar Turbines in 1993, I walked into my first experience of a more neutral workplace environment free from sexually explicit decor. It was a noticeable and welcome cultural shift. I felt like I belonged.

SWE has provided me with an abundance of opportunities. My SWE mentors have given me that extra nudge to try new things, make friends and expand my network. It was the women at General Dynamics Space System Division who invited me to become active in the local SWE San Diego section. I had no clue how much I needed this group. SWE opened up the conversation about work, working environments, and the hope that we too could one day make a difference in the world.

My SWE mentors soon took me far outside of my comfort zone. They had to talk me into running for the 1992 section president, more than once. They talked me into joining a panel discussion at my first SWE Conference (1991 San Diego Convention), where I met Jill Tietjen, SWE president, and Alice May Anderson, convention co-chair. Impressed by the quality of the (other) speakers, I vowed to learn more about public speaking and joined Toastmasters. My journey into running an organization full of unpaid volunteers began. I've now been in SWE for over 35 years and am a huge advocate for engineering society engagement. SWE has opened up several doors for me.

When I became the 1992 SWE San Diego section president, my corporate skip-level manager invited me to conduct a workplace culture study. He authorized a one month half-time workload to complete the study. My final report included interviews and anonymous quotes from women throughout the division. The drafting bullpens were eliminated the following year. I was fortunate to see several female colleagues begin to advance in their careers.

Awed by the research contained in the *SWE Magazine*, I joined the Editorial Board a few years later. There I had the opportunity to review the seminal 2010 American Association of University Women compilation report *Why So Few Women in STEM*. This one report shook the very core of my beliefs. The numbers of women in engineering had barely budged in the last 30 years, doubling to 20% for college graduates and stagnating at only 12% in the workplace. The report raised awareness on the impacts of unconscious gender bias. For the first time, I began to unravel the workplace order and how the unwritten rules for women were simply different. I had direction.

My public speaking career began, primarily addressing unconscious bias. I gave my first solo presentation during the SWE 2011 Conference Invent It Build It Parent Educator program. My skip-level manager invited me to deliver the presentation to his entire directorate: the 120 men and handful of women that comprised his engineering, manufacturing, and technology departments.

The 2016 documentary *The Computer Wore Heels* was another seminal moment. I was gobsmacked at how many women had served in technical roles during World War II. These pioneers were the mathematicians who programmed the first electronic computer. Several pioneering women's documentaries began popping up. There were those who served in construction, technology, and including the NASA "Hidden Figures" who helped launch the first manned rocket to orbit the moon. Who were these pioneers? Would my career path have been more straightforward, or less haphazard, if only I had known about their achievements during high school?

Writing this book on the women who carved out engineering careers during the Great Depression and WWII has been extremely cathartic for me. My first historical research of a SWE Legend began when a founding SWE San Diego member Alice May Anderson (1916–2016) died at age 100. Alice's son helped connect me with her workplace colleagues, whom she had kept in touch with through the decades. You can find her story on the SWE Endowed Scholarship page. I've gained a great deal of respect for what it takes to bring the story of an early pioneer to light.

My passion for learning about the early pioneering engineers has continued to grow over the last decade. Through my local San Diego County Engineering Council, we began looking for pioneers in various engineering disciplines. The challenge has been to identify a diverse array of these pioneering engineering role models and to make their stories more visible. Often, very little information is available. When you do find data, it is frequently missing the photo or significant career details. My latest accomplishment resolved the online content for a founding member of the American Indian Science and Engineering Society (AISES). To date, we have placed more than seventy Engineering Legend profiles on our website; the majority are SWE Legends.

When founding the SWE Late Career and Retiree (LCR) Affinity Group (AG) in 2019, I found my tribe. I've been able to reconnect with SWE friends I've met through the decades and to meet new SWE advocates along the way. I will be forever grateful to the SWE LCR AG for being a welcome home for the SWE Legends project, the project that connected me to SWE fellows who helped launch this very book.

Afterword—Vicki S. Johnson

My Dad wanted sons; he got two daughters. He believed that children who weren't in school or sleeping would get in trouble if they had free time, so my sister and I did a lot of physical work. We did construction cleanup, dug up thistles on our 20 acres, cleaned up a big pile of used lumber, and ran a laundromat. When my fifth-grade math teacher said, "Mathematicians work with their brains, not their bodies," I said, "Sign me up!"

My Mom had two rules for her daughters: (1) get a college degree that would let us get a job and support ourselves; and (2) have that job and be supporting ourselves before we could get married. When my 27-year-old sister came home to talk about getting married, she had met the rules as a Chemical Engineer. Mom was reluctant—no guy was good enough for her daughter. After 36 years, Val and Mike are still married, and Mike became the favorite child of both Mom and Dad before they left this earth. I still haven't met the rules.

Until 10th grade, my goal was to become a high school algebra teacher. Then the Kansas City teachers' strike happened. It did not affect my school district but the publicity influenced me. I learned that teachers had a very tough way to go and didn't make a lot of money. I decided it was time to rethink my career plan.

When I was in junior high school, my Dad supported me joining the Civil Air Patrol (CAP) to learn leadership and flying. My first flight was at a CAP encampment when I was 13; my next flight was my first lesson. I soloed in 10th grade on my 16th birthday. I got my private pilot—single-engine land license in December of my senior year of high school.

I combined my love of math, my like of science, and my love of flying to choose a career in aerospace engineering. My goal was to work for Cessna Aircraft Company because I flew old, beat-up Cessna 172 airplanes and felt very safe in them. Dad argued that I should be a mechanical engineer and work in aerospace so I could get a job in lots of other places when aerospace was down. For me, it had to be aerospace. Dad's backup plan when I got laid off was to set me up with a two-ton truck and backhoe; fortunately, I never was laid off. I was highly motivated not to be!

Somewhere I heard about the cooperative engineering program and wound up at the University of Missouri—Rolla (now Missouri University of Science and Technology) in 1977 because I was eligible for in-state tuition, I got a Curator's scholarship, and they had an engineering co-op program. I was able to get a co-op job at NASA-Langley Research Center in Hampton, Virginia after I completed fall, spring, and summer of my freshman year.

My first co-op assignment was in the Aeronautical Systems Office, where I redesigned the wing thickness distribution for NASA's supersonic transport to minimize the wave drag. At that time, I had not taken any aerodynamics classes and had no experience programming computers. I learned to program an HP-97 calculator to generate the geometry. Then I updated the geometry on 80-column computer cards, ran the deck through the card reader in our office to run the Carlson-Harris Wave Drag Program, and got back a printout from the main computer center four times a day. I would plot the wave drag results by hand and decide what to try next. The result was the NACA 64A Double Flattop wing thickness distribution.

I was the only female engineer in my class of 13 AE students who graduated from Rolla in 1982. I was hired full time by NASA-Langley. There was one other female engineer (an African American) in our office. She was Dr. Christine Darden, who was described in the book *Hidden Figures* as the next generation of African American females who were helped by the original Hidden Figures computer ladies. Soon after I arrived, a female graduate student (Helen Reed) arrived in our office.

Back in those days we wrote our papers by hand and gave them to the branch secretary to get them typed. After I was hired there full-time, the secretaries got word processors with memory so they could save what they typed. My secretary would make the changes I noted, but other errors would creep in. It was a vicious cycle.

In September 1982, I started work on my master's degree from George Washington University (GWU). I put the finishing touches on my thesis as I was moving to Lawrence, Kansas, to do the coursework for a Doctor of Philosophy in Aerospace Engineering at the University of Kansas (KU). From August 1985 until August 1986, I completed the coursework at KU. In the next 2 years at NASA, I did the research and wrote a dissertation "Life Cycle Cost in the Conceptual Design of Subsonic Commercial Aircraft." I typed my dissertation on my home computer. I was the first female to receive a doctorate in aerospace engineering from KU.

When I left NASA in 1990 I had a Sun MicroStation on my desk, could make polished presentation slides, and could run the Carlson-Harris Wave Drag program in about 9 seconds. I also had the ability to save output files and print what was required on a printer in the office.

My next stop was a year at the Aeronautics and Space Engineering Board (ASEB) at the National Research Council in Washington, DC. We traveled quite a bit to hold meetings around the country. Other people were starting to have laptops then, but not us. I took a lot of meeting notes by hand (some meetings lasted 5 days) with a task of typing up the notes after returning to the office.

I left ASEB for Universities Space Research Association (USRA) in a suburb of Houston, Texas. I served as Program Manager for the Advanced Design Program contract which had a requirement that we visit the 43 universities under our purview once every three-year program cycle. I arrived for the third year of the cycle to find that 33 universities still needed to be visited. I was on the road 68% of the time that year. I had a staff of four to manage in Houston along with reports to publish and a summer conference to plan. They found a local homemade laptop for me. Back then (1991), communicating was by dial-up modem. The odds seemed to be about 50/50 that I would get connected at any hotel.

Before leaving USRA, I served as a member of the NASA Federal Laboratory Review Task Force. I was a late addition to try to give the task force a better gender balance. There were 26 people on the Task Force; three of us were female (one of the other females was Helen Reed). We visited all the NASA centers either in person or by video conference.

Embry-Riddle Aeronautical University (ERAU) was my next stop as an associate professor of Aerospace Engineering (AE) with ¾-release time to be Director of Student Success Programs. In my best year, we reduced freshman attrition from 28% to 18%. I taught Introduction to Aerospace Engineering, Fluid Mechanics, and Airplane Stability and Control. As a Welliver Fellow at Boeing in St. Louis during the summer of 1998 I learned what faculty could do to make our students more productive for industry. As a faculty and staff member, I had a computer in both offices. Networks were not available, so I wound up using Zip drives to store files and carry them between computers.

My involvement in the Society of Women Engineers (SWE) started at ERAU. Nancy Sliwa was the Faculty Advisor for the ERAU SWE student section. Nancy asked me to take over as the faculty advisor when the Sliwas left ERAU. I joined SWE in 1997. The professional members at the Cocoa Beach section were very helpful in indoctrinating me in the ways of SWE. They were also very helpful in getting me connected to the Wichita Area Section of SWE when I moved to Wichita three years later.

I left Embry-Riddle for the National Institute of Aviation Research (NIAR) in Wichita, Kansas in 2000. At NIAR I was the Associate Director of Research and Development (R&D) where I did research for the Advanced General Aviation Experiment (AGATE) program, helped with a strategic plan for NIAR, and helped ensure the NIAR research projects for the Federal Aviation Administration were meeting the program requirements.

Next up was a job as the first-level manager for aerial refueling systems on the International Tanker Transport Aircraft (ITTA) Program at Boeing in Wichita just after Italy signed a contract for tankers. This was my first foray into industry. It was very different from my other experiences. As my manager said, they wanted me to get my feet wet headfirst!

As a manager, I had a laptop computer with a docking station in my office and provided my own cell phone. I had a great team of 26 engineers and engineering technologists. The team wrote requirements documents and worked with suppliers to deliver hose drogue systems, and did designs and generated drawings for the aerial refueling boom. When a plan for United States Air Force lease of tankers fell through, I was looking for work again.

I made use of my SWE network to find my next job. Wichita SWE was storing all their supplies for the annual Engineering Expo in my garage. I went to a SWE member who worked at Cessna and told her that I needed to find a job. If I couldn't find one in Wichita, I would need to move, and the Wichita SWE supplies would need to find a new home. Twenty-five years after I embarked on a path to NASA, I finally made it to Cessna. I was a much better engineer when I arrived than I would have been straight out of college. It was nice that I didn't have to move, and the SWE supplies remained in my garage!

My first responsibility at Cessna was on the Research and Advanced Technology (RAT) team. My projects primarily involved electrothermal deicing system design and testing. My iPod was handy for storing and transporting data. The work resulted in two trips to an icing tunnel in downtown Manhattan, New York, and in two patents. During that time, I did a Master of Business Administration with an emphasis in management of technology.

Beginning in 2008, I trained as a Six Sigma Black Belt and Master Black Belt, and then did improvement projects and mentoring. One of the really great things was that Black Belts were issued laptops.

In 2011, I became assistant principal investigator on a NASA contract "N + 3 Small Commercial Efficient and Quiet Transportation for Year 2030–2035." One of our "inventions" to reduce aircraft weight and drag was protective skins for composite airliners. I became the principal investigator for the Phase II contract investigating protective skins. That kept me very busy during a very tough financial time for many companies and people.

I made one last job change before retirement to Spirit AeroSystems in Wichita. My primary job at Spirit was as Lead Engineer for R&D Portfolio Management. That involved helping to determine the major R&D disciplines that would be studied each year and then working with the researchers to define, conduct, report on, and document their projects.

I missed a big retirement party when I retired due to COVID, but my Spirit and Wichita SWE friends did the next best thing. They made a Shutterfly book with pictures of them with me and with notes from them for me. We had the party on a Zoom call. I am touched every time I look at that book and see that I really did make a difference in the lives of the women I touched. I was much more of a role model than I realized.

Freedom from going to the office every day resulted in me buying a 40-ft diesel pusher recreational vehicle, the Queen Bee. I also had time to work more on my hobby of skeet shooting.

Being in Wichita for the last 20 years of my career was excellent for my involvement in SWE. I did about all the jobs in the Wichita Section, including chairing the Engineering Expo Committee one year. I became the first senator from Region i. In 2014, the Wichita Section hosted the Region i conference. I was the Finance Committee Chair. We exceeded our fundraising goal by 17%.

After that, Lee Iacocca's saying, "Lead, follow, or get out of the way" applied perfectly. I didn't want to lead and only wanted to follow occasionally, so I got out of the way. Then in 2019 the Late Career and Retiree Affinity Group (LCR AG) was formed. I got involved at the beginning, serving first as the Local Section Engagement lead, then lead of the group, and now newsletter editor.

The LCR AG has been a lifesaver. I found my new tribe. I am still contributing to SWE. I was recognized with SWE Fellow in 2020 and a Distinguished Service Award in 2024. And being involved has opened up new paths to mentor younger ladies, help share the collective knowledge and wisdom of our group, and write a book!

Afterword—Janet L. Williams

I have been active in the Society of Women Engineers for over 40 years. I led our local section, served on several and chaired some Society committees, gave numerous presentations at conferences, wrote dozens of articles for *SWE Magazine* (the flagship publication of the Society), participated in countless STEM outreach events, and in the last few years have embarked on a new adventure in federal and state legislative advocacy for STEM. What drove me to more than 40 years of commitment to the goals of the Society? The answer lies in my circuitous and serendipitous path to engineering.

I graduated fourth in a high school class of 435 students, that is, the top 1% of my class. I knew I was expected to go to college, and I wanted to go, but I had no idea what major to select. All subjects had come easily to me, and I always earned stellar grades. My parents and older siblings did not attend college, so they could not advise me. We had a high school counseling office, but the counselors would not have mentioned engineering to a woman in those days. I randomly selected mathematics, thinking it would get me into the school of my choice, the University of California, Davis.

At the end of my sophomore year at Davis, my father died. I worried about the loss of financial support. I had already begun to question my decision about the math major. I had to finish college in 4 years, even if I changed my major. How would I survive on odd jobs when my income from Social Security death benefit ceased on my 22nd birthday? My answer was to switch to Spanish, in which I had completed several classes and whose degree requirements I could still complete within my time limit.

Armed with a BA in Spanish, I moved to the Midwest with a friend and found myself in a different world, culturally speaking. There was little call for Spanish majors. I managed to find a job in an administrative position at a local engineering consulting firm. After a year of seeing what engineers do, I said to myself, “I could do this.” I enrolled in the University of Missouri, Kansas City and continued to work part-time as I worked my way through the mechanical engineering curriculum. I was one of two women in that major when I graduated in May 1983.

My story as an “accidental engineer” has propelled me in my commitment to and passion for encouraging young women (and underrepresented minorities) to pursue careers in engineering and related fields. I never had an opportunity to know what engineering is, what engineers do, or how they benefit society. I was never able to develop an identity as a potential engineer because that information was never presented to me through role modeling, mentoring, career guidance, or outreach events. Once I found engineering (the hard way), it was like finding my path home. Engineering has given me a multi-faceted career that was intellectually stimulating and financially rewarding in ways that no one in my immediate family ever experienced. Every young woman (every young person, really) with a creative spark deserves to know the endless possibilities a career in engineering and related fields can provide, and I am more than willing to help provide those opportunities through the Society's many activities and resources.

This inspiring collection of stories about the women who received the Society's first 25 Achievement Awards represents the foundation which made the authors' careers possible. While the very earliest women engineers had to hack the first paths through seemingly impenetrable, forbidden territory, these Achievement Award recipients widened those paths, preparing the way for our generation to build the paths into roads and allow subsequent generations to pave those roads. Their pioneering careers and exceptional lives demonstrate the determination, perseverance, and fortitude that allowed us, the authors, to achieve our dreams. They actively prepared us to traverse our paths by example and through patience and sometimes "tough love" mentorship. They codified our right to be in the engineering workplace by their very presence, their cool competence, and often through direct advocacy for women's rights. We owe everything to these pioneers. For these reasons, when the opportunity arose to help tell their stories, I was beyond eager to participate.

When I interact with the newest generation of women engineers in the Society, I see young women who are so much more confident and self-assured than I was at their age. Many of them are savvy and assertive enough to simply call out discrimination and bad behavior. At their age, I just endured it or ignored it. The fact is, some of these young women don't even understand the focus on discrimination of women in the workplace, because they haven't had to experience discrimination and bad behavior.

I worry that the newest generation of women engineers does not fully appreciate the pioneering shoulders they stand on as they become the first generation to be virtually free of the obstacles these pioneering women faced just to be in the workplace. These young women haven't had to fight to be recognized for their amazing accomplishments like our predecessors have. We, the authors, are proud of our own careers, and we have all contributed to technology and to the fabric of the history of women in engineering, but we know that without the women in this book, our journeys would have been much more difficult, if not impossible. Collectively, we all experienced various forms of prejudice throughout our careers, from veiled innuendo to blatant harassment. We see that unfair treatment of women engineers in the workplace is markedly reduced, and certainly less overt, than it was in our generation's experience. But it still exists in pockets, and manifests in more nuanced ways. Gender parity in engineering is the best strategy to eliminate harassment of women engineers in the workplace and give them equal opportunities to invent, inspire, and shape the future. I hope I live to see it.

I am indebted to my Society colleagues with whom I have collaborated on this project. They have provided an exceptional level of support and encouragement since the day we embarked on this journey. And I should have expected nothing less. As Fellows of the Society of Women Engineers, we have all supported one another through the years. The Society of Women Engineers is our core, our rock, and our nexus. It is the entity that has united us across geographic areas, cultures, beliefs, and family circumstances and has led us to this successful and sublime endeavor.

Afterword—Holly J. Teig

See someone, be someone. It is powerful for a person to see someone who looks like them in a role that they are aspiring to. My journey to becoming an engineer and desire to write this book are aptly described by these words.

I grew up in a small town in the countryside of Minnesota. We didn't have fast food, a stop light, or a movie theater. My Dad worked as a second shift machinist for a commercial airline and farmed part time. Mom was the mother of four children, a homemaker, and a seamstress. Neither went to college, but we were expected to pursue higher education. We were involved in our church, with 4-H, and there was always work on the farm with the animals or out in the fields. Dad taught us kids to run the tractors, bale hay, and perform endless farm chores. I always preferred to work outside chores versus the domestic household chores. My career aspirations growing up ranged from marine biologist to commercial airline pilot. Mostly, I wanted to stretch my wings outside of my small town while supporting myself financially. Money was tight for our family of six and I yearned to have a career that allowed me to live in my own place and buy new clothes rather than dress in the hand-me-downs of my older sister. I recall my mom using her first credit card in the 1970s and instilling in my sister and me that it's a wise idea to be equipped to earn your own living.

In high school, I took pre-college science and math courses and scored well in those subjects on the standardized tests. I was also active in choir, band, and theater, which might be the reason why my high school guidance counselor advised me to go into theater arts. I was flabbergasted! I knew I wasn't talented enough to ever make a decent living doing that. Shocked and determined, I headed to the school library. I drew my finger down the list of starting salaries based upon college degrees, noting that several types of engineering were at the top. Later, I asked my Mom if she knew anyone who was an engineer. She told me that Beth, a girl at my church who was my older brother's age, was studying engineering and I should talk to her. *See someone, be someone.*

Ten minutes into our conversation, Beth assured me I had all the intellect and skills required to pursue engineering. She was soon to be receiving her Civil Engineering degree, and suggested affordable colleges, including one I had researched, the University of North Dakota College of Engineering and Mines (UND). "Be sure to join SWE when you get there."

What was SWE? I wondered.

I am thankful that I followed her lead. I went to my first Society of Women Engineers (SWE) pizza party that fall and signed up as a member. I instantly became part of an incredibly supportive and talented community of women that would shape the rest of my life. Joyce Medalen, the SWE Counselor, and Dr. Donald Naismith, the SWE faculty advisor (a SWE Rodney D. Chipp recipient), were my initial support network when dealing with any college challenges. They helped me through a tough sophomore year and made connections when I wanted to pursue engineering co-op positions (both for experience and financial need). The UND SWE section was well established, and I gained friends and leadership through those experiences. With support from this network, I proudly earned my BS in Mechanical Engineering in 1989.

A younger neighborhood friend from my hometown joined me at UND two years after I began, also joined SWE, and went on to earn her electrical engineering degree. If you do the math, that sums up to three women engineers emerging from my small hometown simply because we were aware of one another's educational paths. Another friend I knew through 4-H also received her electrical engineering degree during this time. *See someone, be someone.*

Computer technology changed rapidly during my time in college. As a freshman, I had an introductory programming assignment using punch cards. The following semester, I spent hours at the computer lab typing FORTRAN programs into a VAX mainframe, compiling jobs, and generating dot matrix printouts of results. Personal computers caught on quickly. For my senior year independent study, I was using a Lotus 123 spreadsheet on a two floppy disk drive PC that my boyfriend Randy (later my husband) had loaned to me. The software program was encoded on one floppy disk and you stored your data file on the second floppy disk. I got very proficient using control function key shortcuts in spreadsheets—which I'm pleased to report still work with the spreadsheet software in use today. Thank you, software engineers!

Somewhere in the middle of my college years the internet took off. I recall typing my first email message on a mainframe server at college to my older brother who was doing research at another college. The subject was about meeting up to drive home for the holidays. I typed a cryptic address to reach him, and his responses would print off on the output server a few minutes later. It wasn't convenient at all, but we still knew it was cutting edge technology.

Upon graduation, I had $21 in my checking account, a rusted out pickup truck, and a top-end job offer at Caterpillar Inc. in Illinois. I felt like I could go anywhere and do anything with that coveted starting engineering salary. I put my mechanical engineering degree to work in manufacturing. Construction equipment was very much a male-dominated environment. Of the roughly 100 engineers hired into the company that year, 11 were women. In our hiring class picture, the women were lined up on the front row of the risers, attired in our 1980s big hair and shoulder-padded suits in various colors. Behind us were multiple rows of men—a sea of dark charcoal suits with white shirts and colored ties. We stood out from the men in so many ways. I learned quickly in my career to use that to my advantage. One manufacturing manager asked me if I was going to cry if he yelled at me. I told him I had grown up with two older brothers, so it would take a lot for me to cry.

The new women engineering recruits were scattered throughout the company and we wanted to have a women's network to exchange information. I befriended others who had been members of SWE student sections and we began meeting up. In 1992, we founded the SWE Central Illinois professional section. We had 15 founding members and just five had been part of SWE in college. I served as treasurer and everyone took turns as officers. The main focus of our fledgling section was to provide engineering outreach to middle and high school students along with professional career support. It was a slow start, but the section grew into a well-run organization now boasting over 320 members, with an endowed scholarship for students and an awards program to recognize SWE members. I sealed my commitment to SWE in 2008 by becoming a life member.

Five years into my career, my business unit was dramatically downsized and I was put on the surplus list. My boss called me into his office and thrust a handful of job postings at me. I naively asked him why I was marked as surplus despite a great performance review. He responded, "You have an engineering degree. You're marketable in lots of other jobs at this company. These other guys, this is all they can do." He swept his hand in a gesture to my much older, male colleagues, many without degrees. I landed an interview for our new logistics organization, but was honest with the interviewer that I had no background in that area and wasn't even sure what the field of logistics was. "Don't worry, we will teach you. And we know you are capable of learning because you have an engineering degree." For the first time, I really appreciated how much of a door opener my degree was. It was providing the financial security and resiliency that I had wanted in my career. I accepted the logistics role and discovered a passion for an industry I didn't know even existed until that moment.

I did a lot of SWE outreach with youngsters through the years. I led experiments during Engineers Week, presented at schools, mentored at a middle school, and worked career fair recruitment at universities and SWE conferences for my company. I had older male colleagues ask me to connect with their high school aged daughters who were interested in math and science. My encouragement led to several of them pursuing engineering. It felt good knowing I was making an impact for women in STEM one person at a time. *See someone, be someone.*

I had the opportunity to travel around the world to work with global colleagues and customers, something I never imagined doing back in high school. My career achievements were validated when I was selected to join a multi-company team that was transforming the supply chain infrastructure—from software to processes to physical layout—creating a legacy of modernization that launched breakthrough technology to the industry. I gravitated to the tough assignments and chipped away at barriers and traditions.

Mid-career, I realized I could have a much greater influence on keeping technical women in the pipeline. I expanded mentoring relationships with younger talent in my company and advocated for promoting technical women. I collaborated with a SWE friend to write submissions for deserving women for SWE awards and for community leadership honors. It was difficult to dig information out of candidates to complete an awards packet. Unfortunately, our society has programmed women to play down their successes. Many are self-conscious about taking credit for what they have achieved, always preferring to credit their team's success. Women engineers are now eligible for recognition during all phases of their career, for technical, leadership, and community accomplishments. For award recipients, this is an independent validation of their success that builds confidence and brings benefits back into their workplace. They are seen and heard. Validated as engineers. I received my own share of honors, including being named a Fellow of SWE in 2016. *See someone, be someone.*

When I announced my retirement after 30 years at Caterpillar, I received a flood of gratitude for my contributions and leadership. One stood out. A young woman from our software team, who wasn't in my direct organization, said "I hope one day to be a great leader just like you." It melted my heart. I didn't realize how far my influence had spread and how many younger women in the company looked up to me as a leadership role model. *See someone, be someone.*

I'm now using my time to continue to pay it forward with SWE. I'm focusing my skills to grow SWE's financial resources to ensure the organization will be around to support future generations of women engineers. I'm serving as the Treasurer of the SWE Board of Trustees overseeing the Society's endowed funds, ensuring women pursuing engineering aspirations have financial support to pursue their education and be successful in the STEM workforce. The financial burden of higher education is a bigger hurdle than ever for students entering college, one I can relate to. To receive SWE scholarship stipends that relieve this worry can be life-changing. A student can exhale and focus on their learning instead of working multiple part-time jobs in order to earn their degree. I'm so proud that SWE gives out over $1.5 million dollars in scholarships every year to deserving recipients. But I want this number to grow.

I believe it is imperative that we, as women engineers, tell our own stories. Not the watered down, *she helped out* versions that exist in so much of the media today. I want to tell the real stories of invention, perseverance, and success that set these women apart. The women engineers that we are featuring have changed our world. They created lasting impacts in their industries and influenced technology, research, and production for upcoming generations. The obstacles in my career were minor compared with the challenges that faced the trailblazers in this book. When I read their stories, I feel excitement, admiration, and pride. I see myself in them. And I want young girls in the future to learn about these trailblazers and know that they can change the world as an engineer. This book is a full circle moment for me. My sincerest hope is that this book of legends inspires the trajectory of someone's life to become an engineer. *See someone, be someone.*

Photo Credits and References for Graphic Timelines

Images for which photo credits are not provided are Public Domain, from either Wikimedia or other sources.

Chapter 2—1952–56 Graphic Timeline

	Early transistors: Courtesy of Alamy Limited
1952	Mária Telkes, PhD—Courtesy Library of Congress
	Grace Murray Hopper, PhD: Courtesy of the Library of Congress
1953	Bram Van Heel: Courtesy of Jeffrey Hecht
1954	First solar cell: Museum of Solar Energy, courtesy of Lynn Salvo
	Edith Clarke: Courtesy of Society of Women Engineers Records, Walter P. Reuther Library, Wayne State University
1955	Margaret H. Hutchison Rousseau, Sc.D.: Courtesy of Society of Women Engineers
	Records, Walter P. Reuther Library, Wayne State University
	Microwave: Courtesy of the National Museum of American History
1956	Elise F. Harmon: Courtesy of Society of Women Engineers Records, Walter P. Reuther Library, Wayne State University
	Video tape recorder: Courtesy of Engineering and Technology History Wiki

Chapter 8—1957–61 Graphic Timeline

1957	Power plant: Courtesy of the Library of Congress
	Rebecca Sparling: Courtesy of Society of Women Engineers Records, Walter P. Reuther Library, Wayne State University
1958	Mabel MacFerran Rockwell: Courtesy of Society of Women Engineers Records, Walter P. Reuther Library, Wayne State University

C. D. Craig et al. (eds.), *Women Engineering Legends 1952-1976: Society of Women Engineers Achievement Award Recipients*, https://doi.org/10.1007/978-3-032-00224-2

	Dr. Joseph Shivers, Jr.: Courtesy of American Association of Fiber Chemists and Colorists
	Pacemaker: Courtesy of Siemens Medmuseum
1959	Désirée Le Beau, PhD: Courtesy of Society of Women Engineers Records, Walter P. Reuther Library, Wayne State University
	Boeing 707–123: Courtesy of Boeing
1960	Esther M. Conwell, PhD: Courtesy of Society of Women Engineers Records, Walter P. Reuther Library, Wayne State University
	Dr. Theodore Maiman: Courtesy of HRL, Inc.
1961	Laurel Van der Wal: Courtesy of Society of Women Engineers Records, Walter P. Reuther Library, Wayne State University
	Discovery of ibuprofen: Courtesy of Boots Archive, Boots UK
	President John F. Kennedy: Courtesy of NASA

Chapter 14—1962–66 Graphic Timeline

1962	Laurence Delisle Pellier: Courtesy of Society of Women Engineers Records, Walter P. Reuther Library, Wayne State University
	John Glenn: Courtesy of the Library of Congress
	AT&T's Telstar: Courtesy of NASA
1963	Beatrice Hicks: Courtesy of Society of Women Engineers Records, Walter P. Reuther Library, Wayne State University
1964	Grace Murray Hopper, PhD: Courtesy of Society of Women Engineers Records, Walter P. Reuther Library, Wayne State University
1965	Astrodome: Courtesy of the Library of Congress
	Martha J. B. Thomas, PhD: Courtesy of Society of Women Engineers Records, Walter P. Reuther Library, Wayne State University
	Stephanie Kwolek: Courtesy of Michael Branscom Photo and Film
1966	CADD drafter: Courtesy of Library of Congress
	Dorothy Martin Simon, PhD: Courtesy of Society of Women Engineers Records, Walter P. Reuther Library, Wayne State University

Chapter 20—1967–71 Graphic Timeline

1967	Marguerite M. Rogers, PhD: Courtesy of Society of Women Engineers Records, Walter P. Reuther Library, Wayne State University
1968	DRAM & Robert Dennard: Courtesy of IBM
	Isabelle L. Karle, PhD: Courtesy of AIP Emilio Segrè Visual Archives
1969	Man on the moon: Courtesy of NASA
	Alice Stoll: Courtesy of Society of Women Engineers Records, Walter P. Reuther Library, Wayne State University
	Invention of Gore-Tex: Courtesy of W. L. Gore & Associates, Inc.

1970	Image symbolizing the first email: Courtesy of geralt
	Irmgard Flügge-Lotz, PhD: Courtesy of Society of Women Engineers Records, Walter P. Reuther Library
1971	Alva T. Matthews Solomon, PhD: Courtesy of Society of Women Engineers Records, Walter P. Reuther Library, Wayne State University

Section 5, Chapter 26—1972–76 Graphic Timeline

1972	CAT scanner: Courtesy of Science Museum Group, UK
	Nancy D. Fitzroy: Courtesy of Society of Women Engineers Records, Walter P. Reuther Library, Wayne State University
1973	SkyLab: Courtesy of NASA
	Irene Carswell Peden, PhD: Courtesy of Society of Women Engineers Records, Walter P. Reuther Library, Wayne State University
1974	Barbara Crawford Johnson: Courtesy of Society of Women Engineers Records, Walter P. Reuther Library, Wayne State University
1975	Sheila E. Widnall, PhD: Courtesy of Society of Women Engineers Records, Walter P. Reuther Library, Wayne State University
	Digital camera: Courtesy of Associated Press
1976	Apple computer & start of Apple: Courtesy of Associated Press
	Seymour Cray and Cray 1: Courtesy of Supercomputing Technology Historical Museum
	Ada I. Pressman: Courtesy of Society of Women Engineers Records, Walter P. Reuther Library, Wayne State University

References (for all Graphic Layouts)

Ali, Ray. 2023. USwitch. History of mobile phones and the first mobile phone. October 3. https://www.uswitch.com/mobiles/guides/history-of-mobile-phones.

American Bar Association. Lawyers by Gender, Demographics: Profile of the Legal Profession 2024. https://www.americanbar.org/news/profile-legal-profession/demographics/. Accessed March 2025.

American Chemical Society. Legacy of Rachel Carson's Silent Spring. American Chemical Society. https://www.acs.org/education/whatischemistry/landmarks/rachel-carson-silent-spring.html. Accessed April 2025.

Attridge, Margaret. 2025. A History of Women in Higher Education. Best Colleges. March 19. https://www.bestcolleges.com/news/analysis/2021/03/21/history-women-higher-education/.

Beck, Adam. 2017. 60 Years of CAD Infographic: The History of CAD since 1957. September 25. https://www.3dfindit.com/en/engiclopedia/60-years-of-cad-infographic-the-history-of-cad-since-1957. Updated Feb 10, 2025.

Bellis, Mary. 2019. How Fiber Optics was invented. ThoughtCo. June 30. https://www.thoughtco.com/birth-of-fiber-optics-4091837.

Bhattacharyya, Kalyan B. 2016. Godfrey Newbold Hounsfield (1919–2004): The man who revolutionized neuroimaging. *Annals of Indian Academy of Neurology.* Oct-Dec. 19(4):448–450. https://pmc.ncbi.nlm.nih.gov/articles/PMC5144463.

Boots UK. Boots Hidden Heroes - Honouring Dr. Stewart Adams. 2019. February 1. https://www.boots-uk.com/newsroom/features/boots-hidden-heroes-honouring-dr-stewart-adams/.

Bruderer, Herbert. 2022. Did Grace Hopper create the first compiler? Communications of the ACM blog. December 21. https://cacm.acm.org/blogcacm/did-grace-hopper-create-the-first-compiler/.

Campbell, Karen E. and McCammon, Holly J. 2005. Elizabeth Blackwell's Heirs: Women as Physicians in the United States, 1880–1920. *Sage Journals* 32(3). August. https://doi.org/10.1177/0730888405277634.

CertiPUR-US Home Sweet Foam blog. Thank you, NASA: The Unusual History of Memory Foam. https://certipur.us/thank-you-nasa-the-unusual-history-of-memory-foam/. Accessed March 2025.

Charles P. Ginsburg, National Inventors Hall of Fame. https://www.invent.org/inductees/charles-p-ginsburg. Accessed February 2025.

Chodos, Alan. 2009. April 25, 1954: Bell Labs Demonstrates the First Practical Silicon Solar Cell. *APS News.* April 1. https://www.aps.org/apsnews/2009/04/bell-labs-silicon-solar-cell.

C. D. Craig et al. (eds.), *Women Engineering Legends 1952-1976: Society of Women Engineers Achievement Award Recipients*,
https://doi.org/10.1007/978-3-032-00224-2

Computer History Museum. The Cray-1 Supercomputer. https://www.computerhistory.org/revolution/supercomputers/10/7. Accessed March 2025.

Computer History Museum. Timeline of Computer History. https://www.computerhistory.org. Accessed February 2025.

Crea. 2024. The Accidental Inventor: Stephanie Kwolek and the Birth of Kevlar®. March 21. https://creaindia.com/blogs/news/the-accidental-inventor-stephanie-kwolek-and-the-birth-of-kevlar®. Accessed March 2025.

Davis, Will. 2017. Shippingport Atomic Power Station: Five Fast Facts. Nuclear Newswire. American Nuclear Society. December 7. https://www.ans.org/news/article-2004/shippingport-atomic-power-station-five-fast-facts/.

Desikan, Anita. 2022. Why Rachel Carson's Silent Spring Still Resonates Today. The Equation, Union of Concerned Scientists. September 22. https://blog.ucs.org/anita-desikan/why-rachel-carsons-silent-spring-still-resonates-today/.

DuPont. What is Kevlar®? https://www.dupont.com/what-is-kevlar.html. Accessed February 10, 2025.

Fay, Bill. 2022. The Equal Credit Opportunity Act (ECOA).March 22. Debt.org. https://www.debt.org/credit/your-consumer-rights/equal-opportunity-act/. Accessed April 2025.

Gore-Tex. Our History. https://www.gore-tex.com/about/history. Accessed March 2025.

Hammar, Peter. Charles P. Ginsburg 1920–1992. *Memorial Tributes of the National Academy of Engineering*, Volume 7. https://www.nae.edu/188541/CHARLES-P-GINSBURG-19201992

Hill, Catherine, Christianne Corbett, and Andresse St. Rose. 2010. Why so few? Women in Science, Technology, Engineering and Mathematics. *American Association of University Women.* February. https://www.aauw.org/app/uploads/2020/03/why-so-few-research.pdf.

History.com editors. 2024. Women's History Milestones: Timeline. History.com. May 2. https://www.history.com/topics/womens-history/womens-history-us-timeline.

History of Women in the U.S. Congress. (n.d.) Center for American Women and Politics, Rutgers-New Brunswick Eagleton Institute of Politics. https://cawp.rutgers.edu/facts/levels-office/congress/history-women-us-congress

Hochheiser, Sheldon. 2015. Electromechanical Telephone-Switching. ETHW. January 9. https://ethw.org/Electromechanical_Telephone-Switching.

Houston Chronicle Staff. 2015. 50 ways the Astrodome was innovative. April 3. https://www.chron.com/news/houston-texas/slideshow/How-the-Astrodome-was-ahead-of-its-time-changed-106803.php. Accessed February 16, 2025.

HRL Laboratories. The First Laser. 2025. https://www.hrl.com/about/laser.

John F. Kennedy Presidential Library and Museum. 2025. United States President's Commission on the Status of Women Records. January 10. https://www.jfklibrary.org/asset-viewer/archives/uspcsw.

Joseph C. Shivers Jr. 2025. National Inventors Hall of Fame. https://www.invent.org/inductees/joseph-c-shivers-jr.

Lemelson-MIT. Erna Hoover: Computerized Telephone Switching System. https://lemelson.mit.edu/resources/erna-hoover. Accessed April 2025.

Library of Congress. The Founding of Apple Computer, Inc. https://guides.loc.gov/this-month-in-business-history/april/apple-computer-founded. Accessed March 2025.

MacNeil, Jessica. 2019. Touch Tone Phones are Invented. November 18. https://www.edn.com/tone-dialing-telephones-are-introduced-november-18-1963/. Accessed February 10, 2025.

Mail.com blog. 2023. The first email message: 52 years of email. May 25. https://www.mail.com/blog/posts/fiftieth-anniversary-of-email/20/.

Marder, Nancy. 2019. Civil Rights Act of 1957: Women's Service on Federal Juries. Wiley Online Library. August 23. https://onlinelibrary.wiley.com/doi/abs/10.1002/9781118929803.ewac0059.

Mattfeld, Jacquelyn A. and Carol E. Van Aken, eds. 1967. The MIT Symposium on American Women in Science and Engineering. June 15. https://mitpress.mit.edu/9780262630122/women-and-the-scientific-professions/.

McLaughlin, Katie. 2014. 5 Things Women Couldn't Do in the 60s. CNN. August 25. https://www.cnn.com/2014/08/07/living/sixties-women-5-things/index.html.

Michals, Debra 2017. Betty Friedan. National Women's History Museum. https://www.womenshistory.org/education-resources/biographies/betty-friedan. Accessed April 2025.

NASA. 2019. July 20, 1969: One Giant Leap For Mankind. July 20. https://www.nasa.gov/history/july-20-1969-one-giant-leap-for-mankind/.

NASA STEM Team. 2016. Was John Glenn? (Grades K-4). NASA. December 8. https://www.nasa.gov/learning-resources/for-kids-and-students/who-was-john-glenn-grades-k-4/. Accessed February 2025.

National Air and Space Museum. Telstar. Smithsonian. https://airandspace.si.edu/collection-objects/communications-satellite-telstar/nasm_A20070113000. Accessed February 10, 2025.

National Inventors Hall of Fame. Erna Schneider Hoover. https://www.invent.org/inductees/erna-schneider-hoover. Accessed April 2025.

Nicol, Katherine. 2010. America's First Civilian Nuclear Plant. Pennsylvania Center for the Book. Spring. https://pabook.libraries.psu.edu/literary-cultural-heritage-map-pa/feature-articles/americas-first-civilian-nuclear-plant.

Norwood, Arlisha. 2017. Rosa Parks (1913–2005). National Women's History Museum. https://www.womenshistory.org/education-resources/biographies/rosa-parks. Accessed March 2025.

Planned Parenthood. 2015. The Birth Control Pill: A History. https://www.plannedparenthood.org/files/1514/3518/7100/Pill_History_FactSheet.pdf.

Robert Dennard. IBM: History. https://www.ibm.com/history/bob-dennard. Accessed March 2025.

Robert Dennard and IBM patent DRAM. Centre for Computing History. https://www.computing-history.org.uk/det/6295/Robert-Dennard-and-IBM-patent-DRAM/. Accessed April 2025.

Ross, Rachel. 2017. Who invented the microwave oven? Live Science. January 5. https://www.livescience.com/57405-who-invented-microwave-oven.html.

Science Museum Group. EMI CT Brain Scanner. https://collection.sciencemuseumgroup.org.uk/objects/co134790/emi-ct-brain-scanner-ct-scanner. Accessed March 2025.

Seven Great Things That Came from 1955. Dayco Inc. 2016. August 15. https://www.daycoinc.com/blog/seven-great-things-that-came-from-1955-bpid_5.aspx.

Staff Writer. 2018 History made when women were allowed to serve on jury. *The Associated Press*. November 16. https://apnews.com/article/50fb651b7fb84221887f8a7534a87fff.

Stevenson, Laney. 2018. Women at Work in the 1950s. The Text Message. National Archives. March 27. https://text-message.blogs.archives.gov/2018/03/27/women-at-work-in-the-1950s/.

SWE Awards: SWE Achievement Award Recipients. 2024. Engineering and Technology History Wiki. October 20. https://ethw.org/SWE_Awards. Accessed February 2025.

Swopes, Brian R. 2023. Tag Archives: Boeing 707–123. This Day in Aviation. https://www.thisdayinaviation.com/tag/boeing-707-123/.

This Month in Physics History April 6, 1938: Discovery of Teflon. APS Archives. https://www.aps.org/archives/publications/apsnews/202104/history.cfm. Accessed March 2025.

Toosi, Mitra. 2002. A Century of Change: the U.S. Labor Force, 1950–2050. *Monthly Labor Review.* May. https://www.bls.gov/opub/mlr/2002/05/art2full.pdf.

Toosi, Mitra and Teresa L. Morisi. 2017. Women in the workforce before, during, and after the Great Recession. July. https://www.bls.gov/spotlight/2017/women-in-the-workforce-before-during-and-after-the-great-recession/

Transistorized. 1999. ScienCentral, Inc, and The American Institute of Physics. https://www.pbs.org/transistor/album1/. Accessed February 2025.

United Nations. World Conference of the International Women's Year 19 June-2 July 1975, Mexico City, Mexico. https://www.un.org/en/conferences/women/mexico-city1975. Accessed April 2025.

U.S. Census Bureau. 2013. Census Bureau Reports Women's Employment in Science, Tech, Engineering and Math Jobs Slowing as Their Share of Computer Employment Falls. US Census Bureau Press Release Number: CB13-162. September 9. https://www.census.gov/newsroom/archives/2013-pr/cb13-162.html.

U.S. Department of Justice. Title IX of The Education Amendments Of 1972. Civil Rights Division. https://www.justice.gov/crt/title-ix-education-amendments-1972. Accessed April 2025.

U.S. Senate. Senate Stories: President John F. Kennedy Address to Congress, May 25, 1961. [VIDEO]. https://www.senate.gov/artandhistory/senate-stories/videos/address-kennedy-moon-landing-1961.htm

University of California Santa Barbara. Executive Order 11375—Amending Executive Order No. 11246, Relating to Equal Employment Opportunity. *The American Presidency Project.* https://www.presidency.ucsb.edu/documents/executive-order-11375-amending-executive-order-no-11246-relating-equal-employment. Accessed April 2025.

Uri, Jon. 2023. 50 Years Ago: The Launch of Skylab, America's First Space Station. NASA. May 14. https://www.nasa.gov/history/50-years-ago-the-launch-of-skylab-americas-first-space-station/.

Voting Rights Act (1965). 2022. Milestone Documents, National Archives. https://www.archives.gov/milestone-documents/voting-rights-act. Accessed April 2025.

Wallace, Dillon. History of Digital Cameras. Analog, a Legacybox Blog. https://legacybox.com/blogs/analog/history-of-digital-cameras. Accessed March 2025.

Wikipedia. Floppy Disk. https://en.wikipedia.org/wiki/Floppy_disk. Accessed April 2025.

Wisys. A Brief History of the Barcode on its 40th Anniversary. https://www.wisys.com/a-brief-history-of-the-barcode-on-its-40th-anniversary. Accessed March 2025.

Women Leaders in Innovation. Erna Schneider Hoover (b 1926). https://njdiversity.com/innovators/41-erna-schneider-hoover-b-1926. Accessed April 2025.

"Women's Lib" is coined. Annenberg Classroom, Annenberg Public Policy Center of the University of Pennsylvania. https://www.annenbergclassroom.org/timeline_event/womens-lib-coined/. Accessed April 2025.

Yellen, Janet. 2020. "The history of women's work and wages and how it has created success for us all." *The Brookings Institution. Gender Equality Series.* May. https://www.brookings.edu/articles/the-history-of-womens-work-and-wages-and-how-it-has-created-success-for-us-all/. Accessed April 2025.

Zenger, Ingo. 2020. A Lifesaver in a Plastic Cup. Siemens MedMuseum. December 1. https://www.medmuseum.siemens-healthineers.com/en/stories-from-the-museum/herzschrittmacher.

Index

C. D. Craig et al. (eds.), *Women Engineering Legends 1952-1976: Society of Women Engineers Achievement Award Recipients*, https://doi.org/10.1007/978-3-032-00224-2

MIX
Papier aus verantwortungsvollen Quellen
Paper from responsible sources
FSC® C105338

If you have any concerns about our products,
you can contact us on
ProductSafety@springernature.com

In case Publisher is established outside the EU,
the EU authorized representative is:
Springer Nature Customer Service Center GmbH
Europaplatz 3, 69115 Heidelberg, Germany

Printed by Libri Plureos GmbH
in Hamburg, Germany